# Conference Proceedings of the Society for Experimental Mechanics Series

*Series Editor*

Tom Proulx
Society for Experimental Mechanics, Inc.
Bethel, CT, USA

AF774546

More information about this series at http://www.springer.com/series/8922

Christopher Niezrecki
Editor

# Structural Health Monitoring and Damage Detection, Volume 7

## Proceedings of the 33rd IMAC, A Conference and Exposition on Structural Dynamics, 2015

Springer

*Editor*
Christopher Niezrecki
Department of Mechanical Engineering
University of Massachusetts Lowell
Lowell, MA, USA

ISSN 2191-5644 ISSN 2191-5652 (electronic)
Conference Proceedings of the Society for Experimental Mechanics Series
ISBN 978-3-319-38603-4 ISBN 978-3-319-15230-1 (eBook)
DOI 10.1007/978-3-319-15230-1

Springer Cham Heidelberg New York Dordrecht London
© The Society for Experimental Mechanics, Inc. 2015
Softcover reprint of the hardcover 1st edition 2015
This work is subject to copyright. All rights are reserved by the Publisher, whether the whole or part of the material is concerned, specifically the rights of translation, reprinting, reuse of illustrations, recitation, broadcasting, reproduction on microfilms or in any other physical way, and transmission or information storage and retrieval, electronic adaptation, computer software, or by similar or dissimilar methodology now known or hereafter developed.
The use of general descriptive names, registered names, trademarks, service marks, etc. in this publication does not imply, even in the absence of a specific statement, that such names are exempt from the relevant protective laws and regulations and therefore free for general use.
The publisher, the authors and the editors are safe to assume that the advice and information in this book are believed to be true and accurate at the date of publication. Neither the publisher nor the authors or the editors give a warranty, express or implied, with respect to the material contained herein or for any errors or omissions that may have been made.

Printed on acid-free paper

Springer International Publishing AG Switzerland is part of Springer Science+Business Media (www.springer.com)

# Preface

*Structural Health Monitoring & Damage Detection* represents one of ten volumes of technical papers presented at the 33rd IMAC, A Conference and Exposition on Structural Dynamics, 2015, organized by the Society for Experimental Mechanics, and held in Orlando, Florida February 2–5, 2015. The full proceedings also include volumes on Nonlinear Dynamics; Dynamics of Civil Structures; Model Validation and Uncertainty Quantification; Dynamics of Coupled Structures; Sensors and Instrumentation; Special Topics in Structural Dynamics; Experimental Techniques, Rotating Machinery & Acoustics; Shock & Vibration Aircraft/Aerospace, Energy Harvesting; and Topics in Modal Analysis.

Each collection presents early findings from experimental and computational investigations on an important area within Structural Dynamics. Structural Health Monitoring is one of these areas. Topics in this volume include:

Structural Health Monitoring
Damage Detection
Numerical Modeling

The organizers would like to thank the authors, presenters, session organizers, and session chairs for their participation in this track.

Virginia Tech, USA

A. Wicks

# Contents

# Chapter 1
# Bearing Faults Simulations Through a Parametric Model of a Gearbox

**S. Cinquemani, F. Rosa, and E. Osto**

**Abstract** The paper deals with a project aimed to improve the reliability of a condition monitoring system applied on gearboxes installed on rolling mills. In this context, to properly set up the algorithm, it is necessary to have measurements associated both to standard operating conditions and to malfunctioning. Since the experimental determination of the latter is obviously not cost and time effective, they can be simulated by means of numerical models of the mechanical system in several operating conditions. The outputs generated, corresponding to different fault conditions of the more critical elements of the system, will provide a useful data base to tune the algorithm of condition monitoring.

**Keywords** Condition monitoring • Bearing • Mechanical transmission • Rolling mill • Parametric model

## 1.1 Introduction

Gearboxes for rolling mills are complex machines that cannot be treated as commodities, since they are parts of a complex drive system that in case of failure could seriously affect the plant productivity [1–3]. The root causes of a geared system failure can sometimes be quite different from the appearing ones. From this point of view, an early detection of improper operating conditions, as condition monitoring can provide, gives a good chance to plan extraordinary maintenance to prevent sudden stop of production and to identify primary failure causes instead of secondary ones. It worth to be mentioned that preventive maintenance to solve primary failure causes implies negligible cost compared to ones related to secondary failure causes (e.g. bearing replacement vs. gear replacement).

Bearings represent a typical source of gearboxes failures or improper operation [4, 5]. Also for bearing, due to the vast number of different failure modes, specific standards have been introduced as ISO 15243. Gradual deterioration of the operating behaviour is normally the first signal of bearing damage. Failures due to poor ordinary maintenance (lack of lubrication, for example) and improper mounting are relatively infrequent, and very often lead quickly to machine downtime. On the other hand, depending on the operating conditions, a few weeks, or under some circumstances, even a few months, may pass from the time damage begins to the moment the bearing actually fails because of the contact fatigue damaging mechanisms. This typical progressive evolution of the damage makes bearings especially suitable for continuous condition monitoring applications. From the point of view of failure detection, the main effects of bearing damage impacts on operating temperatures, lubricant contamination and vibrations. In principle, all these information can be used for condition monitoring application, but the techniques based on the analysis of vibration signals are the most efficient as they can provide an early identification of the specific bearing involved, of the part of the bearing affected by damage and on the degree of the damage itself, thanks to the information coming from frequency and amplitude data. In particular damages related to contact fatigue affecting the races or the rolling elements can be detected and identified [6]. Moreover they are suitable to be modelled by means of models which can provide preliminary information in terms of expected frequencies and amplitudes in the signal analysis.

Condition monitoring algorithms are based on signals in different working conditions. To properly set up the algorithm, it is necessary to have measures associated both to standard operating conditions and to malfunctioning. The latter, not being experimentally determinable, can be simulated by developing numerical models of the machine under varying

S. Cinquemani (✉) • F. Rosa
Department of Mechanical Engineering, Politecnico di Milano, Via La Masa 1, 20156 Milano, MI, Italy
e-mail: simone.cinquemani@polimi.it

E. Osto
Siemens VAI Metals Technologies Srl, Via Luigi Pomini, 92, Marnate, Italy

© The Society for Experimental Mechanics, Inc. 2015

C. Niezrecki (ed.), *Structural Health Monitoring and Damage Detection, Volume 7*, Conference Proceedings of the Society for Experimental Mechanics Series, DOI 10.1007/978-3-319-15230-1_1

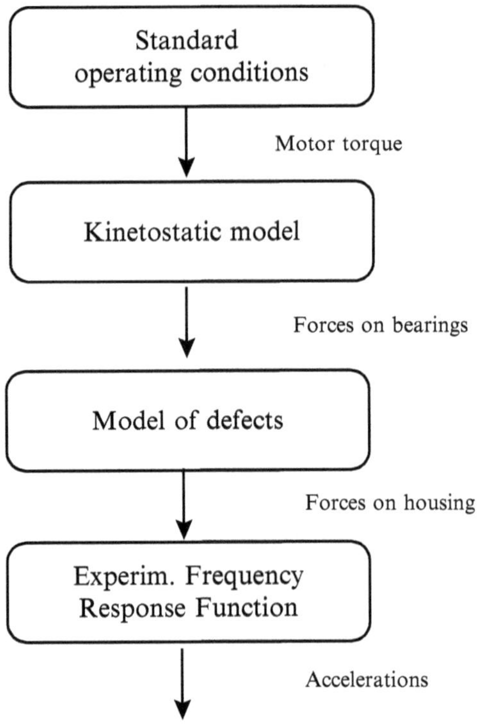


**Fig. 1.1** Overview of the steps to generate virtual signals corresponding to malfunctioning

conditions [7]. The outputs generated, corresponding to different fault conditions associated with the main common failures of the elements that constitute the transmission, will provide a useful data base to properly set the algorithm of condition monitoring.

The paper deals with the description of such approach (Fig. 1.1). Virtual signals related to malfunctioning can be obtained using the real working conditions (torque and motor speed) as the inputs of the model thus implementing the dynamics associated with the malfunctioning of different bearings. The output of the model consists on the harmonic forces acting on the bearing housing. Through experimental transfer functions of the gearbox, virtual accelerations can be calculated.

The paper is structured as follows. Section 1.2 describes the kinetostatic model of the mechanical transmissions. Section 1.3 introduces the mathematical model of main failures of bearings. For the sake of clarity, the presented only cylindrical roller bearings model is here presented, but the reader will understand that this approach can be easily replicated for spherical, tapered and ball bearings. Section 1.4 introduces how to obtain the frequency response function of the gearbox, while Sect. 1.5 shows how the approach can be usefully applied on a real test case. Finally conclusions are drawn in Sect. 1.6.

## 1.2 Kinetostatic Model of the Transmission

Transmissions installed on rolling mills can be generally divided into two groups: with parallel axis (Fig. 1.2) and with orthogonal axis (Fig. 1.3). For each group a set of parameters has been defined to describe the transmission (i.e. mechanical and geometrical features, type of bearings, gears, etc.).

To be able to manage different transmissions, a reduced model with lumped parameters is developed according to the parameterization carried out. Forces on bearings ($F_X$, $F_Y$, $F_z$) can be computed according to [3], as a function of the main features of the transmission and of the operating conditions (i.e. motor torque and angular speed). Then, the displacements of the inner ring of each bearing are calculated with respect to applied forces, as the mathematical models of defects need this input to generate the virtual accelerations.

Unfortunately, while the direct relationship between displacement of the shaft and forces transmitted by the bearing exists [2, 8], the inverse relationship can not be solved analytically. For this reason an iterative procedure has been implemented.

Even if many bearing types (cylindrical rollers, double row tapered rollers, double row self-aligning, and four-point-contact bearings) have been analysed and modelled, only cylindrical roller bearings will be presented hereafter in order to give an overview of the adopted approach. Each model is furthermore capable to simulate the more common bearing defects: inner and outer race pits and cages wear resulting in an uneven roller distribution.

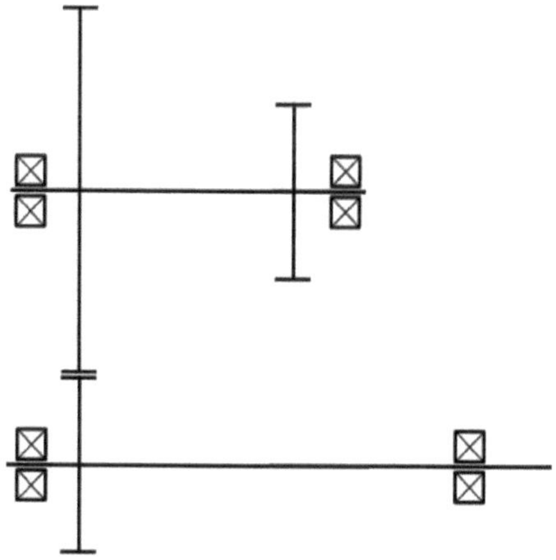


**Fig. 1.2** Parallel axis

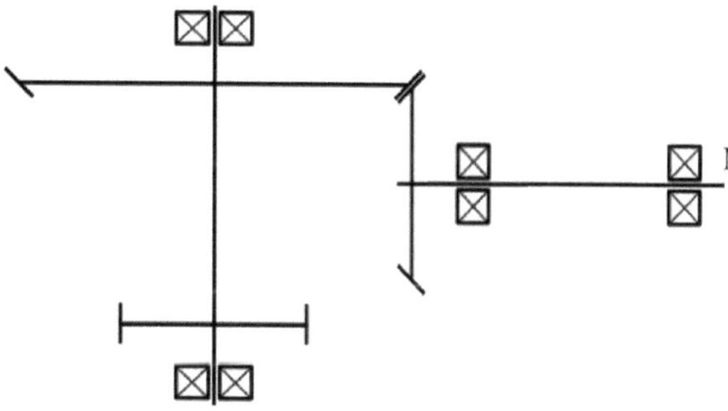


**Fig. 1.3** Orthogonal axis

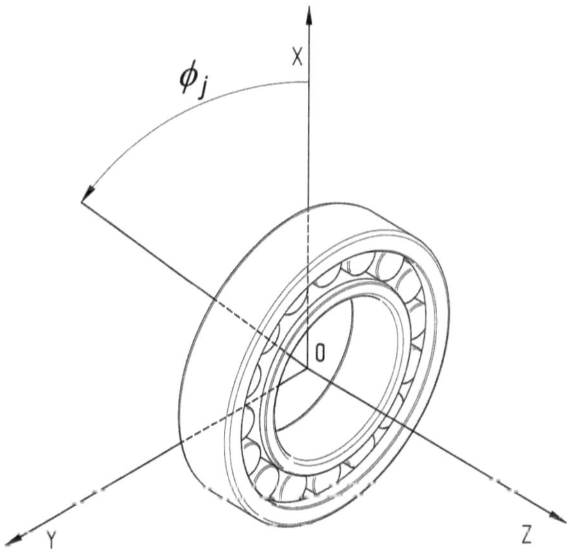


**Fig. 1.4** Cylindrical roller bearing model

In order to ease the integration of these models with the global model of the transmission and to derive an explicit analytical formulation, it was assumed that outer race is fixed, while inner race rotates with the shaft and hence undergoes to the same translational displacements of the shaft section and that bending deformations of inner and outer rings are negligible with respect to roller-race contact deformations, that are hence the only deformations of bearing components considered to determine bearing global stiffness.

Adopting the reference system shown in Fig. 1.4, in a cylindrical roller bearing without any defect, for a given displacement of the inner race $\underline{\Delta} = \Delta_x \, \underline{u}_x + \Delta_y \, \underline{u}_y$, the total radial contact deformation of the generic *j-th* roller at angular position $\phi_j$ is therefore:

$$\delta_j = \Delta_x \, cos \, \phi_j + \Delta_y \, sin \, \phi_j - c$$

where $c$ is the radial clearance of the bearing in operating conditions. It is worth observing that, because of the geometrical nature of this expression, it is an actual contact deformation only if $\delta_j$ is positive. The next step consists in determining the load-deformation relationship.

According to Harris and Kotzalas [4], on the basis of laboratory testing of crowned rollers loaded against raceways, Palmgren [6] experimentally determined the following equation for contact deformation $\delta$ *[mm]*:

$$\delta = 3.84 \cdot 10^{-5} \cdot Q^{0.9} / \, l^{0.8}$$

where $l$ is roller effective length *[mm]*, and $Q$ is the normal total force between a rolling element and a raceway *[N]*.

This expression has been adopted since it allows deriving a direct and explicit relationship between the applied load and the consequent displacements, even if it does not allow taking into account some aspects (i.e. curvature variations and ratios). The numerical constant value for specific roller crowning and/or materials can anyhow be determined by means of experimental tests. By solving this equation with respect to Q and taking into account that each roller is in contact with both races, it is possible to derive an "equivalent stiffness":

$$K_{eq} = k_b \cdot 2^{-n}$$

where $n = 10/9$ for roller bearings and $k_b = 7.7652 \cdot 10^4 \cdot l^{8/9}$ is the "single stiffness".

As a result, the load acting on a single roller and its deformation are related by the following equation:

$$Q_j = \gamma_j \cdot K_{eq} \cdot \delta_j{}^n$$

where $\gamma_j = 1$ if $\delta_j > 0$, and is zero otherwise.

The final step is the summation of components of forces $Q_j$, in order to determine the force exerted by the outer ring on its seat as a consequence of inner race displacement $\underline{\Delta}$. This approach have been further extended to the so called "lamina model" [9], in order to take also into account roller profile modifications [9].

At the end, for each bearing, it is possible to find out a combination of displacements along the three axes (X, Y, Z) and forces transmitted along the same three directions ($F_X$, $F_Y$, $F_Z$). This look-up table can be calculated offline for all the bearings in the database and represents a theoretical stiffness of each bearing, as it relates displacement to forces.

Once the look-up table has been calculated, the displacement of the inner ring of the bearing can be obtained iteratively finding the best combination of forces ($F_X$, $F_Y$, $F_Z$) that matches with the forces applied on the bearing ($\underline{F}_X$, $\underline{F}_Y$, $\underline{F}_z$). It is important to note that the accuracy of the solution is strictly related to the resolution of the look up table. If the combinations are few, the result of the algorithm could be far from the real solution.

## 1.3 Models of Defects on Bearings

Pits presence on a raceway has been introduced as a reduction of roller contact deformation, i.e. as an increase of bearing radial clearance. In practice, this means that the expression of $\delta_j$ modifies as follows

$$\delta_j = \Delta_x \cos \phi_j + \Delta_y \sin \phi_j - c - \beta_j(t) \cdot C_d$$

where $C_d$ is pit depth and $\beta_j(t)$ is a Boolean time dependant function that is equal to 1 only if the theoretical contact point between roller and inner or outer race at time $t$ is within pit circumferential extension. It is worth observing that this function depends also on which race the pit is located, since the inner race rotates, and hence the absolute position of a pit on it is not constant. To reduce the number of variable, a simple relationship between the depth ($C_d$) and the width ($\Delta_d$) of the defect has been imposed as shown in Fig. 1.5

To evaluate the effects of damages on bearings, it is possible to set different values of $C_d$, and then of $\Delta_d$.

It's important to note that models of defects are based on the idea that a roller, when it is over a defect, loses some of the load that can transmit. The limit condition occurs when the roller completely loses the contact with the ring and, therefore, it is no more able to transmit any force. In this condition (easy to be reached for bearing under small a load), even increasing the depth of the defect an increases in force associated with the defect would not be observed.

Coming to cage wear, it was introduced by assuming that its effect on bearing global model can be considered by changing the circumferential positions (with respect to their evenly spaced theoretical positions) of the points where each roller exerts

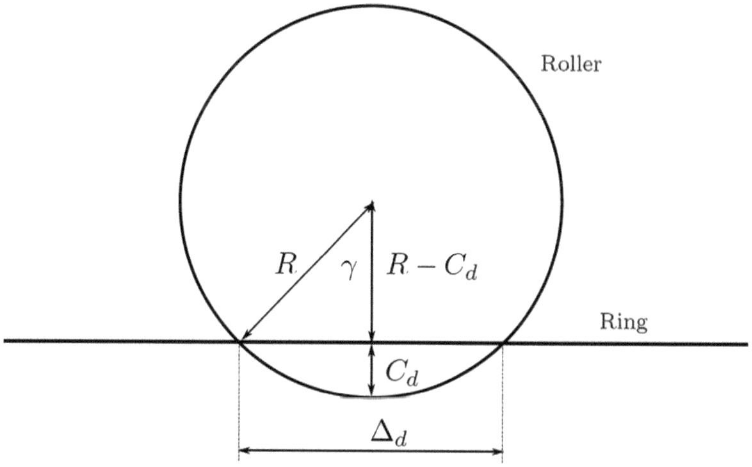


**Fig. 1.5** Geometrical relationship between the depth ($C_d$) and the width ($\Delta_d$) of the defect

its resultant forces on the rings. From a mathematical point of view, after having defined the maximum circumferential displacement ($\varepsilon_{wc}$) of these points, this assumption leads to define a distribution of these displacements in order to locate all of them. If this type of defect is introduced, this implies that angles $\phi_j$ are no longer evenly spaced, as assumed in the model of a "healthy" bearing.

Similar considerations have been made for the other types of bearings taking into account also, if necessary, the relationships between axial and radial components, in order to derive appropriate analytical models.

## 1.4 Experimental Frequency Response Function

The relationship between forces applied on the housing and the corresponding virtual accelerations is obtained experimentally.

Each test consists on hitting the chassis with a dynamometric hammer and measuring the corresponding accelerations through a three axis accelerometer. As a matter of fact, an impulsive test allows driving all the resonances of the system. This means that, with a single test, it is possible to evaluate the dynamics of the system in a wide range of frequency. As the hammer contains a load cell, a measurement of the applied force can be collected. Signal have been acquired using a Daq board Ni 9234 with a sample frequency of 51.2 kHz. Figure 1.6 shows time histories of the force applied to the chassis and the corresponding measured acceleration. The impulsive test is repeated 10 times to improve the quality of the frequency response function, increasing the signal to noise ratio through the so called linear averaging technique.

Knowing the forces applied and the accelerations of the chassis it is possible to estimate the frequency response function of the structure. Figure 1.7 shows, for example, the frequency response function between the force applied on the chassis along the vertical direction, and the corresponding acceleration along the same axis. It is important to highlight that the virtual accelerations resulting from this work, are strictly related to the experimental transfer functions of the transmission.

Information collected in the frequency response function are extremely useful. Once the forces transmitted to the chassis, due to a defect on a bearing, are known from the mathematical model, the corresponding accelerations can be easily obtained simply multiplying the forces (in frequency domain) to the frequency response function of the system. For example, being $TF_{X,B1}(f)$ the frequency response function of the chassis, close to bearing B1, along the X axis and $F_{X,B1}(f)$ the forces along the same direction estimated by the model for a defect on bearing B1, it results:

$$\ddot{x}_{B1}(f) = TF_{X,B1}(f) \cdot F_{X,B1}(f)$$

where $\ddot{x}_{B1}(f)$ is the acceleration measured along x axis due to a defect on bearing B1.

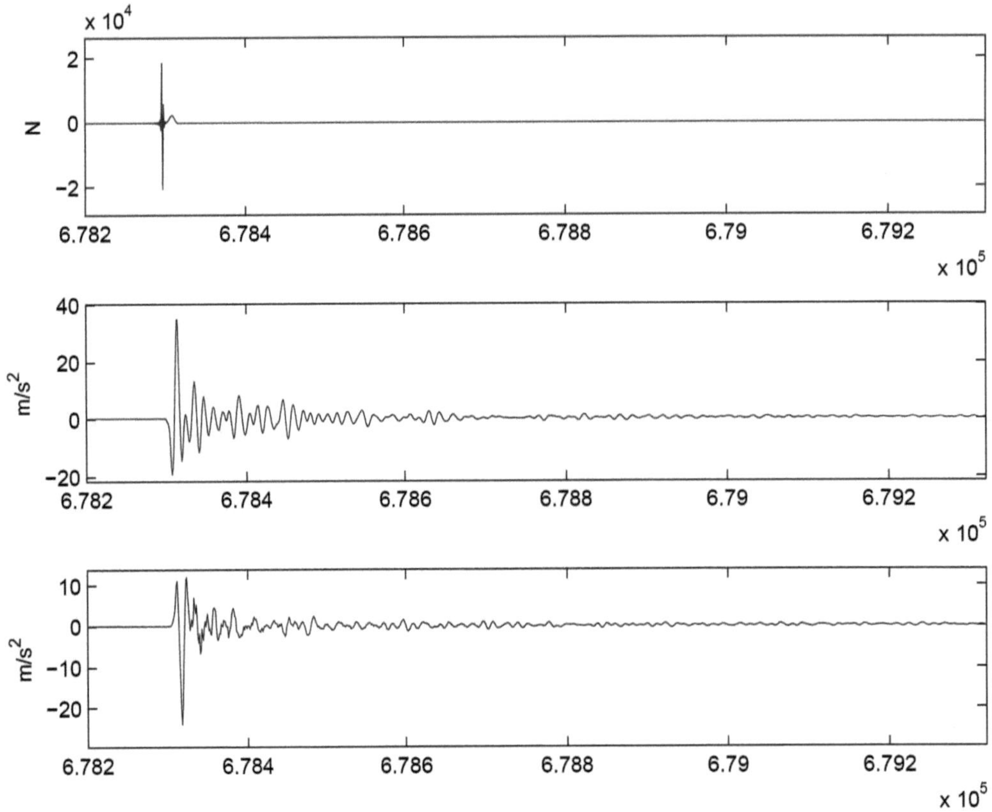


**Fig. 1.6** Force applied to the chassis and corresponding measured acceleration

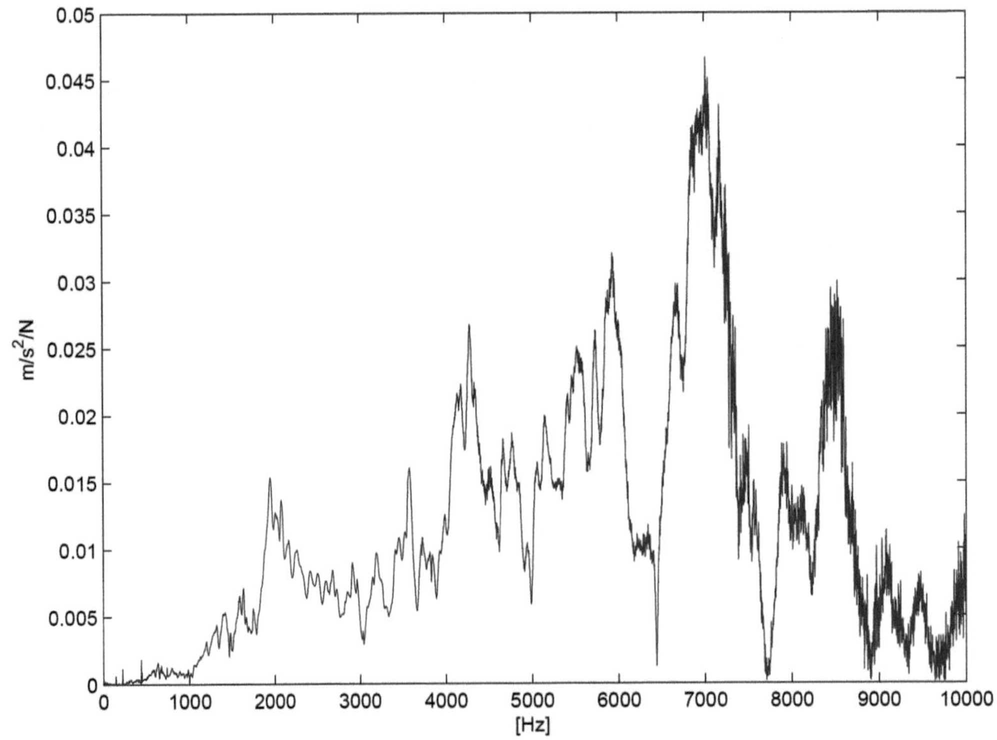


**Fig. 1.7** Frequency response function

## 1.5 Test Case

The parametric model descripted in the paper has been used to simulate failures on bearing of a real mechanical transmission installed on a rolling mill. The layout of the transmission to be analysed is depicted in Fig. 1.8. The transmission is used in a rolling mill plant and it has a speed ratio of 0.072. Shaft 0 is connected to the motor, while shafts 2, 3 are linked by spindles to the rolls of the rolling mill. Speed reduction is obtained from the first two stages of the transmission, while the third one (whose ratio is equal to 1) is used to split the power to the two rolling rolls. Bearings configuration is resumed in Table 1.1.

According to different operating conditions and different kind of defects, the model can evaluate the forces applied to the chassis along X, Y directions and the corresponding accelerations.

Let's consider the tapered roller bearing B1. Figures 1.9 and 1.10 respectively show the forces applied along X direction and the corresponding accelerations. Numerical analysis have been carried out considering motor torque: $T_M = 14kN$, motor angular speed $\omega_M = 500$ rpm, Cd = 0.1 mm for defect on the outer ring, Cd = 0.1 mm for defect on the inner ring, Cd = 3.2 mm for defect on the cage.

Information on virtual accelerations will be useful to train the condition monitoring algorithm.

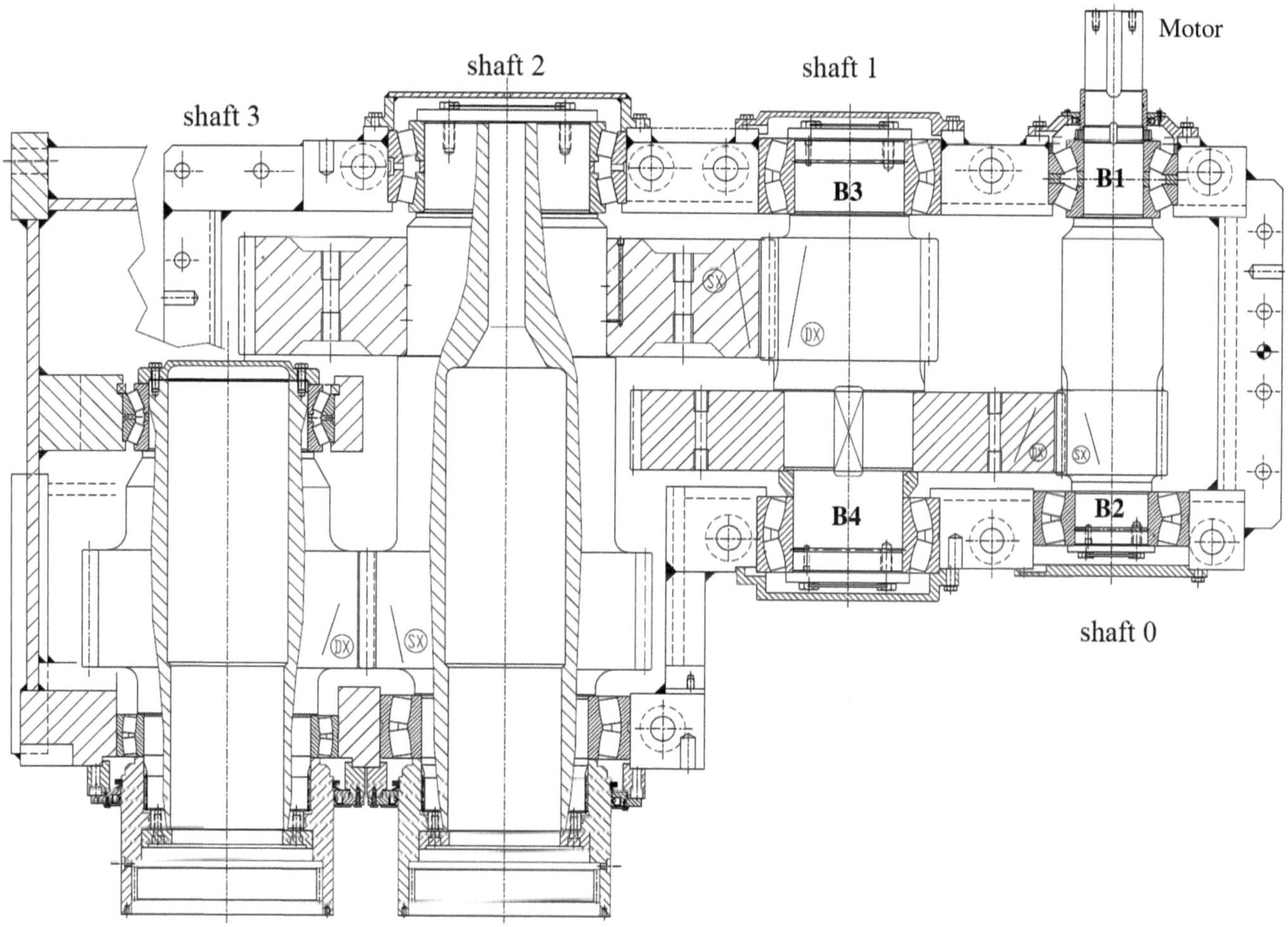


**Fig. 1.8** Layout of the considered mechanical transmission

**Table 1.1** Configuration of bearings

| B1 | B2 | B3 | B4 |
|---|---|---|---|
| Tapered roller b | Spherical b | Spherical b | Spherical b |
| SKF 31326 | SKF 22332 | SKF 24148 | SKF 24148 |

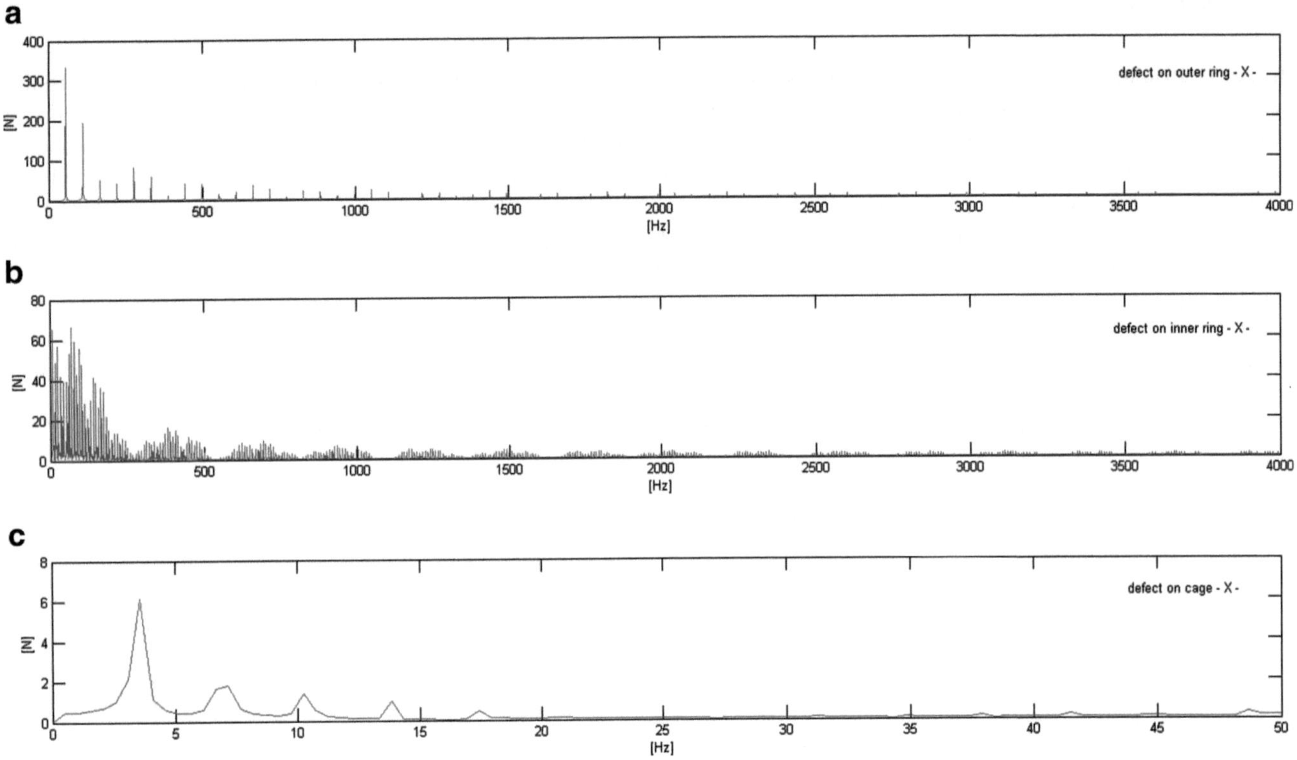


**Fig. 1.9** Forces applied on the chassis along X direction ((**a**) defect on the outer ring, (**b**) defect on the inner ring, (**c**) defect on the cage)

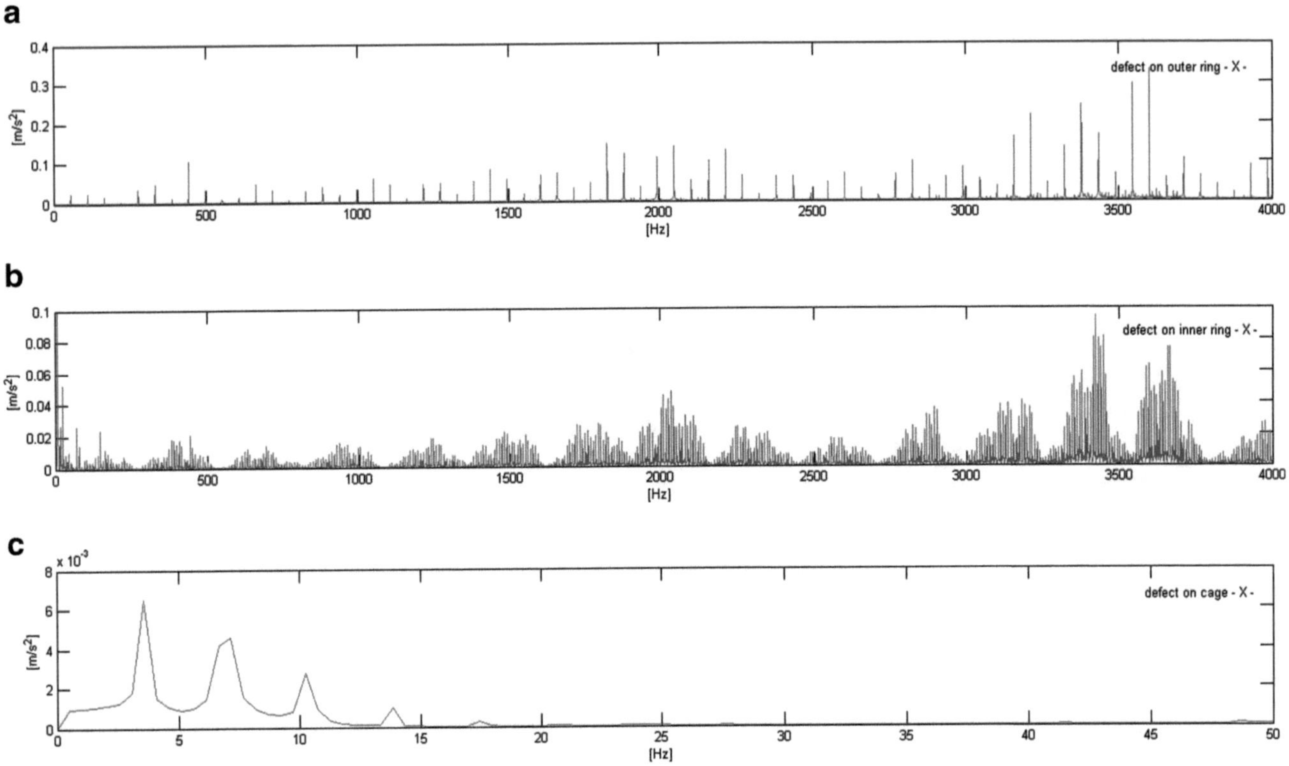


**Fig. 1.10** Virtual accelerations of the chassis along X direction ((**a**) defect on the outer ring, (**b**) defect on the inner ring, (**c**) defect on the cage)

## 1.6 Concluding Remarks

This activity is part of a project aimed to develop a system to monitor the working conditions of mechanical transmissions installed on rolling mills. In this context, to properly set up the algorithm, it is necessary to have measures associated both to standard operating conditions and to malfunctioning. The letters, not being possible to effectively determine them experimentally, have been simulated by developing numerical models of the machine under varying conditions.

The outputs generated, corresponding to different fault conditions of the more critical components of the transmission, provide a useful data base to properly set the condition monitoring algorithm.

## References


1. Brewe D, Hamrock B (1977) Simplified solution for elliptical-contact deformation between two elastic solids. ASME Trans J Lub Tech 101(2):231–239
2. Griseri M (2000) Elementi di meccanica delle trasmissioni – calcolo degli sforzi sui supporti. SERIE SKF – Quaderni di formazione
3. Harris TA, Kotzalas MN (2007) Rolling bearing analysis, vol 1, 5th edn, Essential concepts of bearing technology. Taylor and Francis, Boca Raton
4. ISO (2004) ISO/TS 15243 – Rolling bearing – Damage and failures – Terms, characteristics and causes
5. Palmgren A (1959) Ball and roller bearing engineering, 3rd edn. Burbank, Philadelphia
6. ISO (2007) ISO 281 – Rolling bearings. Dynamic load ratings and rating life
7. Gorla C, Rosa F, Sabbioni E, Osto E, Cinquemani S (2014) A multibody model of a mechanical transmission to simulate failures on bearings. In: Proceedings of the 12th biennal conference on engineering systems design and analysis: ESDA 2014, p 9
8. ISO (2008/2013) ISO/TS 16281 – Rolling bearings – methods for calculating the modified reference rating life for universally loaded bearings
9. Jones A (1946) Analysis of stresses and deflections. New Departure Engineering Data, Bristol

# Chapter 2
# Sensitivity Evaluation of Subspace-Based Damage Detection Method to Different Types of Damage

S. Allahdadian, C.E. Ventura, P. Andersen, L. Mevel, and M. Döhler

**Abstract** In this paper we investigate a damage detection technique based on the subspace method by applying it to an existing bridge structure model. A reference state of the structure is evaluated using this technique and subsequently its modal parameters are indirectly compared to the current state of the structure. There are no modal parameters estimated in this method. A subspace-based residual between the reference and possibly damaged states is defined independently from the input excitations employing a $\chi^2$ test and then is compared to a threshold corresponding to the reference state. This technique is applied to a model of the bridge structure located in Reibersdorf, Austria. The structure is excited randomly with white noise at different locations and the output data is generated at typical locations instrumented and measured in a bridge. Various damages are simulated in different elements and the sensitivity of the method to each type and ratio of damages is assessed. This evaluation is performed by comparing the prediction of the damage state using this technique and the simulated damage of the structure. It can be inferred from the results that in general the statistical subspace-based damage detection technique recognizes most of the damage cases, except the ones with insignificant change in the global dynamic behaviour.

**Keywords** Damage detection • Subspace method • Health monitoring • Bridge damage • Statistical damage detection

## 2.1 Introduction

Structural health monitoring is regarded as the main tool in assessing the functionality of existing structures. The importance of these techniques and researches becomes obvious by considering that failure of a structure can result in catastrophic lost. During past decades extensive researches have been done in the literature in order to investigate an ideal nondestructive damage detection technique. Nondestructive damage detection techniques can be categorized into two groups based on their requirements [1]: (I) local techniques, which need access to all parts of the structure or the location of damage if known, and (II) global damage techniques which use vibration data to evaluate global dynamic characteristics of the structure. In the latter method there is no need to know or have access to the location of damage in priori. The dynamic characteristics used in these methods are usually natural frequencies, mode shapes and damping values.

In order to detect if a damage has occurred in a structure, evaluation of these dynamic characteristics can be avoided by using statistical approaches, e.g. statistical subspace-based damage detection technique (SSDD) [1–5]. The damage can be detected by comparing a statistical model from the possibly damaged structure to the one obtained from a reference state. In other words a subspace based residual function between these states is defined and compared using a $\chi^2$ test. In this way there is no need to estimate the natural frequencies and mode shapes of the structure, making this approach capable of being used in real-time monitoring of structures. In [3] and [6] it is investigated that this approach can also perform robustly under ambient excitations with changing statistics.

S. Allahdadian (✉) • C.E. Ventura
University of British Columbia, Vancouver, Canada
e-mail: saeid@civil.ubc.ca; ventura@civil.ubc.ca

P. Andersen
Structural Vibration Solutions A/S, Aalborg, Denmark
e-mail: pa@svibs.com

L. Mevel • M. Döhler
Inria, Centre Rennes—Bretagne Atlantique, Campus de Beaulieu, 35042 Rennes, France
e-mail: laurent.mevel@inria.fr; michael.doehler@inria.fr

© The Society for Experimental Mechanics, Inc. 2015

C. Niezrecki (ed.), *Structural Health Monitoring and Damage Detection, Volume 7*, Conference Proceedings of the Society for Experimental Mechanics Series, DOI 10.1007/978-3-319-15230-1_2

In this paper the performance of the SSDD technique is assessed for different damage types and ratios. A bridge structure, i.e. S101, located at Reibersdorf, Austria, is investigated and simulated for this purpose. This structure was damaged artificially in a progressive manner and it was continuously measured during each damage level [7]. A finite element model of this structure is created and calibrated using the available measured data. This analytical model is damaged in different locations, for different types of elements with different ratios. Subsequently, the structure is excited using white noise excitations. The simulated ambient vibration data is generated by measuring the acceleration time histories of the nodes typically measured in a bridge structure. The simulated data is consequently processed by ARTeMIS software [8] in order to check with the analytical results obtained from the damaged and reference finite element models.

In order to perform the damage detection, the data acquired from each model is processed using the SSDD technique implemented in ARTeMIS software. A reference state is defined by this technique from the data acquired using the undamaged reference model and then the statistics of the data acquired from the damaged structures are compared to a threshold based on the reference state. From this comparison, the sensitivity of the SSDD technique to different types and states of damage is evaluated.

This paper is organized as follows: in Sect. 2.2 the SSDD technique is described and the formulation of the problem is demonstrated. In Sect. 2.3, the approach of simulating damage in the analytical model and subsequently simulation of ambient vibration test data is described. These techniques are illustrated in a case study of the realistic model of the bridge in Sect. 2.4. The functionality and sensitivity of the SSDD damage detection technique is investigated in that section. In the last section the conclusions are presented.

## 2.2 Statistical Subspace-Based Damage Detection

The statistical subspace-based damage detection (SSDD) technique detects the damage in a structure by using a $\chi^2$ test on a residual function [4–6]. Therefore in this method, there is no need to compute and compare modal parameters of the reference and possibly damaged states of the system. In other words, this residual function represents the changes occurred to the model which can be caused by a damage in structure.

### 2.2.1 *Models and Parameters*

The dynamic system of the model can be considered as a discrete time state space model of

$$\begin{cases} X_{k+1} = AX_k + \varepsilon_k \\ Y_k = CX_k + \vartheta_k \end{cases} \tag{2.1}$$

where, the state is represented by $X \in \mathbb{R}^n$ and the measured output is $Y \in \mathbb{R}^r$ $A$ also represents the state transition matrix and $C$ shows the observation matrix with dimensions $\mathrm{n} \times \mathrm{n}$ and $\mathrm{r} \times \mathrm{n}$, respectively. The state noise, $\varepsilon_k$ and measurement noise $\vartheta_k$ are assumed to be Gaussian unmeasured white noise with zero mean. The covariance of output measurements $Y_k$ can be computed from the state space model (2.1) by

$$R_i = \mathbf{E}\left(Y_{k+i} Y_k^T\right) \tag{2.2}$$

n which operator $\mathbf{E}$ is the expectation function. With choosing parameters $q$ and $p$ such as $q \geq p+1$ the Hankel matrix $\mathbf{H}$ can be written as

$$\mathbf{H} = \begin{pmatrix} R_1 & R_2 & \dots & R_q \\ R_2 & R_3 & & R_{q+1} \\ \vdots & & \ddots & \vdots \\ R_{p+1} & R_{p+2} & \dots & R_{p+q} \end{pmatrix} \tag{2.3}$$

As mentioned earlier the measurements are performed in a reference state and a possibly damaged state. The Hankel matrix of the measurements in reference state, $\mathbf{H}_0$ can then be computed from (2.2) and (2.3). This matrix is then decomposed using singular value decomposition in order to compute the left null space **S**. Defining **H** for the possibly damaged state of the system, the left null space matrix **S** in the reference state is characterized by $\mathbf{S}^T\mathbf{H} = \mathbf{0}$ ([2–4]). Therefore the residual vector $\boldsymbol{\zeta}_n$ can be written as

$$\boldsymbol{\zeta}_n = \sqrt{n}\ \mathrm{vec}\left(\mathbf{S}^T\mathbf{H}\right) \tag{2.4}$$

in which, $n$ represents the number of samples measured for computing **H**. This residual can now be used in order to check if any change is made in the model due to damage. The residual vector $\boldsymbol{\zeta}_n$ is asymptotically Gaussian with zero mean in reference state; significant changes in its mean value indicates the structure is moved from its reference state. In order to check this change from the residual vector mean, the $\chi^2$ test can be performed as following [4–7].

$$\chi^2 = \boldsymbol{\zeta}_n^T \Sigma^{-1} \boldsymbol{\zeta}_n \tag{2.5}$$

Herein, $\Sigma$ represents the covariance matrix of the residual in the reference state, and can be shown as

$$\Sigma = \mathbf{E}\left[\boldsymbol{\zeta}_n\boldsymbol{\zeta}_n^T\right] \tag{2.6}$$

It is worth mentioning that the covariance of the input noise $\varepsilon_k$ is assumed to not change between the reference state and the possibly damaged state. Details of the covariance computation are found in [3].

By monitoring the value of $\chi^2$ and comparing it to a threshold value, the state of the damage of the system can be estimated. This threshold can be simply evaluated using several data sets measured from the structure in its reference state. Subsequently, some other data sets measured from the reference state are used to check the threshold. Then the $\chi^2$ value is computed for the possibly damaged structure. If the computed $\chi^2$ value is higher than the threshold it can be inferred that the structure may be damaged. In other words, damage that provokes a change in the statistics of the measured data leads to an increase in the $\chi^2$ value.

## 2.3 Damage and Data Simulation

The ideal test data that can be used to evaluate a damage detection technique is to damage a real structure progressively and measure its response continuously [7]. Having a clear understanding of the conditions of the structure before damaging plays a critical role in the results. Furthermore, in addition to the cost of the procedure, it is not practical to damage a structure and restore it to its undamaged condition for the next test, especially when different elements of the structure are needed to be damaged separately and to various extents.

Simulating the damage in a structure and subsequently generating data that represents the ambient vibration test data can be a useful approach to evaluate damage detection techniques. This data can be an acceptable benchmark to evaluate the functionality of these techniques by allowing control on the test conditions, e.g. structural properties and damage effects. In order to investigate the effect of noise on these techniques, a predefined amount of white noise can be superposed to the data. For simplicity, in this paper there is no additional noise imposed on the results and its effect on the damage detection technique investigated in future study.

In order to evaluate the functionality of the subspace-based damage detection technique, the ambient vibration test data can be simulated for different damage types and amounts. In order to simulate this data, a finite element model of the structure is created and then calibrated to the real structure. It should be mentioned that calibration of the structure does not have a straight effect on the damage detection technique. In other words, the damage detection technique should be able to detect the damage in any structural model including the uncalibrated one as long as the base of comparison is identical. However in this study, calibration to a real structure is performed to obtain a realistic model and simulate the damage in it. The damage in different elements of the model is simulated by reducing the dimensions of one or some short elements in the intended location of damage. The amount of the damage is defined in terms of ratio of this reduction.

Several points of the structure are excited using white noise excitation in all three directions. Different excitations are imposed on the structure in order to excite the structure as randomly as possible. This excitation can be done by acceleration or load forces in different points of the structure. Subsequently, the simulated data can be obtained by measuring acceleration time histories of the nodes typically measured and instrumented in a bridge.

The simulated data can then be analyzed in order to compute the natural frequencies and their corresponding mode shapes. These can be used to check which mode shapes can be captured by the simulated white noise excitation. Based on the positioning of the sensors and or insufficient excitation of the structure, some mode shapes may not be captured. For the latter, the excitation must be modified to impose to the structure an excitation close to white noise in different points of the structure.

## 2.4 Case Study

Herein, the case study is the model of a bridge structure, namely S101, located in Reibersdorf, Austria. In [7], this structure was progressively damaged and the ambient vibration data was recorded continuously to evaluate statistical damage detection methods. In this study, the finite element model of this structure is used to simulate the damage in more locations of the bridge with various extents. The finite element model is calibrated using the measured data from the bridge to have a precise estimation of the behaviour of structure. The bridge structure and its finite element model are shown in Fig. 2.1. The natural frequencies of the analytical model and the bridge structure are also compared at Table 2.1.

It can be inferred from Table 2.1 that the finite element model of the structure can be a good representative of the dynamic properties of the bridge. As mentioned in previous section, this calibration is to have a realistic model of a bridge and it does not affect the assessment of the functionality of the damage detection technique.

The effect of bearings in simulating the damage in other elements of the bridge is neglected. However, for simulating the damage in bearings the reference structure is equipped by bearings at the supports. Therefore the reference state of the structure with bearings is used to create the threshold of $\chi^2$ and then the damaged models are compared with that reference state.

### 2.4.1 *Damage Simulation*

The damage is modeled in different elements of the structure and in different locations. The damaged element types include girders, cap beams, columns, deck and bearings. Furthermore, since this bridge is composed of three spans the damage for the girder and deck is modeled in two locations, i.e. central span and side span. Girders, cap beams and columns are modeled by a number of finite elements. The damage is simulated only in one of these finite elements by reducing a ratio of its section

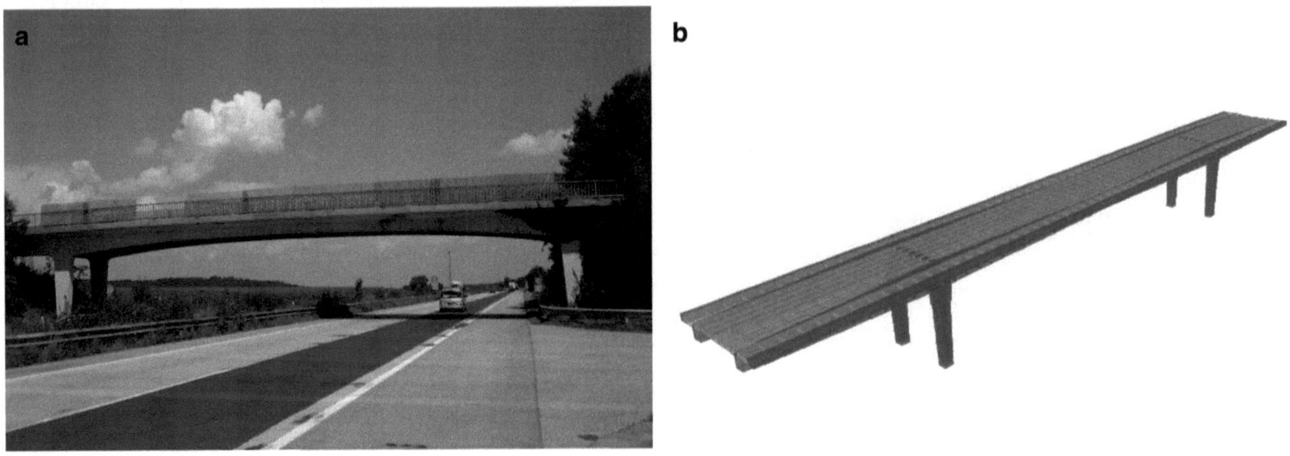


**Fig. 2.1** (**a**) S101 bridge structure, Austria, and (**b**) its calibrated finite element model

**Table 2.1** Natural frequencies of the bridge structure in undamaged condition obtained from the measured data and finite element model

| | Measured data (Hz) | Finite element model (Hz) |
|---|---|---|
| First bending mode | 4.05 | 4.04 |
| First torsional mode | 6.30 | 6.08 |
| Second bending mode | 9.69 | 10.72 |
| Second torsional mode | 13.29 | 12.85 |
| Third bending mode | 15.93 | 19.58 |

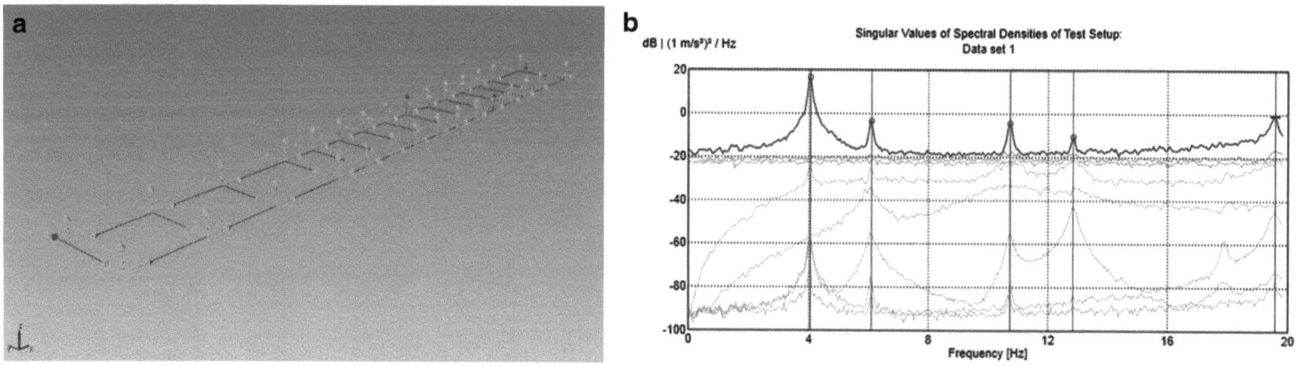


**Fig. 2.2** (**a**) measuring-points corresponding to sensor locations; (**b**) spectral densities of the singular values of the simulated measurement data in undamaged structure

dimension around the strong axis. This damage ratio varies among 20 % (minor damage), 40 % (intermediate damage) and 80 % (severe damage). Damage is simulated in the deck by removing some portions of it and in the bearings by reducing their stiffness.

### 2.4.2 *Data Simulation*

The finite element model of the structure is excited with a white noise excitation as an acceleration time history in three directions. Moreover, the structure is vibrated by different white noise loads in various locations. The measured points to record acceleration time histories are illustrated in Fig. 2.2a. Spectral densities of the simulated data obtained from undamaged reference case are shown in Fig. 2.2b.

It can be seen in Fig. 2.2b that the natural frequencies of the analytical model can be obtained from processing the simulated data accurately. Although, the structure is properly excited by white excitation, but some mode shapes cannot be captured. This stems from the location of the sensors and their resolution. As an example, the mode shapes associated to the longitudinal edges of the bridge cannot be captured by the sensors due to their small accelerations occurring in sensor locations.

### 2.4.3 *Damage Detection*

The undamaged structure is excited for six cases from which four are used to create a threshold for the $\chi^2$ value. The two remaining cases are then used to check the threshold. For each damage type and ratio, the simulated data is created and the $\chi^2$ test is performed. Subsequently, this value is compared to the computed threshold. The reference state for the structure with fixed supports and with bearings are shown in Fig. 2.3a. In order to validate the reference state, the null space of the Hankel matrix is illustrated in Fig. 2.3b, which shows that only a small portion of the singular values are more than the system order. This suggests that the reference state in both cases are reliable.

The threshold is computed for two significance values, namely critical zone for significance level 95 % (shown with yellow line in Fig. 2.3a) and unsafe zone for significance level 99 % (shown with red line in Fig. 2.3b). If the $\chi^2$ test value computed from the structure becomes larger than the yellow line, it suggests that the structure is in critical state. Similarly, if this value passes the red line, then the structure is estimated to be in unsafe conditions.

The $\chi^2$ test values of the simulated data from different damage cases of the fixed support model are illustrated in Fig. 2.4. For each damage ratio, the $\chi^2$ value is computed and compared to the reference state of the structure.

It can be seen from Fig. 2.4a, b, that $\chi^2$ value obtained from the girder damage can be captured for the intermediate and severe damage cases only. Minor damage in girder which involves 20 % damage ratio cannot be captured by SSDD technique. The intermediate damage to the girder in both spans is also recognized as critical condition for the structure. Figure 2.4c, d are also showing that the damage to the deck, which was simulated by removal some plane elements of the deck could be captured in all cases. It is worth mentioning by removal of these elements the stiffness of the structure is not

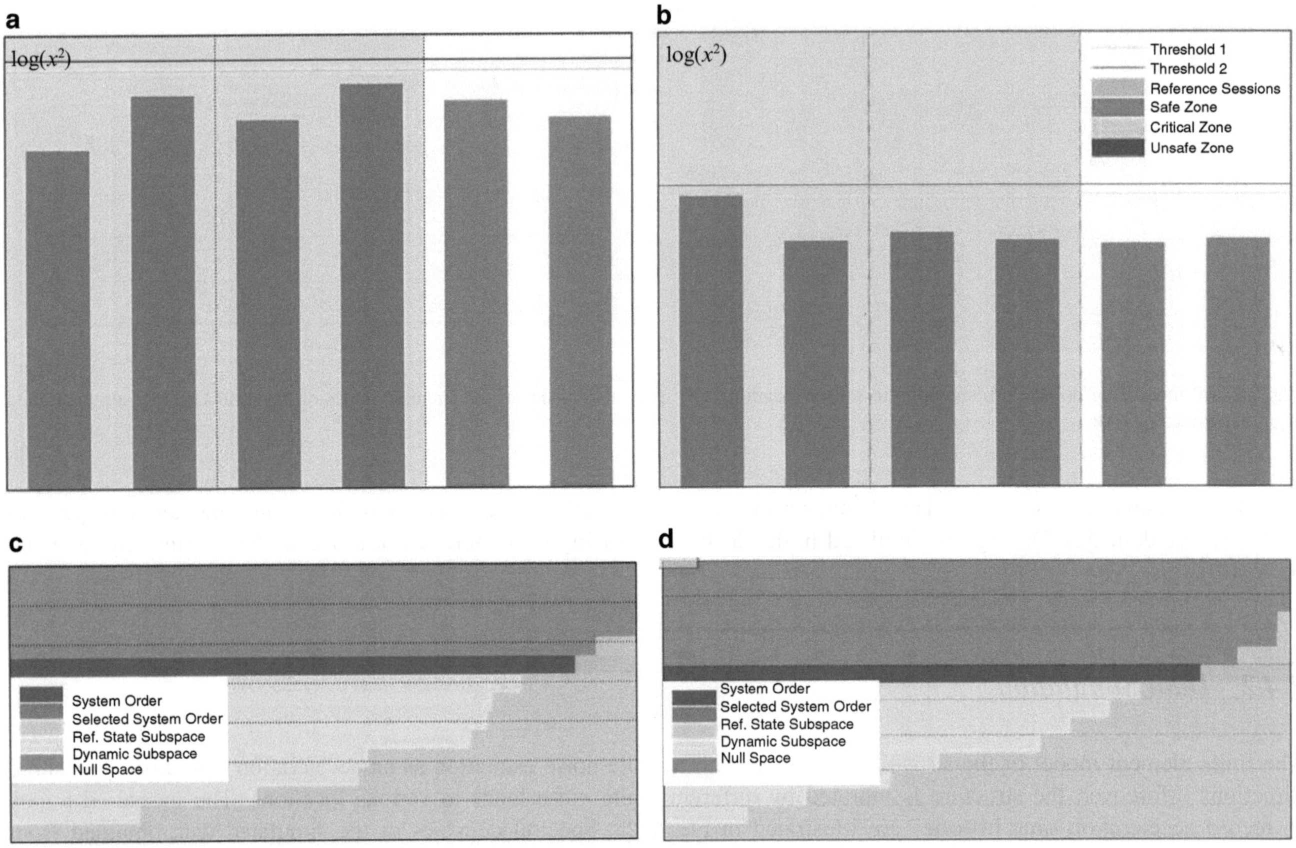


**Fig. 2.3** $\chi^2$ values and thresholds of the reference state in (**a**) structure with fixed supports and (**b**) with bearings; validation of the reference state in (**c**) structure with fixed supports and (**d**) with bearings

changed and the major change happens in mass matrix of the structure, which has a direct effect on dynamic vibration of the structure and can be captured effectively with this technique.

The damage in column can be captured only for the severe case, i.e. damage ratio equal to 80 %, as shown in Fig. 2.4e. The cap beam in the structure is located between the columns and connects the girders by crossing them. However, the girders are directly located on top of the columns and therefore cap beams do not have much effect in dynamic behaviour of the model. In all cases the damage in cap beams is not captured suggesting that damage in the elements not affecting the dynamic behaviour of structure may not be recognized by this monitoring technique.

The SSDD technique is used to evaluate damage in bearings as shown in following figure. The resultant $\chi^2$ values are compared to the reference state computed for the undamaged structure with bearings. It can be seen that in all cases the damage detection technique can successfully identify the damage in the structure (Fig. 2.5).

It can be seen that in all cases except the cap beam case, the severe damage in the structure could be captured by the SSDD technique. In most of the cases the damage for intermediate damage ratio could be also identified using this technique. For the cap beam element type, as this model was investigated, this element has negligible effect on the dynamic response of the structure. For this reason detecting the damage for that types of elements should be investigated in more details.

## 2.5 Conclusion

In this paper an analytical finite element model of a bridge structure namely, S101, located at Reibersdorf, Austria, is created and calibrated. The calibration was performed by measured data from the undamaged bridge. This model was employed as a realistic base model to simulate different damage scenarios. The damage was simulated in different element types, including girder, column, bearing, deck and cap beam. In each element type the damage was simulated separately and for different

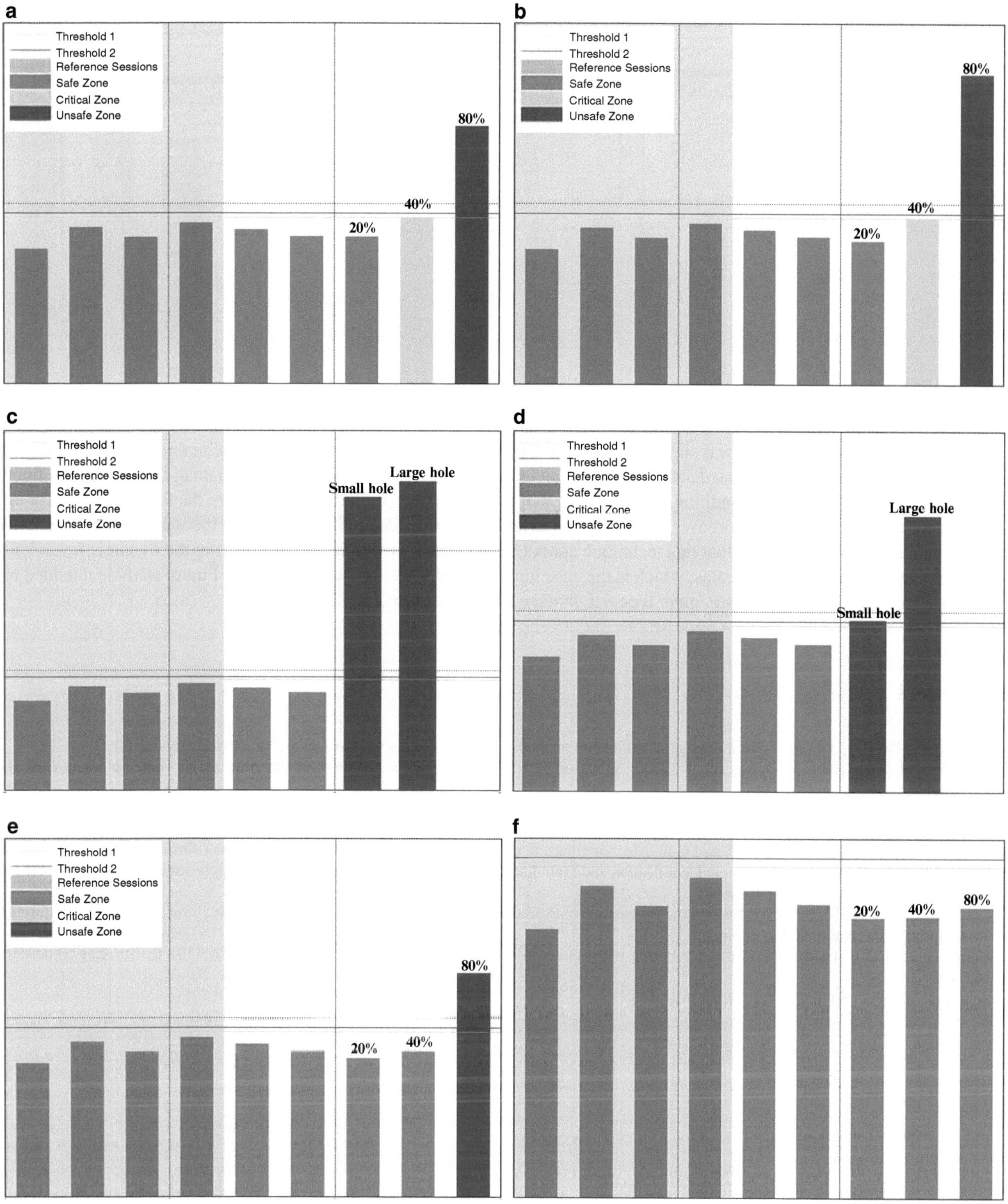


**Fig. 2.4** $\chi^2$ test from SSDD technique for different element types and damage ratios: (**a**) central span girder, (**b**) side span girder, (**c**) central span deck, (**d**) side span deck, (**e**) column and (**f**) cap beam

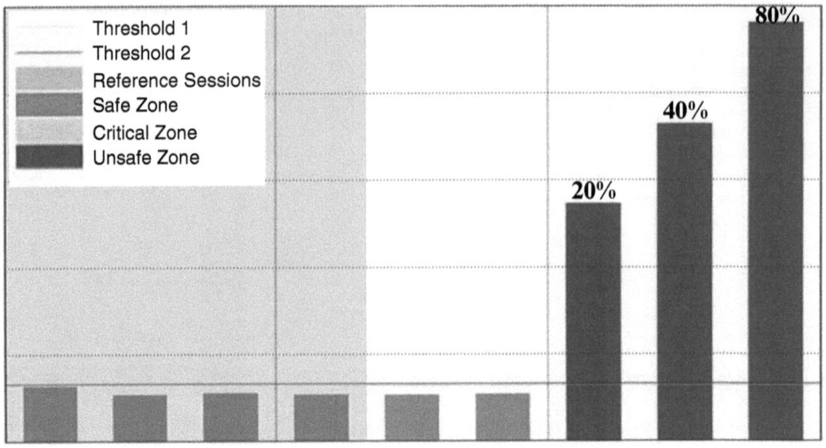


**Fig. 2.5** $\chi^2$ test using SSDD technique for different damage ratios in bearings

ratios. The damage ratio varied among minor (20 %), intermediate (40 %) and severe (80 %) conditions. For each case, a time history analysis was performed with white noise excitation imposed on the structure.

The ambient vibration test data was simulated by recording the acceleration time histories of the nodes typically measured in this type of structure. This simulated data was used as an input to the statistical subspace-based damage detection technique in order to estimate the damage condition. It was shown that this technique could detect the severe damage condition for all cases except cap beam damage. Moreover, in most of the cases the intermediate ratio damage was recognized successfully. It can be confirmed from the results that this technique cannot detect damages which are not affecting the dynamic behaviour of structure, e.g. damages in cap beams, which is the case for all global vibration-based methods. Future study is intended to investigate the effect of noise on data, more types of damage cases and ratios.

## References

1. Fan W, Qiao P (2011) Vibration-based damage identification methods: a review and comparative study. Struct Health Monit 10(1):83–111
2. Basseville M, Abdelghani M, Benveniste A (2000) Subspace-based fault detection algorithms for vibration monitoring. Automatica 36(1): 101–109
3. Döhler M, Mevel L, Hille F (2014) Subspace-based damage detection under changes in the ambient excitation statistics. Mech Syst Signal Process 45(1):207–224
4. Döhler M, Mevel L (2013) Subspace-based fault detection robust to changes in the noise covariances. Automatica 49(9):2734–2743
5. Döhler M (2011) Subspace-based system identification and fault detection: algorithms for large systems and application to structural vibration analysis. Dissertation, Université Rennes 1
6. Döhler M, Hille F (2014) Subspace-based damage detection on steel frame structure under changing excitation. In: Structural health monitoring, vol 5. Springer International Publishing, pp. 167–174
7. Döhler M et al (2014) Structural health monitoring with statistical methods during progressive damage test of S101 Bridge. Eng Struct 69: 183–193
8. Structural Vibration Solutions, http://www.svibs.com/

# Chapter 3
# An Improved Blind Source Separation for Structural Mode Identification Using Fewer Measurements

**Ayan Sadhu and Budhaditya Hazra**

**Abstract** In this paper, we present a novel underdetermined structural mode identification algorithm within the framework of Blind Source Separation (BSS) employing PARAllel FACtor (PARAFAC) decomposition. In the proposed method, BSS is employed on vibration measurements to generate an over-complete dictionary of multi-component bases, which are then used to estimate the modal parameters by utilizing the strong uniqueness property of PARAFAC decomposition. Results show significant computational efficiency of the proposed method over existing BSS algorithms, while substantially improving the upper bound of source separation capability of existing PARAFAC decompositions.

**Keywords** Underdetermined system identification • Blind source separation • Wavelet packets • PARAFAC decomposition • Multi-channel measurements

## 3.1 Introduction

Blind Source Separation (BSS) has recently emerged as a powerful signal processing tool to undertake blind system identification for civil and mechanical structures employing vibration measurements [1–5]. The basic objective in these problems is to estimate the vibration characteristics of flexible systems, which manifest themselves as natural frequencies, damping ratios and mode shapes (called modal parameters), from ambient measurements without the knowledge of input characteristics. Following the estimation of modal parameters, several approaches exist in the literature to undertake the condition assessment of such systems [6–8]. Classic BSS algorithms (e.g., SOBI [9]) primarily rely on classical linear algebra tools where a two-way array (matrix) representation of measurements and their decomposition (e.g. principal component analysis, joint diagonalization) thereof, are utilized. Recently PARAllel FACtor (PARAFAC) decomposition [10, 11], a higher-order signal processing tool based on tensor algebra, has shown significant promise in the area of source separation due to its strong uniqueness and identifiability properties. In this paper, a PARAFAC decomposition-based BSS algorithm is explored towards vibration based structural system identification using multi-channel measurements. Issues related to partial measurements and extraction of system properties are discussed in detail using experiments.

The field of structural system (mode) identification has seen a progressive shift in recent years from centralized batch processing algorithms to de-centralized algorithms [3, 12, 13]. This is because typical civil infrastructure such as multi-storied buildings and bridges require significant sensor and cable deployments in centralized processing, while de-centralized sensing and processing alleviates this need and can operate with significantly limited resources. However, many of the algorithms used in centralized settings cannot be readily ported to decentralized implementations, as the fundamental nature of the system of equations changes from an over-determined set to an underdetermined one [3, 14]. Furthermore, the problem of identifiability under partial sensor measurements (generally the case in decentralized measurements) is exacerbated when the data collected is noisy.

Recently PARAFAC-decomposition is explored for ambient modal identification [5, 15, 16]. In [15], the PARAFAC decomposition is utilized in conjunction with Bayesian model updating to estimate modal parameters. However, unlike the aforementioned study, the authors specifically address the problem of identification in decentralized settings [5] and propose a computationally efficient signal-processing framework for identification. Another key contribution of the paper

A. Sadhu (✉)
Department of Civil Engineering, Lakehead University, Thunder Bay, ON, Canada P7B 5E1
e-mail: ayansadhu.civil@gmail.com; asadhu@lakeheadu.ca

B. Hazra
Director of Research, SpectraQuest Inc., 8227– Hermitage road, Richmond, VA 23228, USA

© The Society for Experimental Mechanics, Inc. 2015

C. Niezrecki (ed.), *Structural Health Monitoring and Damage Detection, Volume 7*, Conference Proceedings of the Society for Experimental Mechanics Series, DOI 10.1007/978-3-319-15230-1_3

is to improve the underdetermined source separation capabilities at zero and finite lags, which cannot be achieved using standard PARAFAC decomposition [11], where the upper bound of source separation is restricted by the number of partial measurements.

Before presenting the details of our algorithm, let us first consider the standard instantaneous mixing BSS model [9]:

$$\mathbf{x}(k) = \mathbf{A}\mathbf{s}(k) \tag{3.1}$$

where, **A** is the mixing matrix and **s** represents the sources. The essence of BSS is to identify both **A** and **s** with only very broad assumptions regarding the statistical nature of **s**. Assuming that the system is linearly time-invariant, and the excitation is wide-band, it can be shown easily that **A** is a matrix of mode shapes and **s** contains the modal responses [17]. BSS tools such as Independent Component Analysis (ICA) and SOBI yield good results for the over-determined case (i.e., **A** is of rank $n$, $n$ being the number of sources to be identified). However, they are not directly applicable to the underdetermined case, which occur in decentralized implementations, where only partial sensor information is available (i.e., does not satisfy the full rank condition).

The essence of the present approach is to transform the data using an orthogonal wavelet basis into wavelet domain and then perform source separation in the transformed space. This approach has two advantages in addition to rendering the signal sparse—an improvement in the identified source quality for noisy signals, and the development of a common framework where a switch from full rank source separation to a rank-deficient one is relatively straightforward and computationally efficient.

## 3.2 Formulation of the Algorithm

Consider a dynamical system with $n_s$ degrees-of-freedom (DOF), subjected to an excitation force, $\mathbf{F}(t)$.

$$\mathbf{M}\ddot{\mathbf{x}}(t) + \mathbf{C}\dot{\mathbf{x}}(t) + \mathbf{K}\mathbf{x}(t) = \mathbf{F}(t) \tag{3.2}$$

where, $\mathbf{x}(t)$ is a vector of displacement coordinates at DOFs. $\mathbf{M}, \mathbf{C}$, and $\mathbf{K}$ are the mass, damping and stiffness matrices of the system. The solution to Eq. (3.2) can be expressed as a superposition of vibration modes (similar to Eq. (3.1) [1]:

$$\mathbf{x}_{n_m \times N} = \boldsymbol{\Phi}_{n_m \times n_s} \mathbf{q}_{n_s \times N} \tag{3.3}$$

where, $\mathbf{q} \in \Re^{n_s \times N}$ is a matrix of the corresponding modal coordinates, $\boldsymbol{\Phi}$ is the modal transformation matrix, $n_m$ is the number of measurements and $N$ is the number of measurement data points.

When all the measurements are available ($n_m \geq n_s$), by comparing Eq. (3.3) with Eq. (3.1), the covariance matrix $\mathbf{R}_{\mathbf{x}}(k)$ evaluated at time-lag $k$ can be written as:

$$\mathbf{R}_{\mathbf{x}}(k) = E\left\{\mathbf{x}(n)\mathbf{x}^T(n-k)\right\} = \mathbf{A}\mathbf{R}_{\mathbf{s}}(k)\mathbf{A}^T \tag{3.4}$$

where $\mathbf{R}_{\mathbf{s}}(k) = E\left\{\mathbf{s}(n)\mathbf{s}^T(n-k)\right\}$. For notational simplicity, we introduce the following definitions:

$$R_{x_i x_j}(k) = R^x_{ijk} \Longleftrightarrow R_{s_l}(k) = R^s_{kl} \tag{3.5}$$

For such a multi-dimensional representation of the system expressed by Eq. (3.2), its covariance matrix in Eq. (3.5) admits a PARAFAC decomposition by virtue of its representation as a three-way array. A three-way tensor is primarily decomposed into a sum of outer products of triple vectors, and can be succinctly expressed as:

$$R^x_{ijk} = \sum_{r=1}^{n_s} a_{ir} b_{jr} R^s_{kr} \Longleftrightarrow \mathbf{R}^{\mathbf{x}} = \sum_{r=1}^{n_s} \mathbf{a}_{\mathbf{r}} \circ \mathbf{b}_{\mathbf{r}} \circ \mathbf{R}^{\mathbf{s}}_{\mathbf{r}} \tag{3.6}$$

where $i \in [1\ \mathrm{I}]$, $j \in [1\ \mathrm{J}]$ and $k \in [1\ \mathrm{K}]$. This is described by a trilinear model of $\mathbf{R}^{\mathbf{x}}$, $\mathbf{R}^{\mathbf{x}} = [[\mathbf{A}, \mathbf{B}, \mathbf{C}]]$, where the matrices are given by $\mathbf{A} = (\mathbf{a}_1, \mathbf{a}_2, \ldots, \mathbf{a}_{\mathbf{R}})$, $\mathbf{B} = (\mathbf{b}_1, \mathbf{b}_2, \ldots, \mathbf{b}_{\mathbf{R}})$, and $\mathbf{R}^{\mathbf{s}}_{\mathbf{r}} = (\mathbf{R}^{\mathbf{s}}_{\mathbf{1}}, \mathbf{R}^{\mathbf{s}}_{\mathbf{2}}, \ldots, \mathbf{R}^{\mathbf{s}}_{\mathbf{R}})$. Due to *symmetric* and *positive definite* nature of $\mathbf{R}^{\mathbf{x}}$, $\mathbf{A} = \mathbf{B}$. Therefore, Eq. (3.6) represents the summation of $R$ rank-one tensors or PARAFAC components that are needed to fit the tensor [18, 19]. PARAFAC components are usually estimated by minimizing the quadratic cost function, achieved using alternating least squares [20] :

$$g(a,b,c) = \|\mathbf{R}^{\mathbf{x}} - \sum_{r=1}^{R} \mathbf{a_r} \circ \mathbf{b_r} \circ \mathbf{R_r^s}\|^2 \tag{3.7}$$

The resulting solutions yield the mixing matrix **A** and the auto-correlation function of $\mathbf{R_r^s}$ from which other modal parameters, e.g., natural frequencies and damping ratios are estimated. Unlike the traditional BSS form, PARAFAC component extracts the auto-correlation of sources explicitly. The PARAFAC decomposition yields a unique decomposition even if the rank is greater than the smallest dimension of the tensor, i.e., $n_m < n_s$ [11]. This property is utilized in the framework of BSS for underdetermined source separation using partial measurements as explained next.

### 3.2.1 *Efficient Underdetermined Source Separation*

Now consider the case when $n_m < n_s$. The measured signal from the $i$th measurement channel of Eq. (3.3) can be expressed as $x_i(t) = \sum_{l=1}^{n_s} A_{il} s_l(t); i = 1, 2, \ldots, n_m$, where, $A_{il}$ is a vector of mixing matrix coefficients corresponding to the $i$th degree of freedom (synonymous with a measurement channel). Considering the wavelet packet coefficients of the sensor responses and sources at a node $(j, v)$ of the wavelet packet tree, and applying the applicable orthogonality condition for the wavelet [3], we get $f_{k,i}^{j,v}(t) = \sum_{l=1}^{n_s} A_{il} e_{k,l}^{j,v}(t)$, for $i = 1, 2, \ldots, n_m$. In matrix vector form, it can be written as:

$$\mathbf{f}_{\alpha}^{j,v} = \mathbf{A_u e}_{\alpha}^{j,v} \tag{3.8}$$

where $j, k, v, i$ and $l$ represents the scale, shift, modulation, sensor and source index, respectively. $f_{k,i}^{j,v}$ and $e_{k,l}^{j,v}$ denote the wavelet packet coefficient of the $i$th channel and the $l$th source at a specific node $(j, v)$, respectively.

By utilizing the *shift-invariance* property of SWPT [3], the covariance matrix of SWPT coefficients of partial measurements $f_{\alpha}^{j,v}$ (i.e., $\mathbf{R}_{\mathbf{f}_{\alpha}^{\mathbf{j,v}}}(k)$) evaluated at time lag $(k)$ can be written as:

$$\mathbf{R}_{\mathbf{f}_{\alpha}^{\mathbf{j,v}}}(k) = \mathbf{A_j R}_{\mathbf{e}_{\alpha}^{\mathbf{j,v}}}(k) \mathbf{A_j}^T \tag{3.9}$$

Equation (3.9) can be further simplified [similar to Eq. (3.4)–(3.6)] to the following PARAFAC models using SWPT coefficients across the scale index and lag respectively, which is discussed in Sects. 3.2.1.1 and 3.2.1.2.

#### 3.2.1.1 Lag Zero PARAFAC Decomposition

The lag-zero PARAFAC model corresponding to Eq. (3.9) can be reformulated in terms of different packets across the scale levels, which results in $\mathbf{R}_{\mathbf{f}_{\alpha}^{\mathbf{j,v}}}(0) = \mathbf{A_j R}_{\mathbf{e}_{\alpha}^{\mathbf{j,v}}}(0) \mathbf{A_j}^T$. Using notational convention described in Eq. (3.5), the PARAFAC model can be expressed as:

$$\mathbf{R}_{ijv}^{\mathbf{f}} = \sum_{r=1}^{n_s} \mathbf{a_r^s} \circ \mathbf{a_r^s} \circ \mathbf{R_r^{e,v}} \tag{3.10}$$

PARAFAC modelling utilizing zero-lag covariance structure is akin to multi-resolution principal component analysis (PCA), where multi-resolution analysis allows for robust de-noising. The difference, however, is that by using PARAFAC the number of free parameters are reduced, contrary to PCA that would require extra degrees of freedom to include the multi-resolution representation. By avoiding other lags, the proposed PARAFAC model is rendered fast and efficient. It yields directly the mixing matrix without using any intermediate concatenation. The sources are estimated using a simple least square estimator on the mixing matrix and measurements. For partial responses, only the mixing matrix can be recovered using the identifiability of the PARAFAC, but sources cannot be recovered using a least squares estimator.

#### 3.2.1.2 Finite-Lag PARAFAC Decomposition

Unlike lag-zero decomposition, here the covariance with finite lags of the SWPT coefficients at a particular node $(j, v)$ are considered. Using notational convention along a scale level $j$ described in Eq. (3.5), the PARAFAC model of Eq. (3.9) can be expressed as [5, 16]:

$$\mathbf{R}^{\mathbf{f}}_{ijk} = \sum_{r=1}^{n_s^j} \mathbf{a}^{\mathbf{s}}_{\mathbf{r}} \circ \mathbf{a}^{\mathbf{s}}_{\mathbf{r}} \circ \mathbf{R}^{\mathbf{e,k}}_{\mathbf{r}} \tag{3.11}$$

where $n_s^j$ is the number of modes mixed in the SPWT coefficient pairs at the scale level $j$, and $n_m \leq n_s^j < n_s$. By comparing Eqs. (3.10) and (3.11), we can decompose $\mathbf{R}^{\mathbf{f}}$ into $n_s^j$ PARAFAC components. However, unlike the former case (i.e., Eq. (3.10)) where $\mathbf{A}_{n_m \times n_s}$ is identified in a single step, here (i.e., Eq. (3.11)) only a subset of the partial mixing matrix (i.e., $\mathbf{A}_{n_m \times n_m^u}$) with $n_s^j$ number of sources is estimated. An additional feature of this PARAFAC model is that it yields the auto-correlation function of $\mathbf{R}^{\mathbf{e}}_{\mathbf{r}}$ for $r = 1, 2, 3, \dots n_m^u$ directly, from which the natural frequency and damping ratios for individual modes can be estimated. Performing similar operations on other SPWT coefficients (i.e., for $v = 0, 1, 2, \dots, (2^j - 1)$) allows us to construct the complete partial mixing matrix (i.e., $\mathbf{A}_{n_m \times n_s}$) for $n_s$ modes. Using this approach, the source separation capability of PARAFAC is significantly improved. In PARAFAC-only method (i.e., Eq. (3.6)), the uniqueness property of PARAFAC applies globally to all the measurement setups. On the other hand, in the proposed method, the same applies only at the individual packet coefficient level. Therefore, while considering all the packet coefficients, the proposed method is able to separate all the structural modes and their associated partial mixing matrices using a limited number of sensors. This results in significantly improving the source separation capability of PARAFAC.

## 3.3 Experimental Results

For demonstration purposes, a simulation model of a 6-DOF building [3] ($n_s = 6$) is considered. We assume that only three measurements (at DOF) ($n_m = 3$) are available. $db5$ is chosen as the wavelet basis. The required scale level, which is a function of the sampling frequency ($f_s = 100\,\text{Hz}$) and the lowest harmonic of the structure (i.e., $f_l = 0.78\,\text{Hz}$), is 7 (i.e., $s = 7$). However, we choose a lower scale level of 4 (i.e., $j = 4$) and apply the proposed method. Obviously, this will result in mode-mixing which the proposed algorithm can handle, where traditional BSS methods will fail to separate the sources. Due to this reduction in the scale level, the number of SWPT coefficients is reduced from 128 to 16, which results in reducing the computational overhead considerably.

Figure 3.1 shows the Fourier transform of the resulting SPWT coefficient pairs, $f_3^{4,1}$, $f_4^{4,1}$ and $f_5^{4,1}$, which indicates that there is a significant amount of mode mixing in the three sources, however, the sources are separated well using Eq. (3.11) under rank-3 ($R = 3$) PARAFAC decomposition as shown in Fig. 3.2 and the sub-set ($\bar{\Phi}_{3\times 3} = A_j$) of the partial mixing matrix ($\bar{\Phi}_{3\times 6}$) is obtained by concatenating the associated $\mathbf{a}_i$'s. When we repeat this exercise for other SWPT coefficients of $i = 5, 4, 3$ floor measurements, the complete partial mixing matrix ($\Phi_{3\times 6}$) is obtained. Figure 3.3 shows the separated sources using four successive measurements with rank-4 ($R = 4$) PARAFAC decomposition. It may be noted that with increasing number of sensors, the scale level for SWPT can be further reduced by utilizing the identifiability of PARAFAC decomposition [11]. In this case, we use scale level only up to $j = 1$ which yields the corresponding partial mixing matrix (i.e., $\Phi_{4\times 6}$). Accuracy of the mixing matrix is evaluated using the modal assurance criteria (i.e., MAC [21]):

$$\text{MAC}_i = \frac{(\psi_i^T \hat{\psi}_i)^2}{(\psi_i^T \psi_i)(\hat{\psi}_i^T \hat{\psi}_i)} \tag{3.12}$$

where $\psi_i$ and $\hat{\psi}_i$ represent $i$th true and estimated mode shape vector respectively. A MAC value of 1 indicates perfect correlation, and the values greater than 0.95 are usually considered good.

Figure 3.4 shows the performance of the proposed method under different measurement cases with various noise environments. While the PARAFAC-only method (i.e., Eq. (3.6)) is able to separate 6 sources using $n_m \geq 4$ [11], it is seen that irrespective of the number of measurement setups, the proposed method yields accurate modal identification (i.e., MAC values of 0.99), even with SNR=10. On the other hand, the performance of PARAFAC-only method is sensitive to noise.

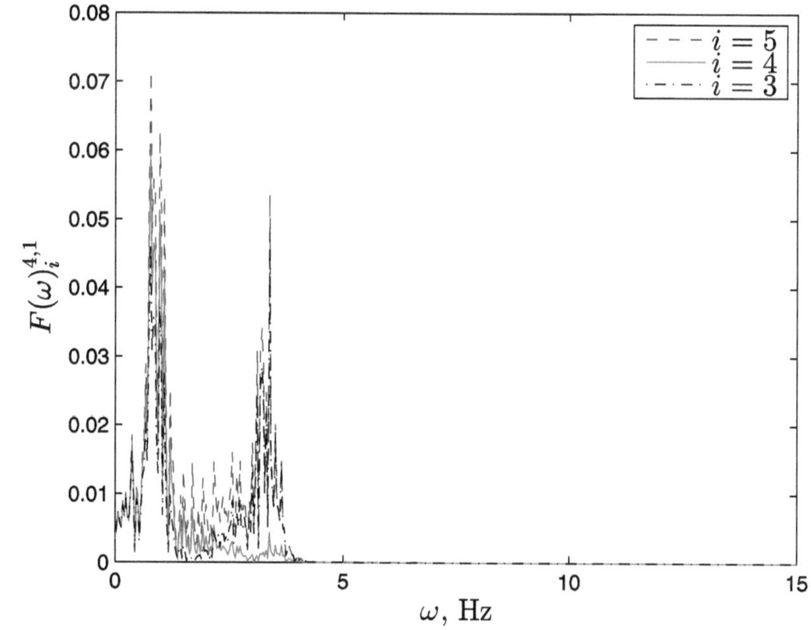


**Fig. 3.1** Fourier transforms of SPWT coefficients

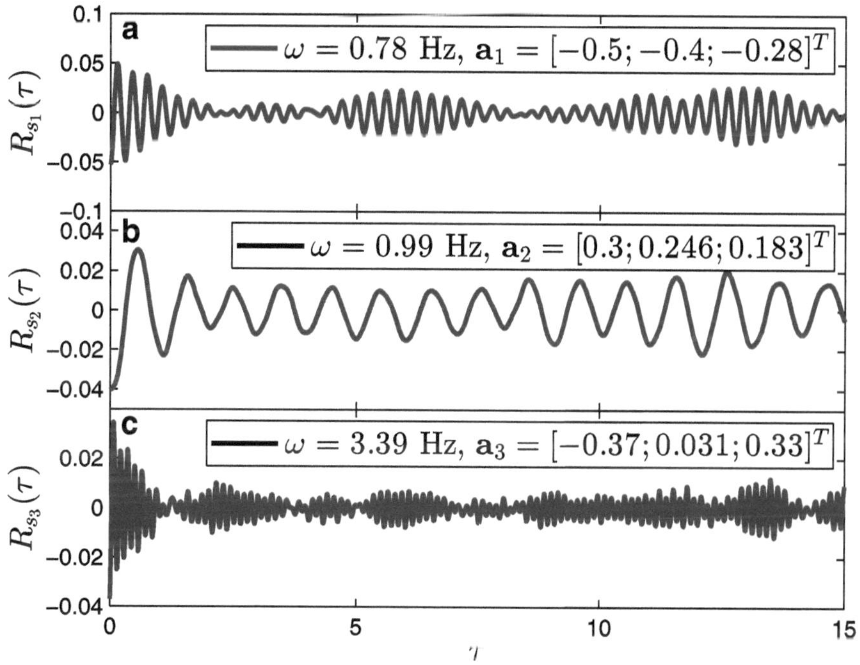


**Fig. 3.2** Source separation using rank-3 PARAFAC decomposition

It should be noted that a maximum of four sources can be extracted from three measurements using PARAFAC-only method [11]. On the other hand, using the proposed method, we implement strong uniqueness conditions only at individual SWPT coefficients, which enables us to identify all six modes using just three sensors. Thus the source separation capability of PARAFC is improved using sparse BSS. Table 3.1 shows the comparison of lag-zero and finite-lag PARAFAC decompositions for SNR=5. The lag-zero method is more computationally efficient than the finite-lag method because for the latter, the main loop has to run $\eta$ times (where $\eta$ is the number of lags). It is also interesting to see that when the number of measurements ($n_m$) is increased, the required scale level (i.e., $j$) as well as the number of required SWPT coefficients is reduced, thereby reducing the computational effort.

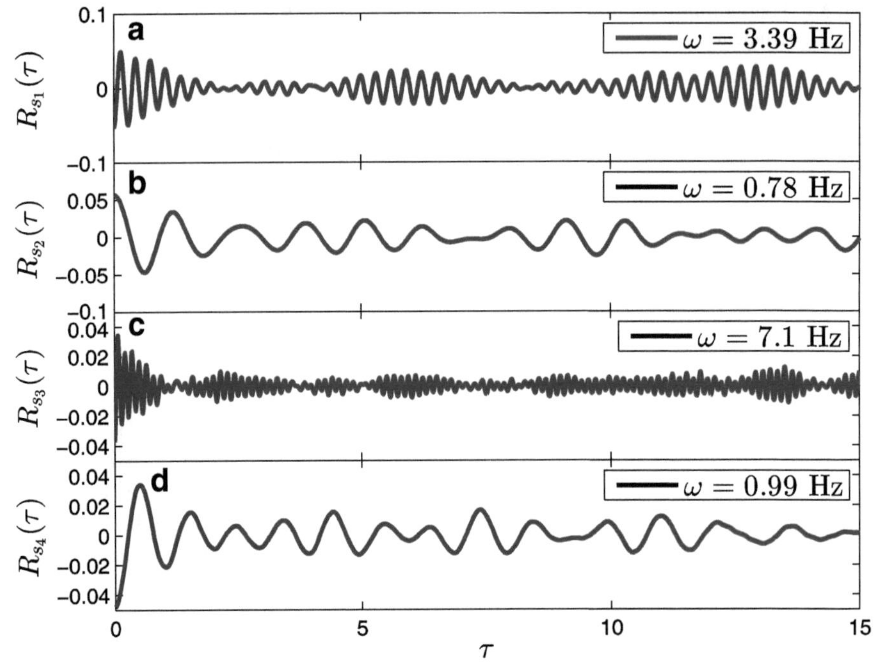


**Fig. 3.3** Source separation using rank-4 PARAFAC decomposition

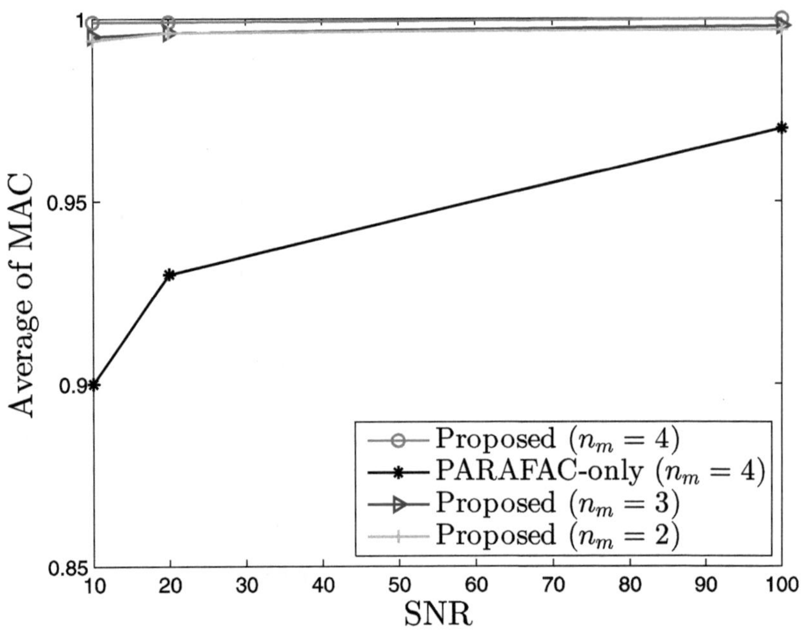


**Fig. 3.4** Performance of the proposed method

**Table 3.1** Relative performance of the proposed methods using SNR=5

| | Finite lag | | | | Zero lag | | | |
|---|---|---|---|---|---|---|---|---|
| $n_m$ | $R$ | $j$ | MAC | Time (s) | $R$ | $j$ | MAC | Time |
| 3 | 3 | 4 | 0.962 | 90 | – | – | – | – |
| 4 | 6 | 1 | 0.955 | 120 | 6 | 3 | 0.99 | 1.25 s |
| 6 | 6 | – | 0.953 | 138 | 6 | 2 | 0.99 | 1.54 s |

## 3.4 Conclusions

An under-determined system identification method employing limited multi-channel measurements is proposed within the framework of BSS in conjunction with PARAFAC decomposition. Results demonstrate significant improvement in reducing the computational burden compared to other popular methods, as a result of utilizing the over-complete dictionary of wavelet packet coefficients, even under the presence of substantial measurement noise and sources with low energy content. It is also observed that the proposed method improves the upper bound of source separation capability of PARAFAC, and hence holds significant promise in the area of ambient modal identification of structures.

## References

1. Antoni J (2005) Blind separation of vibration components: principles and demonstrations. Mech Syst Signal Process 19(6):1166–1180
2. Hazra B, Sadhu A, Roffel AJ, Narasimhan S (2012) Hybrid time-frequency blind source separation towards ambient system identification of structures. Comput Aided Civ Infrastruct Eng 27(5):314–332
3. Sadhu A, Hazra B, Narasimhan S, Pandey, MD (2011) Decentralized modal identification using sparse blind source separation. Smart Mater Struct 20(12):125009
4. Sadhu A (2013) Decentralized ambient modal identification of structures. PhD Thesis, Department of Civil Engineering, University of Waterloo
5. Sadhu A, Hazra B, Narasimhan S (2013) Decentralized modal identification of structures using parallel factor decomposition and sparse blind source separation. Mech Syst Signal Process 41(1–2):396–419
6. Doebling S, Farrar C, Prime M, Shevitz, D (1996) Damage identification and health monitoring of structural and mechanical systems from changes in their vibration characteristics: a literature review. Technical Report LA-13070-MS, Los Alamos National Laboratory, Los Alamos, NM, 1996
7. Carden EP, Fanning P (2004)Vibration based condition monitoring: a review. Struct Health Monit 3(4):355–377
8. Nayeri RD, Masri SF, Ghanem R, Nigbor RL (2008) A novel approach for the structural identification and monitoring of a full-scale 17-story building based on ambient vibration measurements. Smart Mater Struct 17(2):1–19
9. Belouchrani A, Abed-Meraim K, Cardoso J, Moulines E (1997) A blind source separation technique using second-order statistics. IEEE Trans Signal Process 45(2):434–444
10. Lathauwer LD (2006) A link between the canonical decomposition in multilinear algebra and simultaneous matrix diagonalization. Siam J Matrix Anal Appl 28(3):642–666
11. Lathauwer LD, Castaing J (2008) Blind identification of underdetermined mixtures by simultaneous matrix diagonalization. IEEE Trans Signal Process 56(3):1096–1105
12. Magalhaes F, Caetano E, Cunha A (2007) Challenges in the application of stochastic modal identification methods to a cable-stayed bridge. J Bridg Eng ASCE 12(6):746–754
13. Sim SH, Spencer BF, Zhang M, Xie H (2009) Automated decentralized modal analysis using smart sensors. Struct Control Health Monit 17(8):872–894
14. Sadhu A, Narasimhan S (2014) A decentralized blind source separation algorithm for ambient modal identification in the presence of narrowband disturbances. Struct Control Health Monit 21(3):282–302
15. Abazarsa F, Ghahari SF, Nateghi F, Taciroglu E (2013) Response-only modal identification of structures using limited sensors. Struct Control Health Monit 20(6):987–1006
16. Sadhu A, Hazra B, Narasimhan S (2014) Ambient modal identification of structures equipped with tuned mass dampers using parallel factor blind source separation. Smart Struct Syst 13(2):257–280
17. Antoni J, Garibaldi L, Marchesiello S, Sidhamed, M (2004) New separation techniques for output-only modal analysis. Shock Vib 11:227–242
18. Carroll JD, Chang JJ (1970) Analysis of individual differences in multidimensional scaling via an N-way generalization of Eckart-Young decomposition. Psychometrika 35:283–319
19. Harshman RA (1970) Foundations of the PARAFAC procedure: Models and conditions for an explanatory multi-model factor analysis. UCLA working papers in Phonetics 16:1–84
20. Smilde A, Bro R, Geladi P (2004) Multi-way analysis with applications in the chemical sciences. Wiley, West Sussex
21. Maia NMM, Silva JMM (1997) Theoretical and experimental modal analysis. Research Studies Press, UK

# Chapter 4
# Real Time NDE of Cold Spray Processing Using Acoustic Emission

**Patrick L. Clavette, Michael A. Klecka, Aaron T. Nardi, Greg C. Ojard, and Richard S. Gostautas**

**Abstract** As interest continues to grow in Additive Manufacturing (AM) processes, the inspection challenges that arise in this class of material need to be addressed. This must be balanced by considering cost and finding defects as soon as feasible in the process to minimize downstream production impacts. As a means to consider defect detection as early as possible, acoustic emission (AE) was used to monitor the cold spray powder deposition process. The concept was to observe the article being coated for the acoustic signature of internal cracking or interface delamination, resulting in a quick and cost-effective means to detect problems during cold spray parameter development and ultimately during production. Real time monitoring was used to delineate between particle impacts and more significant events such as delamination and internal cracking. The results of this study show that AE can be used to monitor processes in extremely noisy environments. Opportunities for extension to other additive manufacturing processes will be discussed.

**Keywords** Cold spray • Acoustic emission • Additive manufacturing • Process diagnostics • Real time monitoring

## 4.1 Introduction

Additive manufacturing processes are becoming increasingly used in industry. Cold spray (Fig. 4.1) is one process which is beginning to garner attention. It can be used for creating new parts, repairing structures, or coating surfaces of a part. To deposit a layer of material, feedstock powder particles are accelerated in a supersonic jet of compressed cold or pre-heated gas to high velocities, and then directed toward a substrate. Upon impact with the substrate, the powder particles deform plastically and bond to the surface. The supersonic gas jet is produced by using a converging-diverging de Laval nozzle [1]. The expansion of the gas in the divergent section of the nozzle will reduce the gas temperature, resulting in a process where the particles remain in a solid state prior to impacting the substrate, thus reducing oxidation of the deposited layer and percentage of volumetric porosity [2, 3]. Unlike alternative additive manufacturing techniques and thermal spray processes, the solid state cold spray deposition generally retains the chemistry, phase composition, and grain structure of the precursor powder material.

Experimental observations indicate that particle bonding with the substrate occurs only when the particle velocity exceeds a critical value. This value, termed “critical velocity”, has been identified for several materials in the literature [4–7]. The critical velocity is considered to be a function of material properties, particle size, initial impact temperature, and melting temperature [6, 7]. For most of the materials deposited by cold spray to date, the critical velocity is in the range of 150–900 m/s [8].

Acoustic emission (AE) is defined as transient elastic stress waves generated from local internal micro-displacements in a material [9]. These waves can be measured using AE sensors that detect changes in the material. These sources may include crack initiation and/or growth, crack fretting (i.e. rubbing of crack faces as it opens and closes), corrosion, activation of weld discontinuities, and impact. Other potential AE sources can produce “noise” which include friction between parts, leaks, personnel activity, as well as other mechanical operations.

In the case of cold spray material deposition, potential AE sources may include the gas jet, particle impact with substrate, and delamination of the deposition, pitting and chipping of the deposition layer. Regardless of the type of source, AE is

P.L. Clavette (✉) • M.A. Klecka • A.T. Nardi • G.C. Ojard
United Technologies Research Center, 411 Silver Lane, MS 129-73, East Hartford, CT 06108, USA
e-mail: ClavetPL@utrc.utc.com; KleckaMA@utrc.utc.com; NardiAT@utrc.utc.com; OjardGC@utrc.utc.com

R.S. Gostautas
Mistras Group, Inc., 195 Clarksville Road, Princeton, NJ 08550, USA
e-mail: richard.gostautas@mistrasgroup.com

© The Society for Experimental Mechanics, Inc. 2015

C. Niezrecki (ed.), *Structural Health Monitoring and Damage Detection, Volume 7*, Conference Proceedings of the Society for Experimental Mechanics Series, DOI 10.1007/978-3-319-15230-1_4

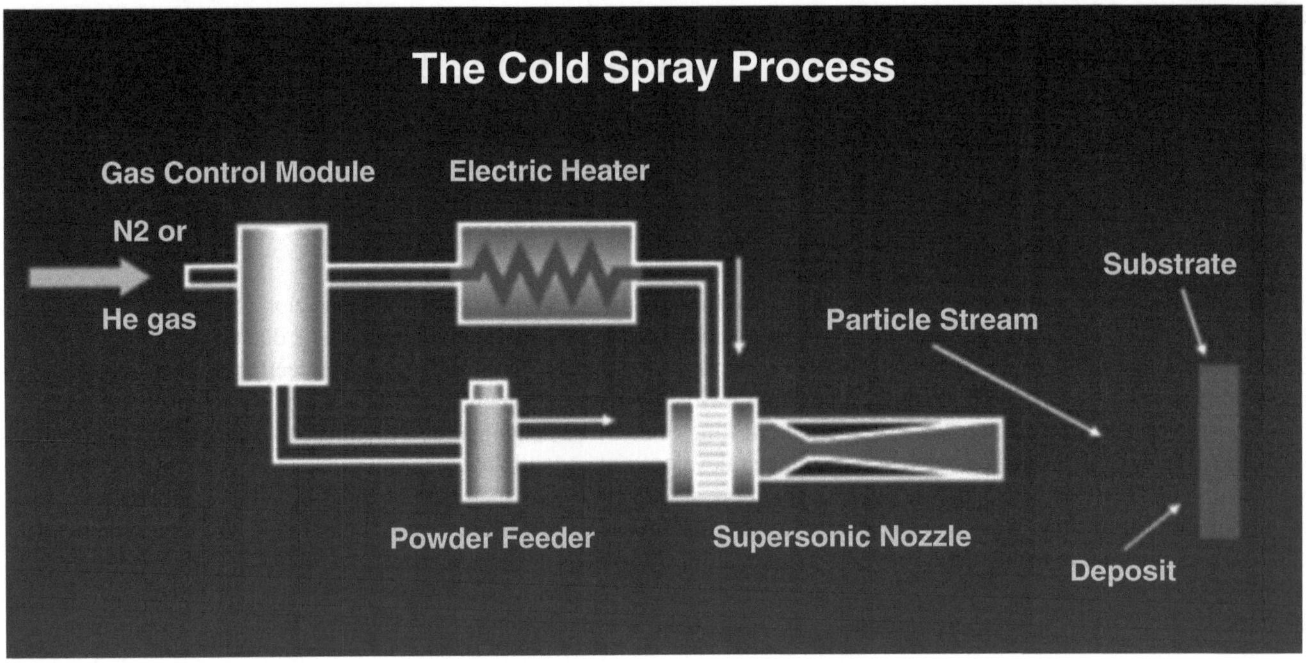


**Fig. 4.1** Schematic of cold spray process, including gas heater, powder feed, and nozzle (www.arl.army.mil)

detected by transducers that are coupled to the material which converts the mechanical energy (e.g. elastic waves) into an electrical signal that is transmitted by transducer cables to the AE data acquisition system (DAQ). Due to the fact that most AE sources are relatively weak, a pre-amplifier boosts the voltage of the signal of the waveform that passes across the transducer. The passing of the waveform is recorded as an AE hit by the DAQ and recorded. The waveform is decomposed into a set of features that include the amplitude, energy, duration, rise time, peak frequency, and frequency centroid.

Successful examples of the uses of AE for subtractive manufacturing processes, forming, and surface modification are documented in the literature [10–23], but at present there exist few if any examples of the technique used for AM processes. Consequently, there is a need for research in the area.

## 4.2 Experimental Approach

AE testing was conducted using the Micro II data acquisition system, from Mistras Group Inc. Two different types of sensors were used for the collection of AE data. These included the WD, wideband (125–1,000 kHz), differential sensor and the μ30 (150–400 kHz), miniature sensor. Each sensor was connected to a 2/4/6 preamplifier that allows the user to boost the AE signal by 20, 40 or 60 dB, selectable by user input. Each substrate sample was approximately $8'' \times 2'' \times 0.25''$ that was constructed of either 1018 steel or 6061 aluminum. The deposition area was approximately $1''$ high with the centerline of the spray area centered along the length of the sample. The sensors were mounted to the backside of the sample approximately $2''$ removed from the center of the spray area with the WD mounted below (Channel 1) and the μ30 located above (Channel 2) (Fig. 4.2). The sensors were coupled to the substrate using a cyanoacrylate adhesive (Vishay Precision Group MBond-200). AE data was collected with AEwin software, also from Mistras Group Inc.

Prior to the start of any AE test, the acoustic coupling of each mounted sensor was verified using the pencil lead break (PLB) test [24]. The PLB test provides a known stress release that generates an acoustic wave in the material. The background noise check was conducted at the beginning of each test in order to identify, and when possible, minimize outside background noise sources either through hardware or software filtering. For AE testing, material deposition represents a high noise environment due not only to the presence of the gas jet used to carry the powder, but the powder particle impacts from the deposition process. Thus the initial testing looked specifically at the AE activity generated by the process in order to lessen some of the activity through the use of sensor orientation, front end filters, analog filters, threshold and preamplifier settings.

**Fig. 4.2** Showing sensor positioning relative to spray band

The cold spray equipment used to produce deposits in the present investigation was a high pressure system designed and constructed by UTRC. The system includes gas heater, gun/nozzle, powder feeder, and controller, all capable of a maximum operating condition of 40 bar pressure and 800 °C temperature. The nozzle-substrate standoff distance for all spraying was in the range of 18–25 mm. In the system, either nitrogen or helium was utilized as the primary accelerating gas and powder carrier gas. For the purposes of the study herein, the powder used was a 90Ta-10W alloy powder specifically designed for the cold spray process.

During acquisition of the AE, initial analysis of the data was performed through the use of comparison graphs, which allowed observation and identification of potential trends in the data in real-time. These trends may have been caused by changes in the test parameters, or more importantly, changes in condition of the material during the AM process. The comparison graphs typically examined a set of AE features and their changes over time or in relation to each other. These graphs included, amplitude vs. time, hits vs. time, frequency centroid vs. time, hits vs. amplitude, among others. Another form of analysis for data discrimination is the evaluation of the waveforms of the detected hits. In most cases mechanical/frictional noise will produce a "continuous" type of AE source emission. This type of emission is normally lower in amplitude and longer in duration. Conversely, emission associated with active cracking (micro/macro) and crack propagation will have a "burst" like emission. Burst emission has a clear and defined initiation, peak and end to the signal. This type of emission is typically higher in amplitude, energy and shorter in duration. Micro-cracking may be lower in amplitude but will still show a "burst" like waveform. Thus, when used with the comparison graphs, waveform discrimination can differentiate between different types of AE sources. However, for material deposition, due to the energy imparted by the jet, there exists a continuous type of emission. As such, waveform discrimination alone is not enough for detecting potential changes in the process.

## 4.3 Experimental Results and Discussion

In the initial stages of testing, the deposition process produced a high noise environment that often saturated the sensors (i.e. constant 100 dB hits). Not only did this make AE difficult for monitoring this type of process, it increased the difficulty in evaluation and analysis of changes in AE response due to variations in the test parameters. However, the following changes allowed testing to move forward: (1) Sensors were reversed – rather than having the crystal face of the sensor bonded to the surface the substrate, each sensor was turned upside down. This had the effect of attenuating the signal by 10–12 dB, thereby reducing saturation of the sensors. (2) Switched the 2/4/6 preamplifier to the lowest gain (20 dB) to reduce input signal, increasing the upper dynamic range. (3) Increasing the software set threshold to above 80 dB. (4) Setting the analog filters at 100–400 kHz band pass.

During the testing it was found that nitrogen (gas jet) produced constant background noise between 70–75 dB (cold), and 77–82 dB (hot) with the nozzle not impinging directly on the substrate. For helium, this constant background noise increased to 85–89 dB (hot). Consequently, the threshold was raised for the helium case.

Due to the high noise and frequent sensor saturation even notwithstanding the measures taken to deal with them, the act of source location, that is the determination of the spatial position of the AE source, was not possible. The continuous emission from the process precluded an accurate determination of the arrival times of the waveforms to the sensor. The small scale of the area sprayed in comparison to the separation of the sensors as well as the fact that the sensors were mounted inverted also would have limited the spatial resolution, due to loss of a clearly defined pulse wave front. The use of two different types of sensors, also made assessment of the zonal source location using relative magnitude of the AE signal difficult.

### *4.3.1 Analysis of a 90Ta-10W on Steel with Parameters Known to Result in Poor Bonding*

For this test, helium gas heated to 585 °C at pressure of 15 bar was used as the spray gas. Steel substrates were known to be difficult to bond for this particular powder and spray conditions and, delamination was expected to occur during the spray process. Fifty grams of powder were placed into the powder feeder and spraying was allowed to continue until all powder was consumed. Referring to Fig. 4.3, numerous distinct periods were visible, corresponding to different stages of the application process. These were as follows: (A) nitrogen purge, (B) helium purge, (C) powder deposition with low amplitude AE, (D) powder deposition with higher amplitude AE, (E) post-delamination powder deposition, (F) movement of the nozzle away from the surface and conclusion of the test. In the course of the test, the nozzle was only directly impinging upon the substrate during periods C, D, and E. Examining the plots, the AE amplitude from the nitrogen in period A was around 81 dB for the WD sensor and around 75 dB for the μ30 sensor. This increased to 92 and 85 dB respectively in period B. Concomitant changes to the frequency centroid were also noted: a decrease from 167 to 159 kHz for the WD sensor and a decrease from 207 to 204 kHz for the μ30 sensor. When the coating was built (period C), the AE amplitudes increased substantially: 95 dB for the WD sensor and 90 dB for the μ30 sensor. Substantial scatter in the distribution of frequency centroids was seen here, indicating multiple likely AE sources which would be expected to include: particle impacts and associated plastic deformation, gas flow, and possibly microcracking. In period D, a sharp increase in the AE amplitude was noted, culminating in the delamination of the coating at the period D – E transitional stage. Also, late in period C and throughout period D, an increase in the frequency centroid was noted, which could indicate cracking with gradually increasing AE amplitude. Period E was characterized as another coating building area, but in this case the frequency centroid showed more scatter. Toward the end of period E, the amplitude increased substantially, but no additional delamination occurred prior to the powder running out at the period E-F transition. In period F, the powder gun was moved away from the substrate and the test concluded.

### *4.3.2 Analysis of a 90Ta-10W on Aluminum with Parameters Known to Result in Good Bonding*

For this test, the same spray conditions as the previous trial were used except that the first pass group was performed with cold He gas. Aluminum was known to be amenable for bonding for this particular powder and spray conditions. Consequently, no delaminations were expected to occur during the spray process. Fifty grams of powder were placed into the powder feeder and spraying was allowed to continue until all powder was consumed. Referring to Fig. 4.4, numerous distinct periods were visible, these were as follows: (A) nitrogen purge, (B) helium purge, (C) powder deposition (with cold He gas) with fairly high amplitude AE, (D) movement of the nozzle away from the surface followed by switching from helium to nitrogen, (E) helium purge in preparation for second run, (F) second powder deposition using hot He gas, (G) a third powder deposition period using hot He gas, (H) movement of the nozzle away from the surface and the conclusion of the test. Examining the plots, the AE amplitude from the nitrogen in period A was around 75 dB for the WD sensor and around 67 dB for the μ30 sensor. This increased to 85 and 76 dB respectively in period B. Concomitant changes to the frequency centroid were also noted: a decrease from 152 to 142 kHz for the WD sensor and a decrease from 183 to 170 kHz for the μ30 sensor. When the coating was built (period C), the AE amplitudes increased substantially: 100 dB for the WD sensor and 96 dB for the μ30 sensor. Substantial scatter in the distribution of frequency centroids was seen here, indicating that there were likely multiple AE sources which were expected to include: particle impacts and associated plastic deformation and gas flow. In period D, a sharp decrease in the AE amplitude and frequency centroid was noted with final values comparable to period A. Period E showed similar behavior to period B as expected. Period F was characterized as another coating building area with similar characteristics to period C, except in this case the frequency centroid shows variation, being much higher than the prior application, but then settling to a range seen in period C. A subsequent group of passes in period G showed behavior which returned to the levels seen in period C and finally, period H showed behavior similar to that at the end of period D. It should

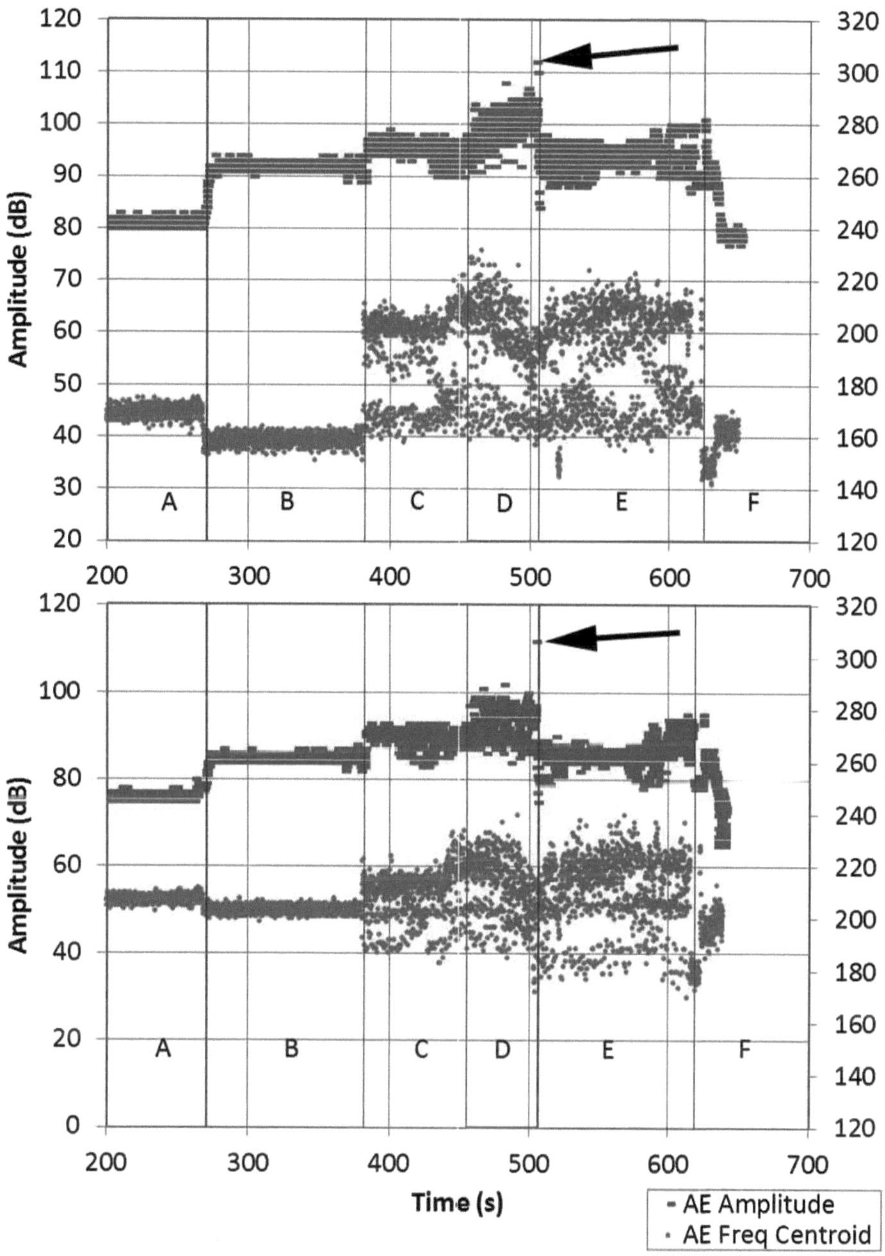


**Fig. 4.3** AE for 90Ta-10W sprayed on mild steel. *Top plot* is the WD sensor, *bottom* is the μ30 sensor. Periods correspond to following (*A*) nitrogen purge, (*B*) helium purge, (*C*) powder deposition with low amplitude AE, (*D*) powder deposition with higher amplitude AE (delaminating at the end of this period as shown by the *arrow*), (*E*) post-delamination powder deposition, (*F*) movement of the nozzle away from the surface and conclusion of the test

be stressed here, that throughout the course of application of the three different coating pass groups, that the overall AE amplitude did not change appreciably and there was no delamination. It is also noted that the overall AE amplitude during the spray process (at a time when no cracking is occurring) was higher in the aluminum case as compared to the steel case. This was considered a direct consequence of the higher transmissibility of the energy from the impacts into the aluminum substrate, which has considerably lower acoustic impedance than the steel of the prior experiment.

### *4.3.3 Analysis of a 90Ta-10W on Aluminum with Powder Which Contained Large Agglomerates*

For this test, the exact same parameters were used with only hot He gas. One major difference was the use of powder containing large particles, as might be expected in a low quality powder. Past experience indicates that the presence of large particles results in pitting of the sample surface, resulting in lower deposit quality. The consideration was to assess if AE

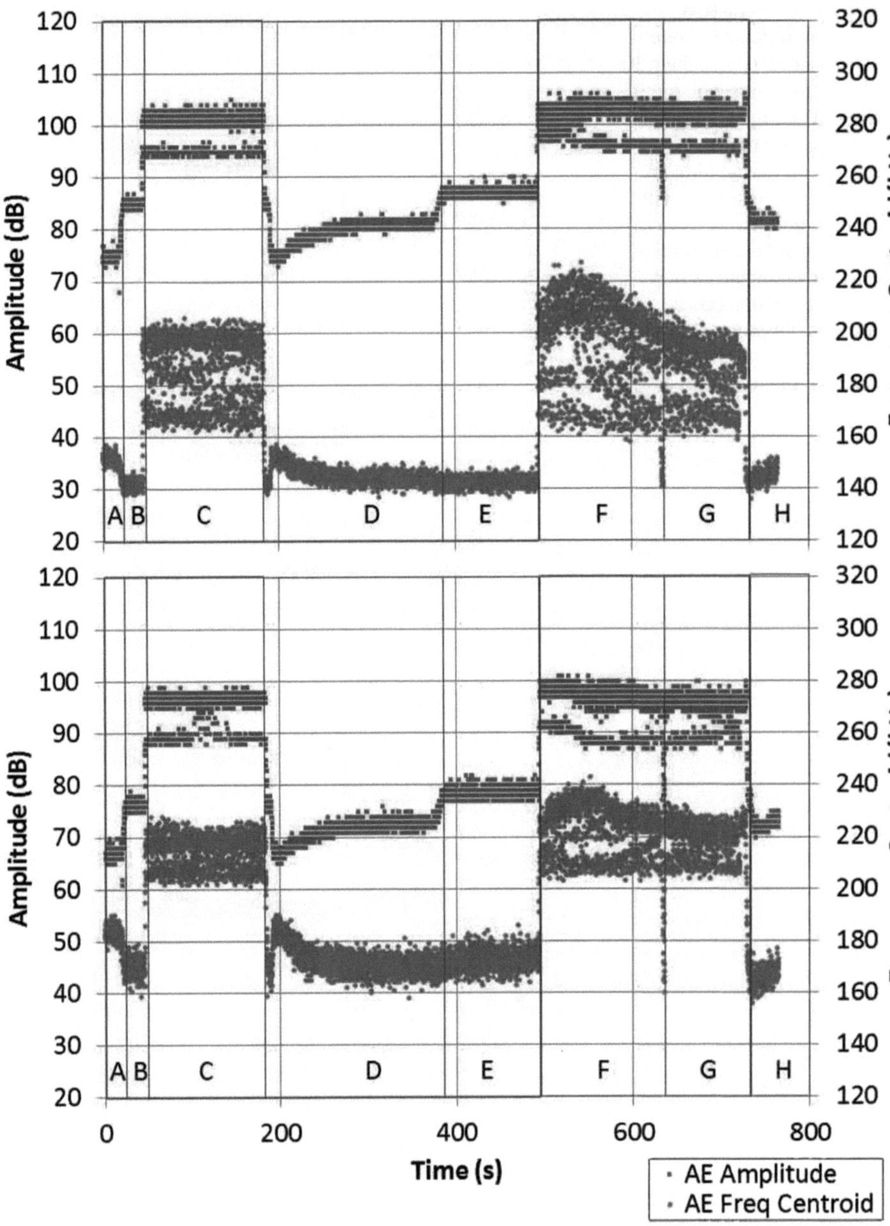


**Fig. 4.4** AE for 90Ta-10W sprayed on aluminum. *Top plot* is the WD sensor, *bottom* is the μ30 sensor. Periods correspond to following: (*A*) nitrogen purge, (*B*) helium purge, (*C*) powder deposition using cold He with fairly high amplitude AE, (*D*) movement of the nozzle away from the surface followed by switching from helium to nitrogen and addition of another 50 g of powder, (*E*) helium purge in preparation for second run, (*F*) second powder deposition with hot He, (*G*) a third powder deposition (hot He) period with no inter-layer purges, (*H*) movement of the nozzle away from the surface from the surface and conclusion of the test

could be used to detect in real time, potential issues related to feedstock powder. Fifty grams of powder were placed into the powder feeder and spraying was allowed to continue until all powder was consumed. Referring to Fig. 4.5, numerous distinct periods were visible, these were as follows: (A) nitrogen purge, (B) helium purge, (C) powder deposition with fairly high amplitude AE, (D) a second period of powder deposition, and (E) movement of the nozzle away from the surface and conclusion of the test. Examining the plots, the AE amplitude from the nitrogen in period A was around 79 dB for the WD sensor and around 72 dB for the μ30 sensor. This increased to 86 and 79 dB respectively in period B. For this gas switch, changes to the frequency centroid were seen, but significantly less than the prior run. The reasons for this were unclear. When the coating was built (period C), the AE amplitudes increased substantially: 101 dB for the WD sensor and 98 dB for the μ30 sensor. The frequency centroid again exhibited a significant shift lower as the coating built. What should be noted here though are some outlier events with amplitudes in the range 105–112 dB, which contrasts with the prior run without large agglomerates. This was better seen in Fig. 4.6. The outlier events were associated with the impacts of the large agglomerates with the surface. It should be stressed here, that throughout the course of application of the two different coating pass groups, that the overall AE amplitude did not change appreciably. It is also noted that the AE amplitude during the spray process again was higher in the aluminum case as compared to the steel case.

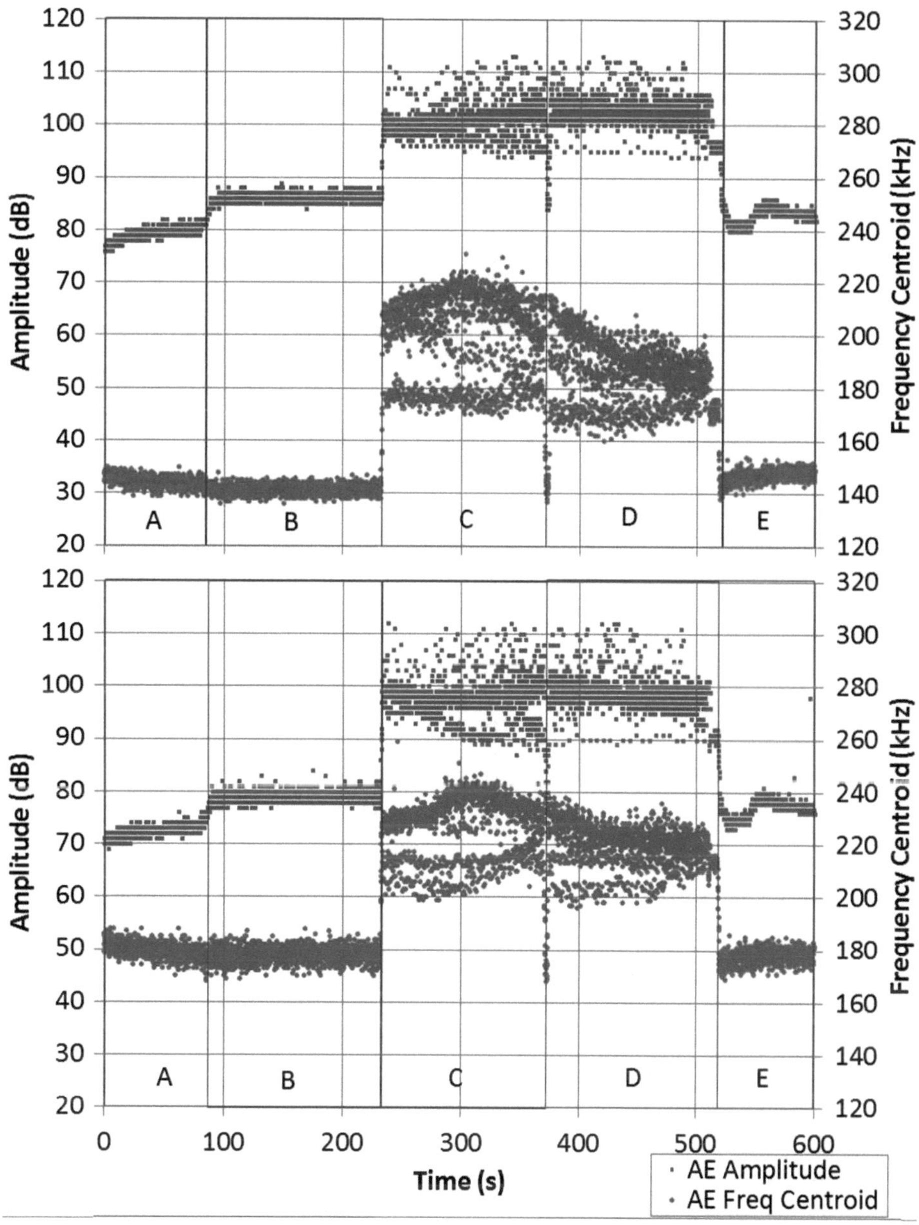


**Fig. 4.5** AE for 90Ta-10W containing large agglomerates sprayed on aluminum. *Top plot* is the WD sensor, *bottom* is the μ30 sensor. Periods correspond to following: (*A*) nitrogen purge, (*B*) helium purge, (*C*) powder deposition with fairly high amplitude AE, (*D*) a second period of powder deposition, and (*E*) movement of the nozzle away from the surface and conclusion of the test

### 4.3.4 Analysis of a 90Ta-10W on Aluminum with Powder Which Contained Large Agglomerates and Delaminated

In a test essentially identical to the previous experiment, a delamination occurred late in the deposition process. Referring to Fig. 4.7, numerous distinct periods were visible, these were as follows: (A) nitrogen purge, (B) helium purge, (C) powder deposition with fairly high amplitude AE, (D) a second period of powder deposition leading up to delamination, and (E) a period of poor coating (there were small fragments spalling from the deposit) leading up to the powder running out, (F) a period when the nozzle was still making passes on the substrate with no powder in the feeder, and (G) movement of the nozzle away from the surface and conclusion of the test. Examining the plots, the AE amplitude from the nitrogen in period A was around 79 dB for the WD sensor and around 73 dB for the μ30 sensor. This increased to 87 and 80 dB respectively in period B. For this gas switch, small changes to the frequency centroid were seen. When the coating was built (period C –first pass group and D – second pass group), the AE amplitudes increased substantially: 103 dB for the WD sensor and 98 dB for the μ30 sensor. The frequency centroid again exhibited a significant shift lower as the coating built. The coating delamination occurred at the D-E transition with a substantial reset of the frequency centroid higher. What should be noted

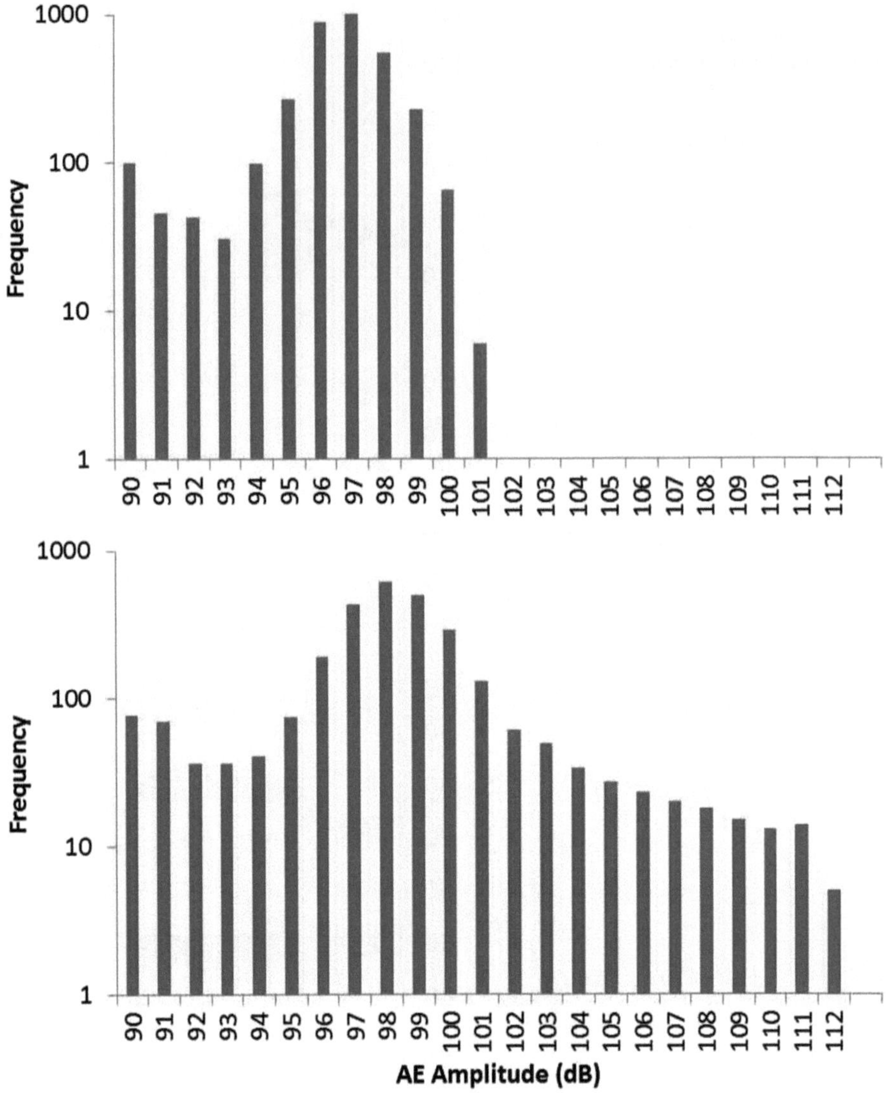


**Fig. 4.6** Comparing the histograms during the application process from Figs. 4.4 to 4.5 for the μ30 sensor. Hits greater than or equal to 90 dB are shown only on a log scale to accentuate the high magnitude impacts

here though are some outlier events with amplitudes in the range 105–112 dB, which was seen with the prior run with large agglomerates. The signature of the gun on the substrate is well demonstrated in period F after the powder was consumed, in this case the amplitude was around 96 dB for the WD sensor and around 89 dB for the μ30 sensor, with the frequency centroid 177 and 213 kHz respectively.

## 4.4 Discussion

In examining the previous data, changes to the spray process were clearly indicated by changes to the amplitude and frequency centroid of the resultant waveforms. Attempts were made to glean more subtle details during the powder application period of the process. NOESIS, a pattern recognition software from Envirocoustics S.A. was used to further analysis the data and identify potential trends, but unfortunately this did not provide additional insight above that directly visible with the standard analysis. It is probable that with changes to the test setup (e.g. using a 0 dB preamplifier and mounting the sensors in the conventional fashion), that the output waveforms might be better defined, allowing more in-depth analysis and insight.

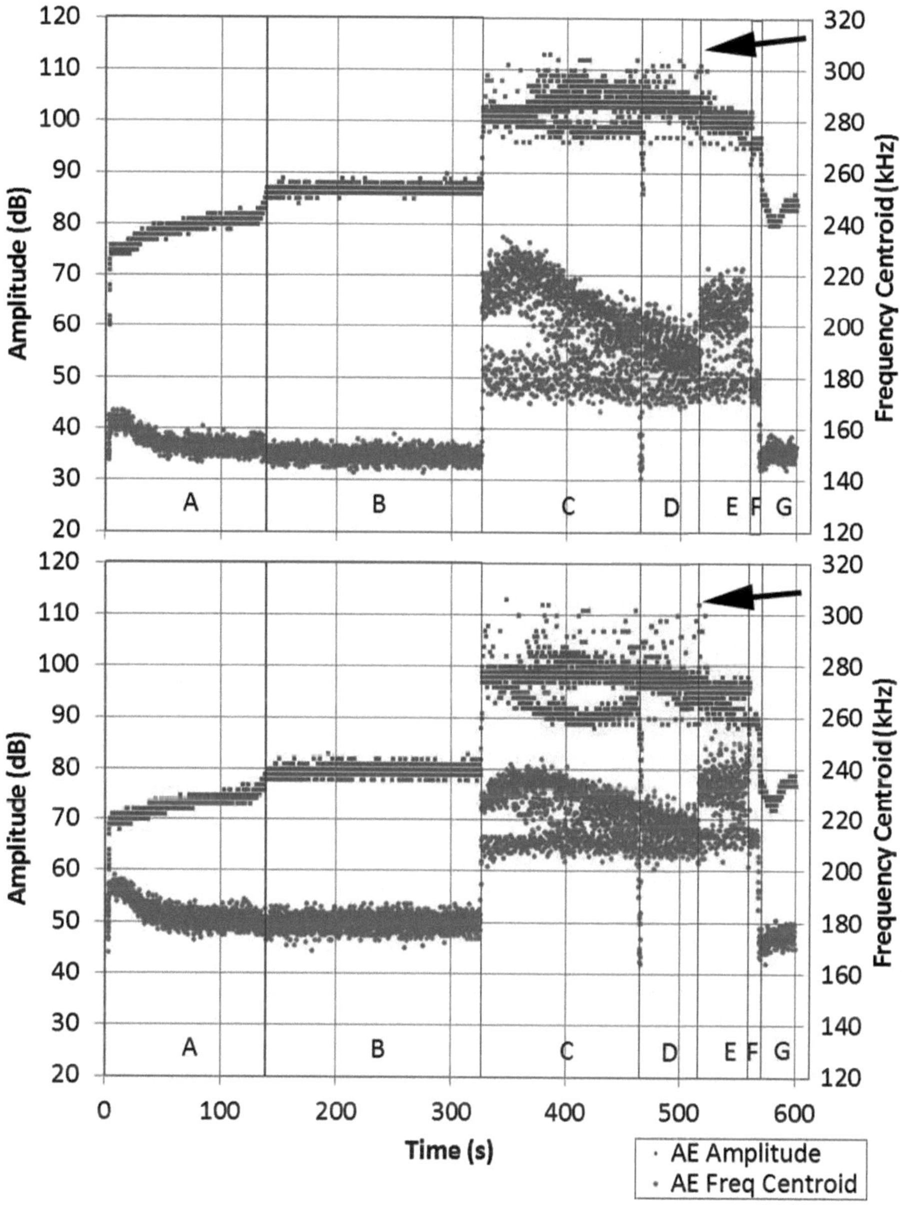


**Fig. 4.7** AE for 90Ta-10W containing large agglomerates sprayed on aluminum, where a delamination occurred. *Top plot* is the WD sensor, *bottom* is the µ30 sensor. Periods correspond to following: (*A*) nitrogen purge, (*B*) helium purge, (*C*) powder deposition with fairly high amplitude AE, (*D*) a second period of powder deposition leading up to delamination as shown by the *arrow*, and (*E*) a period of poor coating (there were small fragments spalling from the deposit) leading up to the powder running out, (*F*) a period when the nozzle was still making passes on the substrate with no powder in the hopper, and (*G*) movement of the nozzle away from the surface and conclusion of the test

## 4.5 Conclusion

This work has shown that AE is a promising technology to monitor the cold spray process in real time. Specifically, AE was used to detect changes to the gas flow, impacts from large particles which could be present in feedstock, delamination of coating, and also to determine when powder is exhausted. Hence, the use of real time AE diagnostics would also be beneficial for other types of additive manufacturing processes (i.e. high velocity oxy fuel, plasma spray deposition, laser sintering). However, much work remains to be done in the area, which will need to focus upon optimizing the system parameters and instrumentation, tailoring the signal conditioning system to address the high noise, performing source location analysis to assist in defect isolation, and working more with NOESIS to better characterize the multiple sources of AE in this challenging application of the technology.

**Acknowledgements** The authors thank the United Technologies Research Center for support of this study and permission for publication.

## References


1. Dykhuizen RC, Smith MF (1998) Gas dynamic principles of cold spray. J Therm Spray Technol 7:205
2. Smith MF (2007) Comparing cold spray with thermal spray coating technologies. In: Champagne VK (ed) The cold spray materials deposition process – fundamentals and applications. Woodhead Publishing Limited, Cambridge, UK, pp 43–61
3. Karthikeyan J (2007) The advantages and disadvantages of the cold spray coating process. In: Champagne VK (ed) The cold spray materials deposition process – fundamentals and applications. Woodhead Publishing Limited, Cambridge, UK, pp 62–71
4. Stoltenhoff T, Kreye H, Richter HJ (2002) An analysis of the cold spray process and its coatings. J Therm Spray Technol 11:542–550
5. Grujicic M, Saylor JR, Beasley DE, DeRosset WS, Helfritch D (2003) Computational analysis of the interfacial bonding between feed-powder particles and the substrate in the cold-gas dynamic spray process. Appl Surf Sci 219:211
6. Assadi H, Gärtner F, Stoltenhoff T, Kreye H (2003) Bonding mechanisms in cold gas spraying. Acta Mater 51:4379–4394
7. Schmidt T, Gartner F, Assadi H, Kreye H (2006) Development of a generalized parameter window for cold spray deposition. Acta Mater 54:729–742
8. Hussain T, McCartney DG, Shipway PH, Zhang D (2009) Bonding between aluminum and copper in cold spraying: story of asymmetry. J Therm Spray Technol 18:364–379
9. Miller RK, Hill EK, Moore PO (eds) (2005) Nondestructive testing handbook, vol 6, Acoustic emission testing. American Society for Nondestructive Testing, Columbus, p 428
10. Jayakumar T, Mukhopadhyay CK, Venugopal S, Mannan SL, Raj B (2005) A review of the application of acoustic emission techniques for monitoring forming and grinding processes. J Mater Process Technol 159:48–61
11. Bordatchev EV, Nikumb SK (2006) Effect of focus position on informational properties of acoustic emission generated by laser–material interactions. Appl Surf Sci 253:1122–1129
12. Li L (2002) A comparative study of ultrasound emission characteristics in laser processing. J Appl Surf Sci 186:604–610
13. Kong LX, Nahavandi S (2002) On-line tool condition monitoring and control system in forging processes. J Mater Process Technol 125–126:464–470
14. Arul S, Vijayaraghavan L, Malhotra SK (2007) Online monitoring of acoustic emission for quality control in drilling of polymeric composites. J Mater Process Technol 185:184–190
15. Smith C, Koshy P (2013) Applications of acoustic mapping in electrical discharge machining. CIRP Ann Manuf Technol 62:171–174
16. Kang MC, Kim JS, Kim JH (2001) A monitoring technique using a multi-sensor in high speed machining. J Mater Process Technol 113: 331–336
17. Kwak J-S, Ha M-K (2004) Neural network approach for diagnosis of grinding operation by acoustic emission and power signals. J Mater Process Technol 147:65–71
18. Schulze V, Weber P, Ruhs C (2012) Increase of process reliability in the micro-machining processes EDM-milling and laser ablation using on-machine sensors. J Mater Process Technol 212:625–632
19. Hao S, Ramalingam S, Klamecki BE (2000) Acoustic emission monitoring of sheet metal forming: characterization of the transducer, the work material and the process. J Mater Process Technol 101:124–136
20. Binsaeid S, Asfour S, Cho S, Onar A (2009) Machine ensemble approach for simultaneous detection of transient and gradual abnormalities in end milling using multisensor fusion. J Mater Process Technol 209:4728–4738
21. Kim HY, Kim SR, Ahn JH, Kim SH (2001) Process monitoring of centerless grinding using acoustic emission. J Mater Process Technol 111:273–278
22. Chung KT, Geddam A (2003) A multi-sensor approach to the monitoring of end milling operations. J Mater Process Technol 139:15–20
23. Yang M, Manabe K, Hayashi K, Miyazaki M, Aikawa N (2003) Data fusion of distributed AE sensors for the detection of friction sources during press forming. J Mater Process Technol 139:368–372
24. ASTM E2374-10, Standard guide for acoustic emission system performance verification, ASTM International, West Conshohocken, 2010, www.astm.org

# Chapter 5
# Prototyping and Testing of a Graphene-Oxide Tamper Evident Seal

**Jason A. Bossert, Christian Guzman, Axel W. Haaker, Joseph H. Dumont, Gautam Gupta, Aditya Mohite, Karen A. Miller, David D.L. Mascarenas, and Alessandro Cattaneo**

**Abstract** Structural Health Monitoring (SHM) technology has great unexplored potential in security applications. Specifically breakthroughs in graphene-oxide (GO) damage-detecting skins coupled with nonlinear, sparse signal processing techniques being used for SHM lend themselves to addressing the need for low-power remotely-readable tamper-evident seals. Assessing the integrity of a tamper-evident seal is inherently an SHM problem. In this case damage is caused by a human adversary, not the environment. This paper presents a novel architecture that leverages the tunable electrical properties of a GO-paper-based seal with a compressed-sensing (CS) acquisition protocol. This architecture allows the seal to characterize its integrity, while simultaneously providing an encrypted authentication feature making the seal difficult to counterfeit and/or spoof. The electrical properties of GO are sensitive to the traditional methods used to attack paper-based seals (mechanical lifting, solvents, heat/cold, steam). This property of GO allows us to determine if a seal has been tampered with simply by measuring its electrical properties. Specific areas of focus addressed by this work include the quantitative analysis of the encryption/authentication capabilities provided by CS, and methods for enhancing the detection of cracks/cuts propagating through the sensitive GO paper.

**Keywords** Tamper-evident seal • Crack-detection • Compressed sensing • Distributed sensors network • Graphene-oxide

## 5.1 Introduction

Tamper evident seals are widely used by both the government and industry in order to easily recognize the unauthorized access to a protected article. The seals are not designed to offer physical security, and instead act simply to identify an intrusion into a secure unit [1]. Tamper evident seals are used by the International Atomic Energy Agency (IAEA) to protect nuclear material. However, current designs of tamper evident seals require specialized teams to be onsite for installation and removal, and after removal, the seals must be processed by multiple labs to ensure accuracy of the seal's assessment [2]. Furthermore, a LANL vulnerability assessment team found that a well-practiced attacker could trivially defeat most tamper evident seals [3]. The tamper evident seal presented in this paper is motivated by the need for a remotely readable seal that can reduce the required man hours to operate, while also achieving a high level of security.

The tamper evident seal presented in this paper combines the novel signal processing technique of compressed sensing (CS) with the unique material properties of graphene oxide to produce a remotely readable tamper evident seal. Compressed

J.A. Bossert
Department of Physics, University of California at Santa Barbara, Broida Hall, Santa Barbara, CA 93106-9530, USA

C. Guzman
Thermal Science Research Center, Prairie View A&M University, MS-2525, P.O. Box 519, Prairie View, TX 77446, USA

A.W. Haaker
Mechanical Engineering Department, University of New Mexico, MSC01 1150, Albuquerque, NM 87131, USA

J.H. Dumont • G. Gupta • A. Mohite
Center for Integrated Nanotechnologies, Los Alamos National Laboratory, MS-K763, P.O. Box 1663, Los Alamos, NM 87545, USA

K.A. Miller
Safeguards Science and Technology, Los Alamos National Laboratory, MS-E540, P.O. Box 1663, Los Alamos, NM 87545, USA

D.D.L. Mascarenas • A. Cattaneo (✉)
Engineering Institute, Los Alamos National Laboratory, MS-T001, P.O. Box 1663, Los Alamos, NM 87545, USA
e-mail: cattaneo@lanl.gov

© The Society for Experimental Mechanics, Inc. 2015

C. Niezrecki (ed.), *Structural Health Monitoring and Damage Detection, Volume 7*, Conference Proceedings of the Society for Experimental Mechanics Series, DOI 10.1007/978-3-319-15230-1_5

sensing is an exciting new field that allows the reconstruction of sparse signals that have been sampled below the Nyquist frequency [4]. This allows measurements to be made quickly and with very little power; a key feature for remotely deployed seals. Furthermore, CS compresses and encrypts the signal that the seal returns to the reader [5, 6], ensuring both a low data transmission volume and secrecy from eavesdroppers. Finally, the encryption that CS provides is not bit sensitive like other common encryption systems. This means that in a real world setting, where there will be small variations in parameters that the seal uses to construct a 'key' for encryption, there will not be spurious readings indicating a seal has been tampered with [6]. For the purposes of a tamper evident seal, CS offers a robust encryption solution.

The heart of the tamper evident seal is made out of graphene oxide (GO); a promising new material that has unique electrical and mechanical properties. GO is typically prepared in solution, and then deposited onto a substrate as a thin film [7, 8]. These GO films are perfect analogs to commonly used tamper evident tapes. GO is also electrically tunable through reduction [8]. Naturally a strong insulator, reduced GO (rGO) is electrically conductive, and the degree of conductivity can be precisely controlled by the amount of reduction. This allows the construction of primitive resistive and capacitive circuits. Furthermore, GO is highly sensitive to many physical stressors. GO dissolves in water and some solvents; completely destroying it's structure [9]. GO reduces when exposed to strong acids or high temperatures, drastically altering its electrical properties [7, 9, 10]. Finally, GO films are thin and highly susceptible to damage from mechanical stresses [11]. This means that GO is highly susceptible to many common forms of attack on tamper evident seals.

In this paper, we describe how a remotely-readable, low-power, tamper-evident seal was prototyped by creating a 'key' for CS out of an array of resistors printed onto GO film. We then present simulation and experimental results that demonstrate the robustness of CS for assessing the structural health of the GO seal.

## 5.2 Theoretical Background

Compressed sensing (CS), also called compressive sampling, is a signal acquisition protocol that allows one with knowledge of the underlying structure of a general signal to simultaneously sample, compress, and encrypt it prior to storage or transmission. What makes CS exceptional is that, by leveraging the principles of sparsity and incoherence, one can use it to recover signals sampled below the Nyquist rate [4–6].

Briefly, a signal is considered sparse with respect to some basis if it has a concise representation in that basis, i.e., if there are few nonzero expansion coefficients relative to the dimension of the signal [4, 5]. Simply put, a signal is sparse if the information it carries is, in a sense, predominantly redundant. Conversely, a signal is said to be incoherent if it has few vanishing expansion coefficients relative to its dimension [4, 5]. For a much more detailed discussion of the protocol, the reader is directed to [4].

To outline the CS protocol, start with a discrete signal $x \in \mathbb{R}^N$ and a sparsifying basis $\Psi \in \mathbb{R}^{M \times N}$, or more generally a dictionary. The expansion coefficients $s_i \in \mathbb{C}$ are given by the decomposition $x = \Psi s$. The compressed signal $y \in \mathbb{R}^M$ is acquired in the linear measurement step $y = \Phi x$, where $\Phi \in \mathbb{R}^{M \times N}$ is known as the measurement, or sensing, matrix. The columns of $\Phi$ are populated by waveforms that are incoherent with respect to $\Psi$, i.e., waveforms that do not have sparse representations in $\Psi$. It has been shown that matrices populated with Gaussian, and Bernoulli, random variables, among others, make good choices for $\Phi$ given any fixed $\Psi$ [4, 5]. The signal is recovered by solving the convex optimization problem below for the approximate solution $s'$ using an algorithm such as Yall1 [4, 5, 12].

$$\min_{s' \in \mathbb{R}^N} \|s'\|_{L_1} \text{ subject to } y = \Phi \Psi s' \tag{5.1}$$

As noted previously, CS also functions as a shared, or secret, key encryption algorithm. In this light, the shared key is the measurement matrix $\Phi$, the plaintext is the sampled signal $x$ [13–16], and the ciphertext is the compressed signal $y$. As is customary in security analyses of cryptosystems, assume that the key is shared across a secure channel between two friends, Alice and Bob, and that an eavesdropper, Eve, cannot access it [5]. It has been shown that CS is computationally secure in response to ciphertext-only attacks, i.e., no practical means by which Eve can recover the plaintext from the cyphertext alone are currently known [5, 6]. Furthermore, CS is computationally secure with regard to known plaintext attacks provided one refrains from reusing measurement matrices $\Phi$ [6]. Moreover, CS differs from many other popular cryptosystems in that it is not bit-sensitive. Rather, CS is considered to be robust in that small perturbations of the encryption key manifest as small changes in the ciphertext [6]. This is an unusual property for a cryptosystem; indeed, most popular systems are bit-sensitive and consequently the integrity of the transmitted information is very much dependent on the integrity of the key [6].

These properties of CS make it particularly attractive in embedded systems and structural health monitoring applications. Additionally, the fact that signal acquisition/encryption is performed in a single linear measurement means that it can be physically implemented using a microcontroller and elementary circuit elements. In this work, $\Phi$ is realized by a shunt circuit containing a set of resistances engraved onto GO and a de-multiplexer. A thin GO film is a good choice for a tamper-indicating element in that it is an electrically tunable, hydrophilic, carbon-based material that is highly sensitive to physical attacks [7–10]. A fundamental aspect in designing the seal consists in providing guarantees that the system cannot easily be exploited. In order to achieve this goal, the solution proposed by the authors takes in due consideration the limits of CS used as encryption/authentication technique. In addition, a reconfigurable architecture is considered to strengthen the overall security of system.

## 5.3 Prototype Description

The present work develops a working prototype of a tamper evident seal which physically implements the measurement matrix used by the compressive sampling procedure. The completion of this task is achieved by leveraging both the tunable electrical properties of GO and the reprogrammability of a microcontroller unit (mCU). Before entering the details of the architecture adopted in this work, it is worth noticing the scheme proposed in this work exploits CS to encrypt the messages sent from the seal back to the remote reader. Although not directly explored in this work, the mCU provides means for securely change both the measurement matrix used to perform the compressive sampling procedure (i.e. the encryption key) and the plaintext message encrypted by the seal. Examples of approaches that can be implemented to reconfigure the seal for increased security are provided in [13–16]. The seal architecture adopted in this work is shown in Fig. 5.1.

The seal architecture provides an easy query method to obtain the status of the seal. For a status query, the signal temporarily saved on the tamper evident tag is parsed by the mCU, after which a voltage signal corresponding to the signal saved in the memory of the microcontroller is sent out by the DAC unit (element #3, Fig. 5.1) to the MUX (element #4, Fig. 5.1). The MUX controls the particular resistor (element #5, Fig. 5.1) which the voltage will be applied across. The MUX uses a switching sequence corresponding to particular resistors to determine where to route the voltage. The voltage drop across the shunt resistance (element #6, Fig. 5.1) is then measured by the ADC unit (element #7, Fig. 5.1). The voltage across the shunt resistance is used to deduce the voltage across the seal resistor using the voltage divider shown in Fig. 5.2.

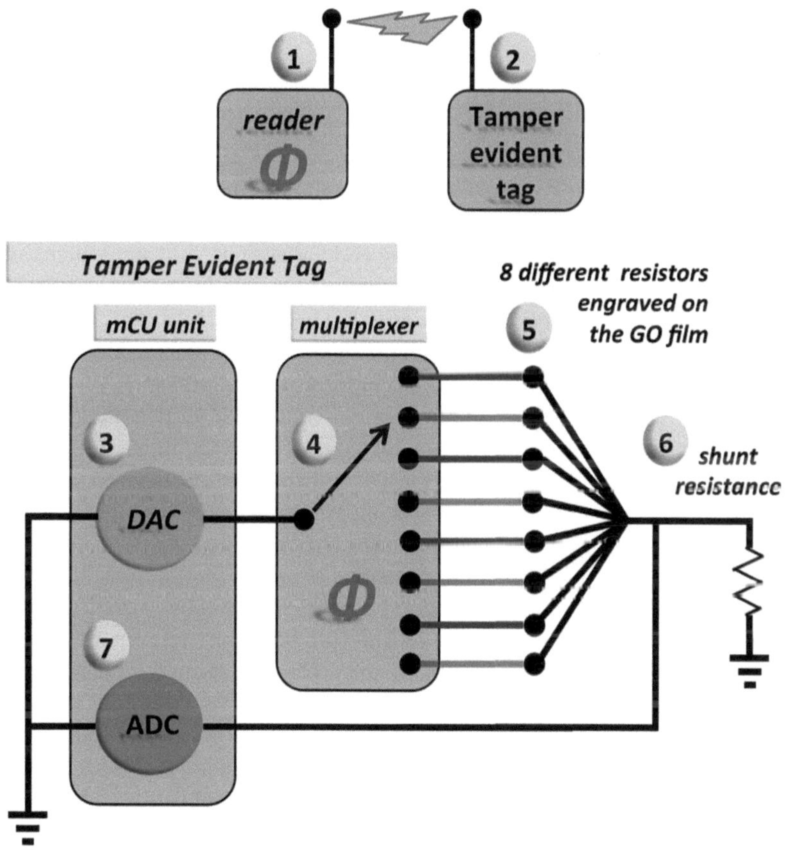


**Fig. 5.1** The communication between the reader (*1*) and the seal (*2*) is operated over an unsecure channel. A plaintext message saved in the memory of the microcontroller is converted into an analog voltage via the DAC (*3*). The DAC sends the signal to the multiplexer (*4*), which duplicates the switching sequence, Φ, by switching to the correct resistor (*5*) at the correct time. The voltage drop across the resistor is measured across the shunt resistor (*6*) and converted to a digital sample at the ADC (*7*). The microcontroller emulates the compressed sensing procedure by keeping a running sum of the digital samples collected for the analog signal generated from the DAC and sent through the multiplexer. The compressed samples are sent back from the tamper evident tag (*2*) to the reader (*1*)

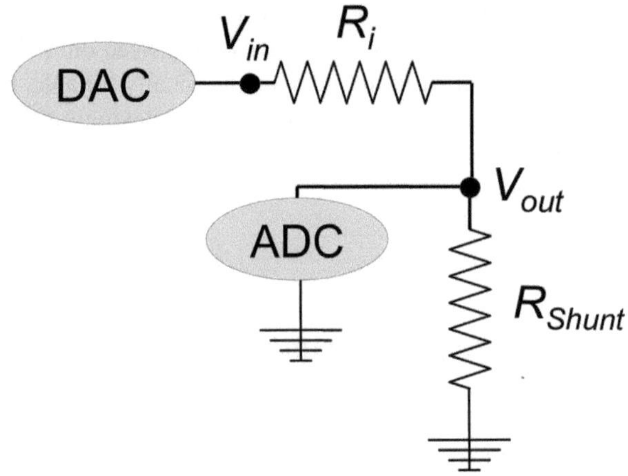


**Fig. 5.2** The voltage division circuit. The signal, represented as voltage in, is dropped across both the seal resistor and the shunt resistance. The voltage out, representing one multiplication in the inner product of the signal and one row of the Φ matrix, is measured by the ADC between the two resistors

In particular the relationship between the voltage drop $V_{ADC}$ measured by the ADC unit across the shunt resistance and the voltage $V_{DAC}$ generated by the DAC unit for a given signal sample can be expressed as follows:

$$V_{ADC} = \frac{R_{shunt}}{R_x + R_{shunt}} V_{DAC} \tag{5.2}$$

Following the completion of the measurement of the voltage drop, the process is repeated for every element in the signal saved in the memory of the microcontroller. A given element of the voltage signal will be applied across several different resistors. The mCU keeps a running sum of the measured voltage drops until M samples are collected. This process emulates the matrix multiplication between the sparse signal and the measurement matrix resulting in a compressed signal. Specifically, the generic element $\phi_{ij}$ of the equivalent measurement matrix associated with the linear projections performed by the seal is equal to

$$\phi_{ij} = \frac{R_{shunt}}{R_{x,ij} + R_{shunt}} \tag{5.3}$$

where $R_{x,ij}$ is the value of one of the eight resistors connected to the multiplexer drawn at random from a uniform distribution for the generic $ij$ element of the measurement matrix $\Phi$, with $i \epsilon [1, M]$ and $j \epsilon [1, N]$.

This compressed signal is then sent back to the remote reader for reconstruction. Tampering with the seal will result in a change in the seal resistances, which will change the effective measurement matrix that is being applied to compress the signal. This implies that a particular measurement matrix is associated with a unique switching sequence. Additionally, the seal architecture provides flexibility and variation in the components' characteristics. The architecture offers the ability to dynamically select a particular measurement matrix by setting a unique switching sequence. A new switching sequence may be selected at every query if desired. The seal architecture also is not constrained to a predetermined sparse signal. The remote reader may use a new sparse signal at every query. The tunable properties of GO provide an additional layer of flexibility within the architecture. This allows for a large range of possible measurement matrices available for selection.

The GO seal is constructed on a 36 mm diameter by 15 μm thick GO film. This film is then reduced using a 40 W $CO_2$ laser made by Full Spectrum Laser, which outputs infrared light at 10.4 μm. The reduction of the GO film produces reduced GO (rGO), which is conductive to electricity [8], while regular GO is highly insulating. Using a space-filling, self-avoiding random walk algorithm with eight starting points, a design for eight resistors can be made and printed onto the seal, as seen in Fig. 5.3a. This algorithm creates a random path for each resistor, allowing unique seal to be easily fabricated. The random path is space filling, and is therefor sensitive to cracks or other mechanical damage across its entire exposed surface.

Similar to the GO seal, a collection of traditional resistors was also used as described in the "Model results" and "Testing results" sections. This was done in order to provide a baseline for comparison to the GO seal.

The GO seal is held in an ABS plastic container, printed on a MakerBot, and seen in Fig. 5.3b. The plastic container consists of two identical rings, each 70 mm in diameter with a 28 mm diameter hole in the center, and with eight evenly spaced holes around the perimeter so that they can be secured together with bolts. The upper container element has copper tape that extends from the underside – where it makes contact with the GO seal – to the top, where it is soldered to wires that connect it to the rest of the seal system. The GO seal is placed between the two container halves, where its outer 4 mm is compressed between the two halves by the securing bolts. This keeps the seal in place and ensures proper contact between the seal and the copper tape.

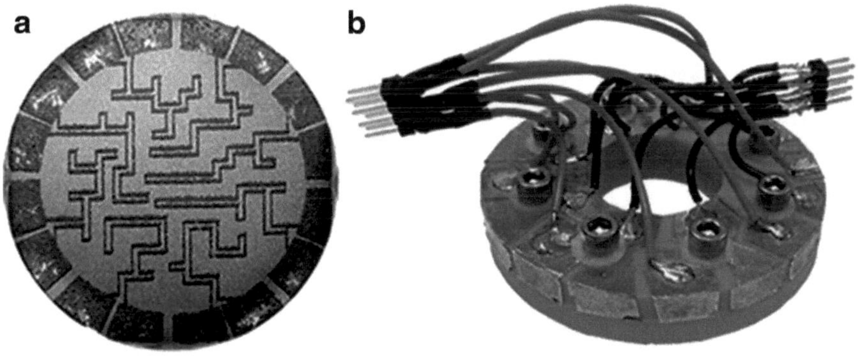

**Fig. 5.3** (**a**) The printed GO seal. The contact pads rim the outside, while the resistive paths meander on the inside. The *light gray* is indicative of insulating GO, while the *darker gray* and more textured parts are conductive rGO. (**b**) The seal container. The GO seal is placed in the central hole, and in the space (not visible) between the two layers of ABS plastic. The copper tape wraps over the outer edge of the upper ABS piece, and down to its underside where it makes contact with the GO seal

Once the seal has been designed and printed, the shunt resistance must be chosen to match the resistance values of the seal. The shunt resistance acts as the second resistance in the voltage divider, and as such, it controls the voltage dropped over the measurement resistor. If this voltage is too small, than if that measurement resistor is destroyed, it's inability to pass voltage will only have a minor effect on the overall compressed sample. However, if the shunt resistance is too high, the majority of the voltage will be dropped across the shunt resistance, and the ADC's readings will all be very similar, making each compressed sample too similar for a proper reconstruction. In practice, a shunt resistance that is between the highest resistance and twice the highest resistance is the best option.

The microcontroller was programmed in C++ using object-oriented techniques. A series of classes were created to facilitate communication between the remote reader and the seal. The classes grant the ability to query the seal, set switching sequences, as well as setting properties of the microcontroller remotely. The behavior of the microcontroller classes are largely dynamic and do not require a hard-reset for unit tests. Additionally, MATLAB scripts were created to simulate the functions of the remote reader. The seal is connected to the MATLAB running computer via a serial port.

## 5.4 Model Results

Modeling was performed in MATLAB in order to analyze the appropriateness of compressed sensing for application in a tamper evident seal. Four aspects of compressed sensing were analyzed: the reliability of the signal reconstruction process, and the effects of environmental variation, measurement error, and mechanical failure. The CS protocol has been implemented making use of the dictionary $\Psi$, the sparse signal $x$ and the measurement matrix $\Phi$ described as follow. The dictionary $\Psi$ is populated with the elements belonging to the Fourier basis as described in [17]. The Fourier basis is an $N \times N$ orthogonal matrix, where the $k$th column is given by

$$\psi_k(n) = \frac{1}{\sqrt{N}} e^{-2\pi i k n/N} \quad 0 \leq n, k \leq N-1 \tag{5.4}$$

where N is the length of the signal vector indexed by $n$.

In this basis, the sparse signal used for all of the modeling is given by $x(t) = 0.9\sin(3t)$. The amplitude is scaled such that it does not cause the DAC to saturate, and three periods of the sine wave are used because, in our experience, the Yall1 algorithm is better able to perform successful recoveries of the compressed sample when more than one period is used. The resistances used for modeling were 1.9, 4.9, 8.9, 20.7, 34.6, 58.6, 67.2, and 83.0 kΩ. With a shunt resistance of 99.7 kΩ, these resistance values give possible $\Phi$ matrix values of 0.981, 0.953, 0.918, 0.828, 0.742, 0.62, 0.597, and 0.545. Therefor, each element of the $\Phi$ matrix is chosen randomly from the possible values. When perturbations are applied in the models, they are applied as a percentage wise change to the value of the resistance, thereby mimicking the physical effects of damage or disturbance to the seal.

The error between the reconstructed signal and the original signal is used as the primary metric to determine whether or not there has been physical damage to the seal. The error metric is calculated as $\varepsilon = \|x - \widetilde{x}\|_2 / \|x\|_2$, where $\widetilde{x}$ is the reconstructed signal. In order to calculate the error, the Yall1 reconstruction algorithm needs to perform in a predictable manner.

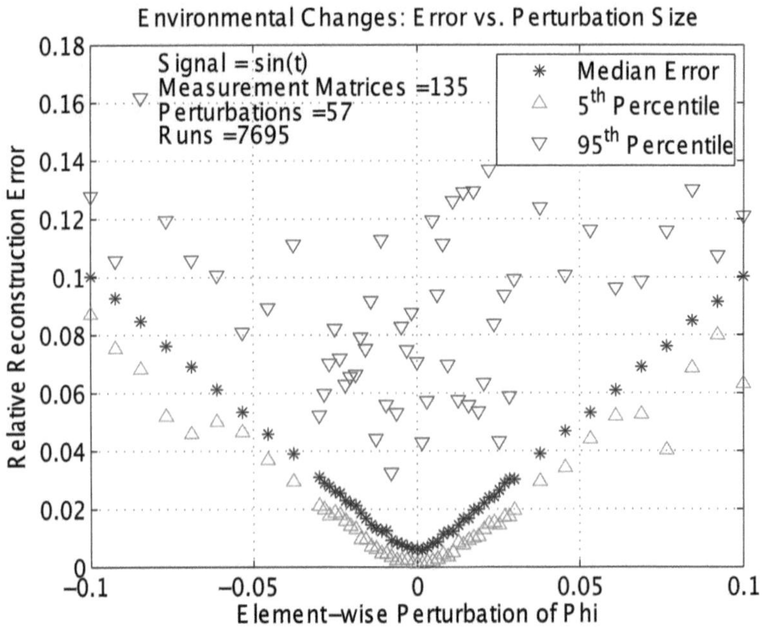


**Fig. 5.4** The mean and lowest percentage error data suggest a linear relationship between perturbation and error, while the highest percentage error data are still generally under 14 % for all simulations. This indicated that CS is not bit sensitive and capable of handling environmental variations

The expected outcomes of the Yall1 algorithm are either a pass or a fail, but occasionally the Yall1 algorithm fails to process the compressed sample in any meaningful way, and a null recovery (a reconstructed signal of zero) is returned [12]. To quantify the rate at which the Yall1 algorithm returns a null recovery the null rate for all simulations and experimental tests were recorded and analyzed. Throughout all simulations, the null recovery rate was found to be 30 %, with little variation due to different perturbations in resistance values as described below.

The sensitivity of GO and rGO to moisture, temperature, and solvents are one of the primary reasons it was chosen as a material to create the tamper evident seal [7, 9, 10]. However, for the seal to be usable without returning false-positives, the Yall1 algorithm needs to be able to reconstruct the original signal with low error, even with changes in environmental conditions. Under the assumption that changes in temperature, humidity, and of trace gasses and salts in the air affect all resistors on the seal equally, simulations were run where all resistances changed by an equivalent percentage. In total, 135 measurement matrices were tested with 57 perturbations ranging from −10 % to 10 % being applied to all resistors equally. The results are presented in Fig. 5.4, and suggest that there is a linear trend between the size of the perturbation and the error introduced into the reconstruction. This indicates that CS is robust with regards to homogeneous changes in resistance, and will not trigger false positives for minor environmental variations.

The effects of measurement error were analyzed by applying a different, stochastic, perturbation to each resistor to emulate the random error introduced into the system by the DAC and the ADC. Again, 135 matrices were analyzed, each being perturbed 57 times. Each perturbation altered every element of the measurement matrix by a percentage randomly chosen from a normal distribution centered at zero, with a standard deviation of 2 %. This value for the standard deviation was chosen to fit well with the error observed in a closed DAC to ADC loop. The closed DAC to ADC loop connects the input of the ADC directly to the output of the DAC, allowing the creation of a mapping between the DAC output and ADC input. The results indicate that for the vast majority of the measurement matrices analyzed, small random perturbations have very little effect on the reconstruction error.

Under the assumption that mechanical failure would introduce a large change in mechanical and electrical properties to one or more resistors [11], we simulated mechanical damage by introducing a large perturbation on a single resistor at a time. Eighty-five matrices were analyzed, and each of the eight resistors were perturbed by −75 %, −50 %, 0 %, 50 %, 75 %, and finally by setting the resistance equal to infinity. The results, presented in Fig. 5.5, show that for small changes, CS is relatively robust, and the reconstruction error increases slightly. However, for large changes such as ±75 % the reconstruction error becomes much larger. Finally, by setting a resistance equal to infinity and opening the circuit, the reconstruction error is often above 100 %, which correctly indicates a seriously damaged seal. To ensure a correct identification of a damaged seal, only measurement matrices that exclude the possibility of false negatives (i.e. reconstruction error below 100 % when the seal is damaged) must be used to effectively run the seal.

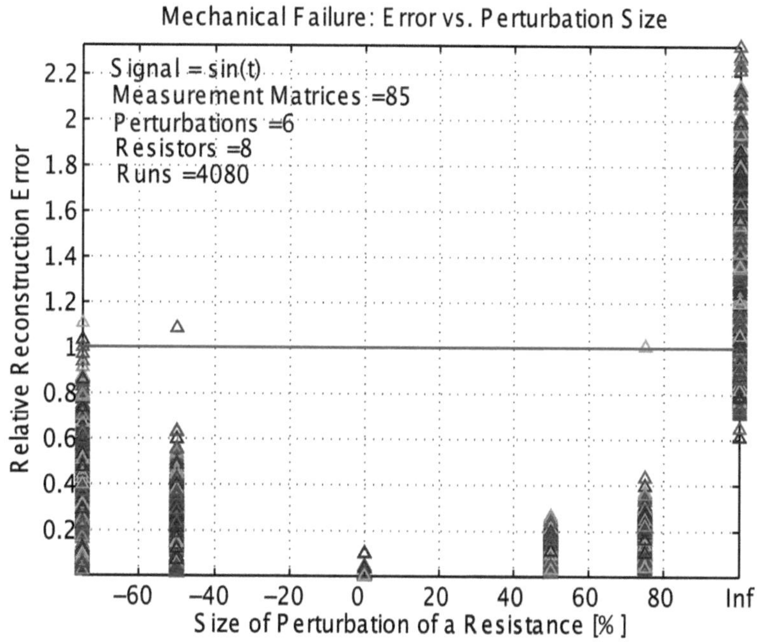


**Fig. 5.5** Larger damage to a single resistor results in a larger reconstruction error. Destroying a resistor entirely drastically increases the reconstruction error. Among the tested measurement matrices, the ones that ensure a zero false negative rate must be chosen to ensure a correct identification of a damaged seal

## 5.5 Testing Results

Initial testing and verification of the microcontroller based seal was performed on the traditional resistor bank. A switching sequence that determined the order of switching of the demultiplexer was hardcoded into the microcontroller. Resistor values were chosen to be spread out across a range that was similar to the expected resistance values of the seal. The nominal values were 5, 10, 15, 22, 30, 39, 46, and 61 kΩ, with a shunt resistance of 91 kΩ. The resistances were then measured by the microcontroller in order to negate the effects of the de-multiplexer and other circuitries, and these resistance values, along with the shunt resistance, were used to calculate each element in the $\Phi$ matrix. The switching sequence remained unchanged for all trials, and was chosen for its high compatibility with the Yall1 algorithm. The input signal was a three period sine wave with maximum amplitude of 0.9 V. Nine tests of 100 compressed signals each were taken. The first test was with all the resistors intact, and each subsequent test removed one of the eight resistors to simulate the cutting of the resistor. The results are summarized in Table 5.1.

Testing on the GO seal followed the same procedures as listed above, and the results are summarized in Table 5.2. In order to preserve the seal, no physical damage was inflicted. Instead, the wires leading into the enclosure were disconnected one at a time to prevent the signal from reaching each resistor on the GO film.

The data from the tests on both the resistor bank and the GO seal indicate that the presented design for a GO based tamper evident seal is feasible. Disregarding null recoveries, both the resistor bank and the GO show a false-positive and false-negative rate of zero. This indicates that the tamper evident seal is capable of robustly detecting severe mechanical damage without generating frequent false alarms. Presented in Fig. 5.6 is a graphical representation of the recovery rates of the simulated destructive testing on the GO seal. All of the undamaged recoveries have an error that is less than the 100 % error threshold, agreeing with the model in that an undamaged seal will not produce false positives. When the connection to a resistor is broken, the reconstruction always returns an error of greater than 100 %, as shown in Fig. 5.6. This result was obtained by properly selecting a measurement matrix that eliminates the possibility of false negatives.

## 5.6 Summary/Conclusion

In this paper we have presented a design for a remotely readable tamper-evident seal. Utilizing the tunable electric properties of graphene oxide we have designed a seal that is sensitive to mechanical damage as well as other attacks that are commonly used to defeat tamper evident seals. We have introduced compressed sensing as a novel and robust method to detect damage to the graphene oxide seal, and we have simulated the robustness of compressed sensing to changes in the seal due to the environment and measurement error. The simulation results indicate that compressed sensing is a viable and robust method

**Table 5.1** Results obtained by computing the error produced by removing one resistor at a time from the resistor bank

| Cutting resistors from the resistor bank | | | | | | | | | |
|---|---|---|---|---|---|---|---|---|---|
| Trial | 1 | 2 | 3 | 4 | 5 | 6 | 7 | 8 | 9 |
| Resistor removed | None | 5,080 Ω | 9,868 Ω | 14,837 Ω | 21,849 Ω | 29,814 Ω | 38,545 Ω | 46,469 Ω | 61,310 Ω |
| Percent successful | 83 | 0 | 0 | 0 | 0 | 0 | 0 | 0 | 0 |
| Percent failed | 0 | 84 | 79 | 73 | 59 | 32 | 51 | 80 | 13 |
| Percent null | 17 | 16 | 21 | 27 | 41 | 68 | 49 | 20 | 87 |

**Table 5.2** Results obtained by computing the error produced by disconnecting one resistor from the GO seal at a time

| Cutting resistors from the GO seal | | | | | | | | | |
|---|---|---|---|---|---|---|---|---|---|
| Trial | 1 | 2 | 3 | 4 | 5 | 6 | 7 | 8 | 9 |
| Resistor removed | None | 13,880 Ω | 10,821 Ω | 24,493 Ω | 21,053 Ω | 5,772 Ω | 49,656 Ω | 40,150 Ω | 27,687 Ω |
| Percent successful | 45 | 0 | 0 | 0 | 0 | 0 | 0 | 0 | 0 |
| Percent failed | 0 | 82 | 89 | 54 | 25 | 76 | 52 | 76 | 83 |
| Percent null | 55 | 18 | 11 | 46 | 75 | 24 | 48 | 24 | 17 |

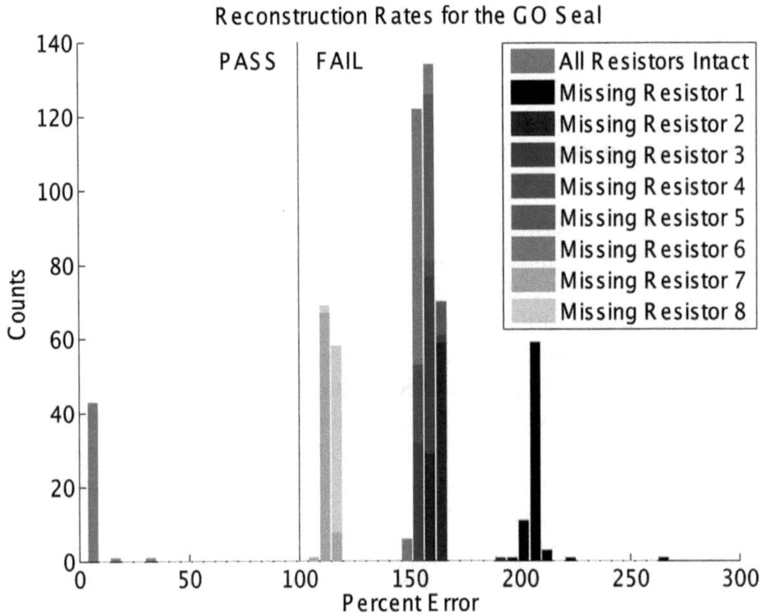


**Fig. 5.6** On the GO seal, with a proper shunt resistance, destroying a resistor always results in a failed reconstruction. Furthermore, the intact seal always reconstructs successfully. This indicates the GO seal suffers from no false-positives nor false-negatives (Null recoveries are not shown)

to query the status of the seal. Furthermore, we have designed and prototyped a graphene oxide seal, and initial tests indicate that the presented work and methodologies show great promise for application as a tamper evident seal.

**Acknowledgements** The authors would like to acknowledge the support of the Los Alamos National Laboratory – Laboratory Directed Research and Development program. Grant number 20130527ER. Any opinions, findings, and conclusions or recommendations expressed in this material are those of the authors and do not necessarily reflect the views of the particular funding agency.

## References

1. Johnston RG (2008) Tamper detection for safeguards and treaty monitoring: fantasies, realities, and potentials. Nonprolif Rev 8(1):102–115
2. Doyle JE (2008) Nuclear safeguards, security, and nonproliferation: achieving security with technology and policy. Butterworth-Heinemann, Amsterdam
3. Johnston RG (2003) Tamper indicating seals: practices, problems, and standards. In: World Customs Organization Security Meeting, Brussels
4. Candes E, Wakin M (2008) An introduction to compressive sampling. Signal Process Mag IEEE 25(2):21–30
5. Rachlin Y, Baron D (2008) The secrecy of compressed sensing measurements. In: 46th annual Allerton conference on communication, control, and computing, Urbana-Champaign

6. Orsdemir A, Altun H, Sharma G, Bocko M (2008) On the security and robustness of encryption via compressed sensing. In: Military Communications Conference, San Diego
7. Marcano DC, Kosynkin DVBJM, Sinitskii A, Sun Z, Slesarev A, Alemany LB, Lu W, Tour JM (2010) Improved synthesis of graphene oxide. ACS Nano 4(8):4806–4814
8. Gao W, Singh N, Song L, Reddy ALM, Ci L, Vajtai R, Zhang Q, Wei B, Ajayan PM (2011) Direct laser writing of micro-supercapacitors on hydrated graphite oxide films. Nat Nanotechnol 6:496–500
9. Gein AK, Novoselov KS (2007) The rise of graphene. Nat Mater 6:183–191
10. Gómez-Navarro C, Weitz TR, Bittner AM, Scolari M, Mews A, Burghard M, Kern K (2007) Electronic transport properties of individual chemically reduced graphene oxide sheets. Nano Lett 7(11):3499–3503
11. Sharp N, Kuntz A, Brubaker C, Amos S, Gao W, Gupta G, Mohite A, Farrar C, Mascareñas D (2014) A bio-inspired asynchronous skin system for crack detection applications. Smart Mater Struct 23(5):055020
12. Yang J, Zhang Y (2011) Alternating direction algorithms for L1-problems in compressive sensing. SIAM J Sci Comput 33(1):250–278
13. Dautov R, Tsouri G (2013) Establishing secure measurement matrix for compressed sensing using wireless physical layer security. In: 2013 international conference on computing, networking and communications (ICNC), San Diego
14. Lin G-S, Chang HT, Lie W-N, Chuang C-H (2003) Public-key-based optical image cryptosystem based on data embedding techniques. Opt Eng 42(8):2331–2339
15. Sreedhanya AV, Soman K (2013) Ensuring security to the compressed sensing data using a steganographic approach. Bonfring Int J Adv Image Process 3(1):1–7
16. Abhishek O, George S, Deepthi P (2013) PWLCM based image encryption through compressive sensing. In: 2013 IEEE recent advances in intelligent computational systems (RAICS), Trivandrum
17. Candés EJ, Eldar YC, Needell D, Randall P (2011) Compressed sensing with coherent and redundant dictionaries. Appl Comput Harmon Anal 31(1):59–73

# Chapter 6
# Solitary Waves to Infer Axial Stress in Slender Structures: A Numerical Model

**Abdollah Bagheri, Piervincenzo Rizzo, and Leith Al-Nazer**

**Abstract** *In-situ* testing of beams subjected to axial stress may prevent structural anomalies such as buckling. We describe the coupling mechanism between highly nonlinear solitary waves (HNSWs) propagating along a granular system and a beam in contact with the granular medium with the aim of assessing the ability of HNSWs at measuring axial stress. Nonlinear solitary waves are compact non-dispersive waves that can form and travel in nonlinear systems such as one-dimensional chains of particles. In the study presented in this article we investigated numerically straight chains of particles in contact with a prismatic beam subjected to axial stress. Two configurations were considered: one or two chains. The effect of the particles' diameter and material was studied to find those optimal conditions that maximize the sensitivity to axial stress variation. We found that the configuration with one chain is preferable whereas certain particles' materials and diameter are better at sensing the variation of the axial stress.

**Keywords** Highly nonlinear solitary waves • Discrete particle model • Nondestructive testing • Buckling • Thermal stress

## 6.1 Introduction

Slender prismatic columns or beams subjected to thermally axial stress are common in civil and mechanical engineering structures. For example, continuous welded rails (CWR) which are typically hundreds of meters long, are prone to buckling in hot weather and to breakage or pulling apart in cold weather.

In the study presented in this paper, we propose to exploit the propagation of highly nonlinear solitary waves (HNSWs) to measure thermal stress indirectly. HNSWs are compact non-dispersive waves that can form and travel in nonlinear systems such as one-dimensional chains of particles [1–3], where they can be generated by the mechanical impact of a striker, or by using a transducer [1] or laser pulses [2]. In chains of particles, the nonlinearity arises from a Hertzian-type contact between two adjacent spheres subjected to compressive force. Recent studies on the propagation of nonlinear solitary waves in one-dimensional chains of spherical elastic beads have demonstrated that these waves can be used as vibration absorbers [4], impurity detectors [5], acoustic diodes [6], and for the nondestructive testing of structural and biological materials [7–11]. One peculiar feature of the HNSWs is that their amplitude, speed, and the number of pulses can be tuned by engineering the granular system, by changing the bead's diameter, or by adding static pre-compression.

We investigated the effect of the thermally axial stress acting in a beam on the propagation of HNSWs along one or two chains of particles in contact with the beam. The objective of the study was to: (1) evaluate the chain configuration, one or two chains, most sensitive to the variation of stress; (2) determine the optimal combination of particles' diameter and material. A discrete particle model [3] was employed to predict the propagation of the wave along the chain and to derive the shape and amplitude of the solitary pulse at the chain-beam interface. A continuous beam theory was then used to estimate the response of the structure subjected to the forcing function equal to the solitary pulse. The response served to predict the characteristics of the solitary waves generated by the impact of the vibrating beam to the chain.

A. Bagheri • P. Rizzo (✉)
University of Pittsburgh, 3700 O'Hara Street, Pittsburgh, PA 15261, USA
e-mail: pir3@pitt.edu

L. Al-Nazer
Office of Research and Development, Federal Railroad Administration, 1200 New Jersey Avenue, SE, Washington, DC 20590, USA

© The Society for Experimental Mechanics, Inc. 2015

C. Niezrecki (ed.), *Structural Health Monitoring and Damage Detection, Volume 7*, Conference Proceedings of the Society for Experimental Mechanics Series, DOI 10.1007/978-3-319-15230-1_6

## 6.2 One-Chain Configuration

In this section, we describe the case of a beam in contact with a single chain of particles. We considered the effect of the beam's axial stress on the values of certain features extracted from the force profile of the pulses.

### 6.2.1 *Numerical Formulation*

Figure 6.1a shows the top view of the numerical simulation model. The beam is clamped at two ends, and one chain of particles is in contact with the beam at its mid-span.

When a solitary pulse propagates along a chain of $N$ identical beads having radius $R$, mass $m$, and subjected to a static pre-compression force $F_0$ and initial approach $\delta_0$, Newton's second law relative to the *i-th* particle can be written as [3]:

$$\ddot{u}_i = \eta_c[(\delta_0 - u_i + u_{i-1})]_+^{3/2} - \eta_c[(\delta_0 - u_{i+1} + u_i)]_+^{3/2}\ , 2 \leq i \leq N-1 \tag{6.1}$$

where $u_i$ denotes the displacement of the $i$-th particle from its equilibrium position, and $\eta_c$ is the stiffness constant normalized with respect to the mass $m$, namely $\eta_c = E\sqrt{2R}/3m\left(1-\nu^2\right)$, where $E$ and $\nu$ represent the Young's modulus and the Poisson's ratio of the particles, respectively. The bracket $[s]_+$ denotes *max*($s$,0) which identifies a system unable to support a tensile force between particles.

For the particles located at the two ends of the chain, i.e. when $i = 1$ and $i = N$, Eq. (6.1) becomes:

$$\ddot{u}_1 = \eta_c\delta_0^{3/2} - \eta_c[(\delta_0 - u_2 + u_1)]_+^{3/2} \tag{6.2a}$$

$$\ddot{u}_N = \eta_c[(\delta_0 - u_N + u_{N-1})]_+^{3/2} - \eta_b[(\delta_{b0} - u_b + u_N)]_+^{3/2} \tag{6.2b}$$

where $\delta_{b0}$ represents the initial approach between the last bead and the beam, $u_b$ is the displacement of the beam at its mid-span, and $\eta_b$ is the normalized stiffness constant between a particle and the beam.

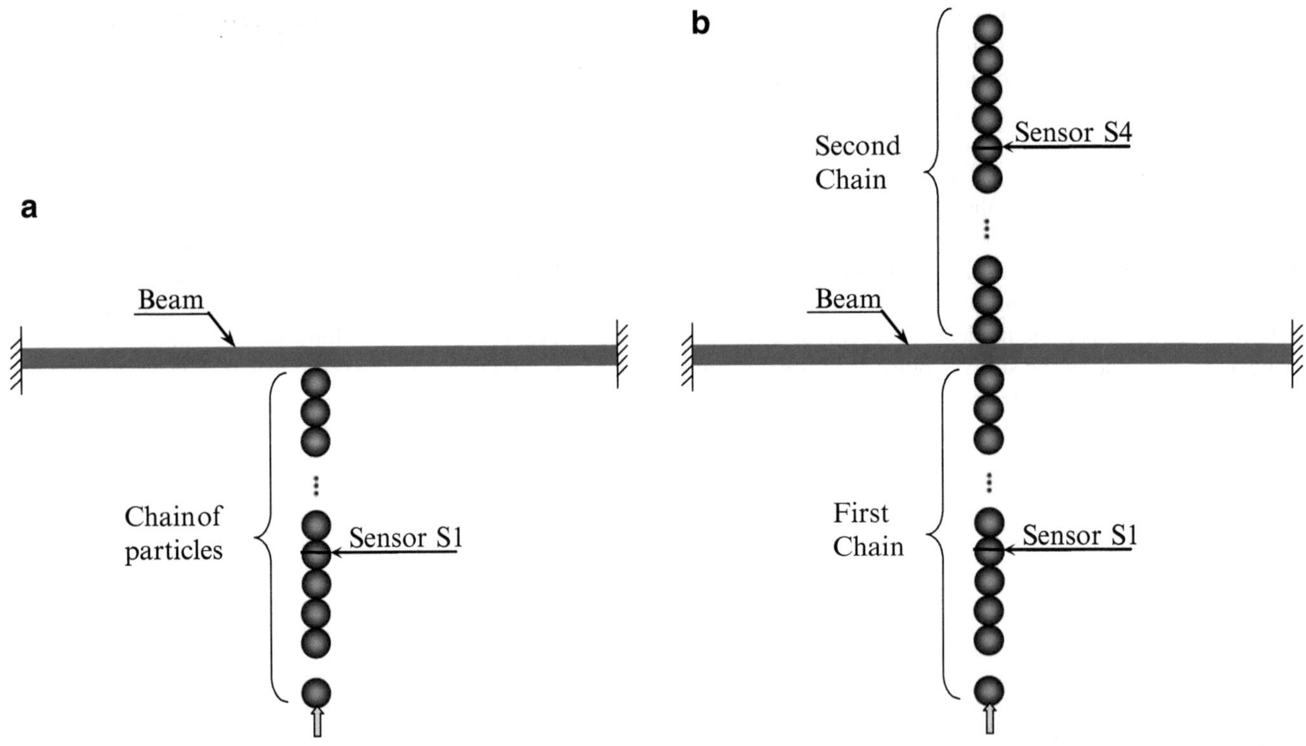


**Fig. 6.1** *Top view* of the numerical simulation setup. The beam is clamped at two ends. (**a**) One and (**b**) two chains of particles are in contact with the beam at its mid-span. Each chain consists of $N = 14$ particles (Drawing not to scale)

In our simulations, the beam in contact with the granular medium was subjected to an impulse force equal to the solitary wave arriving at the chain/beam interface. To predict the response of the beam, we used the continuous beam theory. The equation of the beam's motion can be written:

$$\ddot{u}_b(t) = \sum_{n=1}^{\infty} \phi_n(x) \left[ \frac{1}{M_n} \int_0^L F(t)\phi_n(x)dx - \omega_n{}^2 q_n \right] \tag{6.3}$$

where $\phi_n(x)$, $q_n(t)$, and $M_n$ are the $n$-th normal mode, normalized coordinate, and generalized (modal) mass, respectively; $F(t)$ is the force applied to the beam by the solitary pulse, and $L$ is the beam's length. For more details the reader is referred to Cai et al. [10].

The second order differential equations in (6.1), (6.2) and (6.3) were solved using the 4th order Runge-Kutta method implemented in MATLAB software.

### 6.2.2 Numerical Setup

A 914 mm long stainless steel beam, clamped at both ends, was considered. The beam was 9.525 mm wide and 19.05 mm deep. For stainless steel we assumed $\rho = 7{,}800$ kg/m$^3$, $E = 200$ GPa, $\nu = 0.28$, and yielding stress $\sigma_Y = 206.8$ MPa. The geometric and mechanical properties of the beam were such that its Euler buckling load $F_{cr}$ and the corresponding buckling stress $\sigma_{cr}$ were equal to $-12.95$ kN and $-71.395$ MPa, respectively.

The single chain consisted of fourteen 19.05 mm in diameter particles. Each particle weighed 29 g and was made of stainless steel for which the Young's modulus and the Poisson's ratio were considered equal to 200 GPa and 0.28, respectively. Figure 6.1a shows the top view of the numerical setup. The force profile associated with the propagation of the solitary pulses was measured at the center of the 5th and 10th particles away from the interface. These particles are hereafter indicated as sensor beads 2 (S2) and sensor bead 1 (S1), respectively. The pre-compression force was set to zero, and the effect of a striker was simulated by setting the initial velocity $\dot{u}_0$ of the first particle at 0.3018 m/s. This value was chosen in agreement with some experimental studies recently conducted in our laboratory.

### 6.2.3 Numerical Results

Figure 6.2 shows the dynamic force measured at the two sensors site when the axial stress in the beam is zero. The first pulse is the incident solitary wave (ISW) traveling from the first particle of the chain towards the beam. The next pulse is the reflected solitary wave (RSW) which is generated by the reflection of the ISW at the beam-chain interface. The third and fourth pulses are induced by the vibration of the beam: as the incident pulse reaches the interface part of its energy is reflected to become the RSW and part triggers the motion of the beam. When the beam bounces back toward the chain these pulses, named vibration-based solitary wave (VSW), are generated. Beyond the VSWs, there are a few smaller pulses which are likely generated by the beam's higher modes vibration. Figure 6.3 compares the force profiles measured at sensor S1 at the largest compressive and tensile load. In Fig. 6.3a, there is a few changes in the RSW for different axial loads that is due to changing the beam's stiffness. From Fig. 6.3b, it can be seen that the arrival time of the VSW for two axial loads has a significant difference, and it increases with increasing compressive load. At the largest tensile load, the VSW arrives at ~5 ms whereas it occurs at ~145 ms when there is the largest compressive load. This is due to increasing the beam's natural frequency at higher compressive loads.

To quantify the effect of the stress on the propagation of the HNSWs, the four features summarized in Table 6.1 were extracted from the dynamic force measured at the sensing sites S1 and S2. The first two features are related to the amplitudes of the pulses, whereas the remaining features are related to the time-of-flight (TOF) which is the difference of the time of arrival at the same sensor site.

Figure 6.4 displays the wave features measured at sensor S1 as a function of the beam's stress. The features relative to the reflected wave denote a monotonous trend above $-40$ MPa. Below that mark, a change in the gradient's sign is visible; this variation is associated with the change of the value of the coefficient $\phi_n{}^2/M_n$, associated with the $n$-th normal mode and the corresponding generalized mass. Owing to this higher stiffness, the amplitude of the reflected wave increases and its traveling time decreases. The feature VSW/ISW instead does not show any trend. This is likely due to the combination of the vibrational modes that change with the variation of the axial stress. However, the time of flight of the VSW denotes a sharp

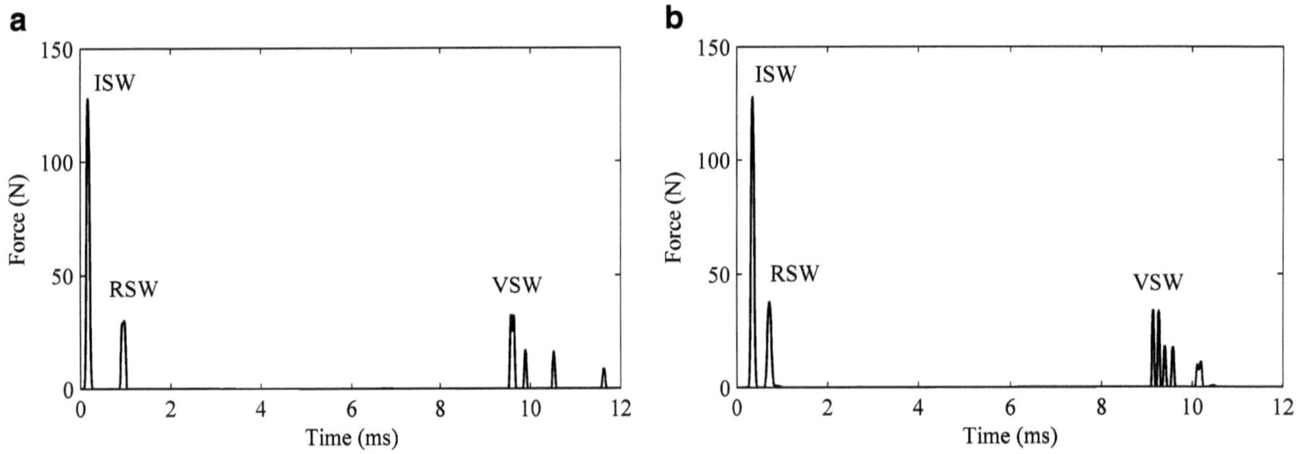


**Fig. 6.2** One chain configuration. Force profile measured at: (**a**) sensor S1, and (**b**) sensor S2, when the axial stress in the beam is zero

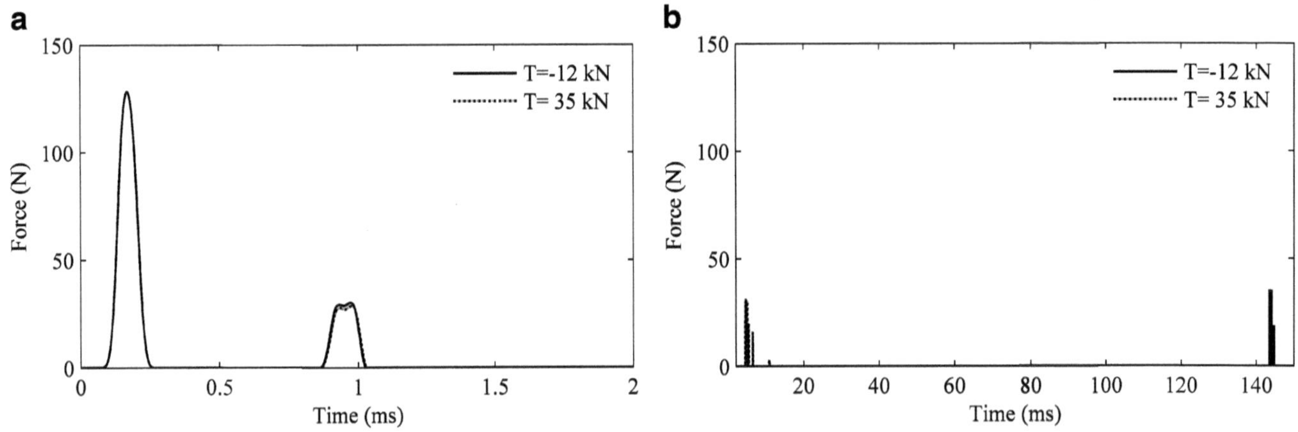


**Fig. 6.3** One chain configuration. Comparing force profiles measured at sensor S1 at largest compressive and tensile load. (**a**) The reflected and incident waves, and (**b**) the vibration-induced wave

**Table 6.1** Features selected in this study for one chain configuration

| Feature | Description |
|---|---|
| RSW/ISW | Ratio of the RSW amplitude divided by the ISW amplitude |
| VSW/ISW | Ratio of the VSW amplitude divided by the ISW amplitude |
| TOF RSW | Transit time between the incident and the reflected wave |
| TOF VSW | Transit time between the incident and the vibration-induced wave |

increase as the stress approaches buckling. By comparing the time of flight associated with both the RSW and the VSW, it can be seen that the latter is more sensitive to the axial stress. This is related with the period of vibration of the beam, which can be expressed in terms of the angular natural frequency $\omega_n$ as a function of the axial load [12]:

$$\omega_n = \frac{\beta_n}{\sqrt{m_b}}\sqrt{T + \beta_n^2 E_b I} \tag{6.4}$$

where $T$ is the axial load, $\beta_n$ is the $n$-th root of the characteristic equation, and subscript $b$ refers to the beam.

To compare directly the sensitivity of the four features to the change of stress, we normalized the values plotted in Fig. 6.4 with respect to their corresponding maximum value. The results are presented in Fig. 6.5. The feature associated with the amplitude of the reflected wave decreases by 6 % when $T$ spans from buckling to yielding. Across the same range, the time of flight relative to the vibration-induced wave decreases by 95 %.

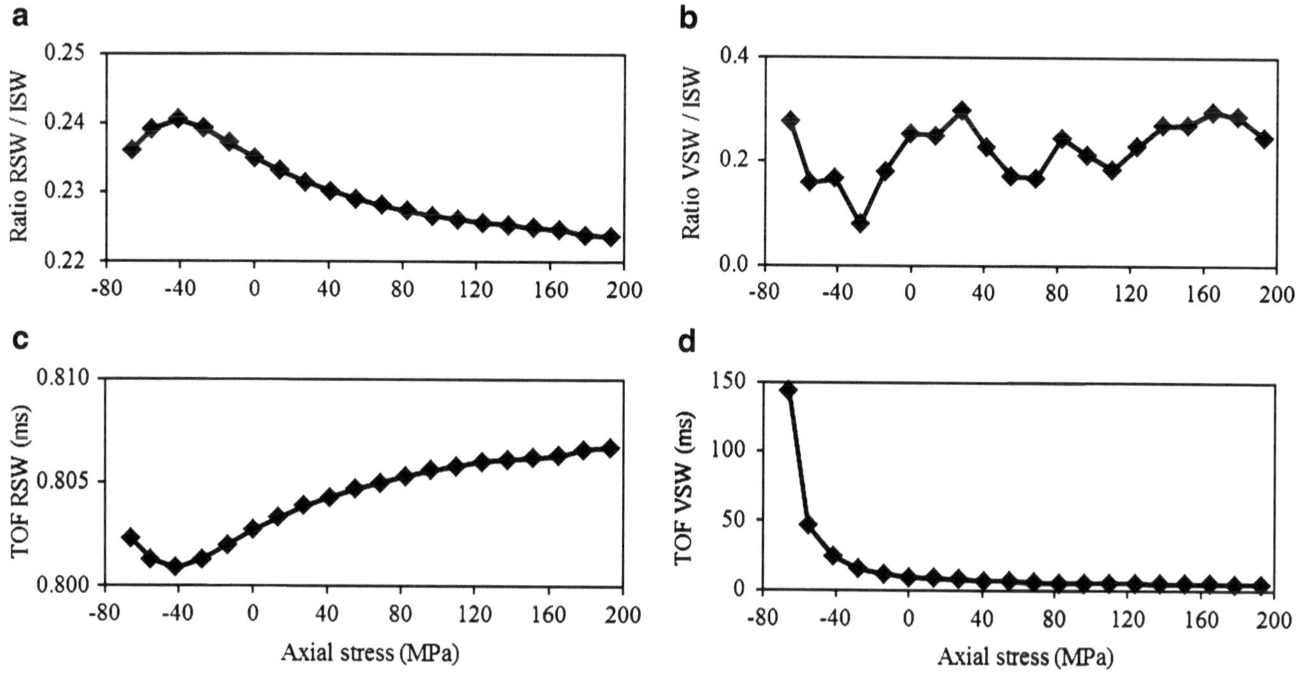


**Fig. 6.4** One chain configuration. Solitary wave features extracted from the force profiles measured at sensor S1 as a function of the axial stress. (**a**) Ratio associated with the amplitude of the reflected and incident waves. (**b**) Ratio associated with the amplitude of the vibration-induced and incident waves. (**c**) Time of flight of the reflected wave. (**d**) Time of flight of the vibration-induced wave

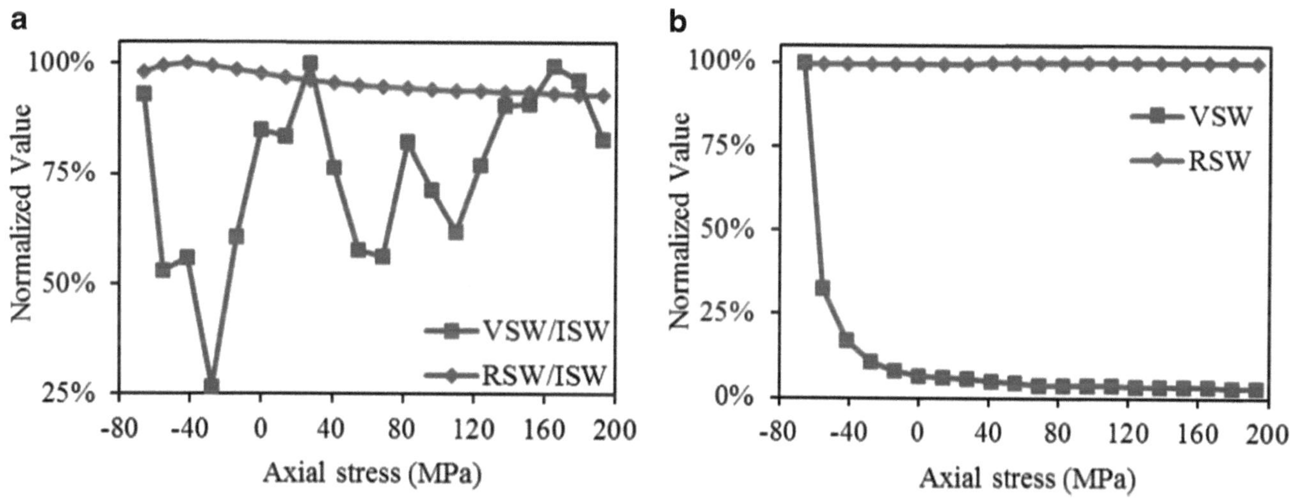


**Fig. 6.5** One chain configuration. Normalized features extracted from the force profiles measured at sensor S1 as a function of the axial stress. (**a**) Ratio associated with the amplitude of the reflected and incident waves, and ratio relative to the amplitude of the vibration induced and incident waves. (**b**) Time of flight of the reflected wave and the vibration-induced wave

## 6.3 Two-Chain Configuration

The use of a single granular medium, although effective, may result unpractical as the contact between the chain and the structure being inspected must be guaranteed. Thus, we also modeled the presence of a second chain of particles located at the opposite face of the beam's mid-span according to the scheme shown in Fig. 6.1b. The objective of the study was to determine whether this second configuration offer the same sensitivity to the presence of axial stress.

### 6.3.1 Numerical Formulation

For the sake of clarity we define as chain 1, the granules subjected to the action of the striker, i.e. the granules where the solitary wave pulse is induced. Therefore, chain 2 indicates the chain where the solitary wave is transmitted after traveling through the cross-section of the beam.

For the particles located at the two ends of the first chain, i.e. when $i = 1$ and $i = N$, we can use Eq. (6.2). For the particles located at the two ends of the second chain, i.e. when $i = N + 1$ and $i = 2N$, the equation of motion becomes:

$$\ddot{u}_{2N-1} = \eta_c[(\delta_0 - u_{2N-1} + u_{2N-2})]_+^{3/2} - \eta_c[(\delta_0 + u_{2N-1})]_+^{3/2} \tag{6.5a}$$

$$\ddot{u}_{2N} = \eta_c[(\delta_0 + u_{2N-1})]_+^{3/2} - \eta_c(\delta_0)^{3/2} \tag{6.5b}$$

This formulation assumes that the last particle of chain 2 is fixed. The differential equation of motion for other particles is same as Eq. (6.1). The discrete particle model is then coupled the continuous beam theory as described before. In the simulation, the effect of the last particle in the first chain was simulated by applying a force to the beam and vice versa, and the influence of the beam on the first particle in the second chain was simulated by applying a force on the first particle in the second chain and vice versa. The force applied to the beam by the solitary pulse is:

$$F(t) = A_b\left[[(\delta_{b0} + u_N - u_b)]_+^{3/2} - [(\delta_{b0} + u_b - u_{N+1})]_+^{3/2}\right] \tag{6.6}$$

By using this force, the beam acceleration $\ddot{u}_b$ can be computed by Eq. (6.3).

### 6.3.2 Numerical Setup

The beam and the granules were the same as in Sect. 6.2.2. The dynamic force associated with the propagation of the solitary pulses was measured at the four sensor sites, indicated as S1, …, S4, shown in Fig. 6.1b. Sensors S3 and S4 were located at the 5th and 10th particle location, respectively, in chain 2. For generation of the HNSW the same an initial velocity of 0.3018 m/s was applied to the first particle.

### 6.3.3 Numerical Results

Figure 6.6 shows the force profile measured at the sensing sites S2 and S3 when the axial stress is null. Three main pulses are visible in the first chain (Fig. 6.6a). They are the ISW, the RSW, and the vibration-based solitary wave which we indicate here as (FVSW), where the first letter in the acronym identifies the fact that this pulse is measured in the first chain. Figure 6.6b shows the force profile measured in the second chain. The first pulse is generated by the transmission of the ISW through the beam's thickness and we named it the transmitted solitary wave (TSW). The next few pulses are generated by the vibration of the beam and they are indicated as SVSW, where the first letter of the acronym denotes the second chain. The third pulse is generated by the reflection of the transmitted pulse from the last particle of chain 2 which is fixed; we named this pulse as the boundary reflected solitary wave (BRSW). The figure shows that the BRSW is higher than the transmitted wave; we hypothesize that this is due to constructive interference of the SVSW and the TSW at the end of the second chain; in fact, it can be seen from Fig. 6.6b that the amplitude of the BRSW is about the sum of the amplitudes of the TSW and the SVSW.

Similar to one-chain configuration study, we extracted a few features from the solitary waves that are listed in Table 6.2.

The normalized values of the solitary wave features are presented in Fig. 6.7. The ratios TSW/ISW and TSW/RSW show a few percent variation across the whole range of linear stress, and the reason is that the reflected and transmitted pulse are minimally affected by the beam's stiffness and axial load. Similar conclusion can be drawn for the time of flight associated with these two pulses. The TOF of the FVSW smoothly decreases with the increase of the axial stress, and it reduces by ~15 % within the compression range. We believe that this low variation is related to the presence of the second chain with fixed boundary condition. Finally, the largest variation with respect to the axial stress is visible at the feature associated

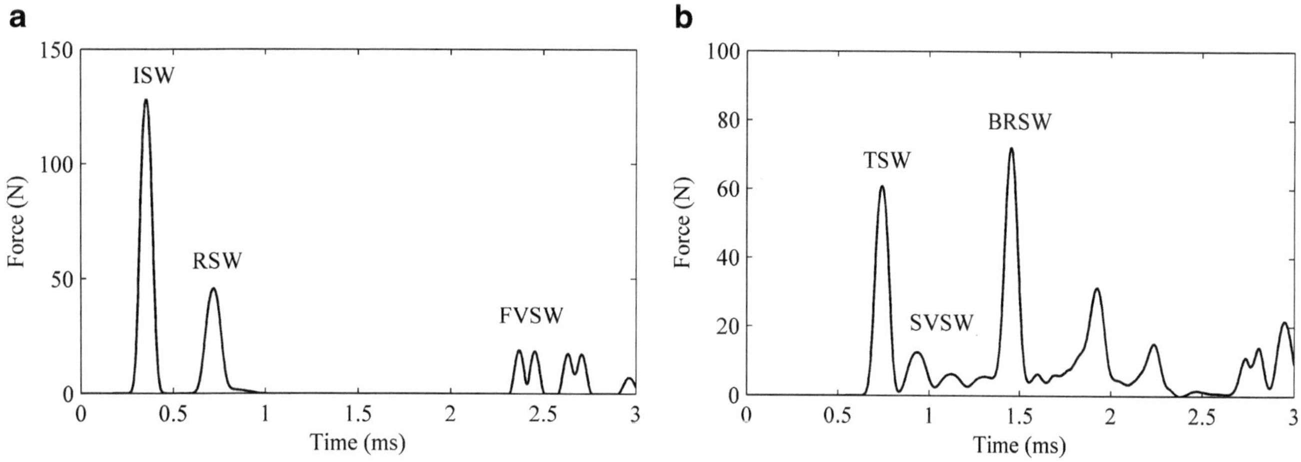


**Fig. 6.6** Two chains configuration. Force profile measured at: (**a**) sensor S2, and (**b**) sensor S3, when the axial stress in the beam is zero

**Table 6.2** Features selected in this study for two chains configuration

| Feature | Description |
|---|---|
| TSW/ISW | Ratio of the TSW amplitude divided by the ISW amplitude |
| TSW/RSW | Ratio of the TSW amplitude divided by the RSW amplitude |
| FVSW/ISW | Ratio of the FVSW amplitude divided by the ISW amplitude |
| SVSW/TSW | Ratio of the SVSW amplitude divided by the TSW amplitude |
| TOF RSW | Transit time between the incident and the reflected wave |
| TOF TSW | Transit time between the reflected wave and the transmitted wave |
| TOF FVSW | Transit time between the incident and the FVSW |
| TOF SVSW | Transit time between the transmitted wave and the SVSW |

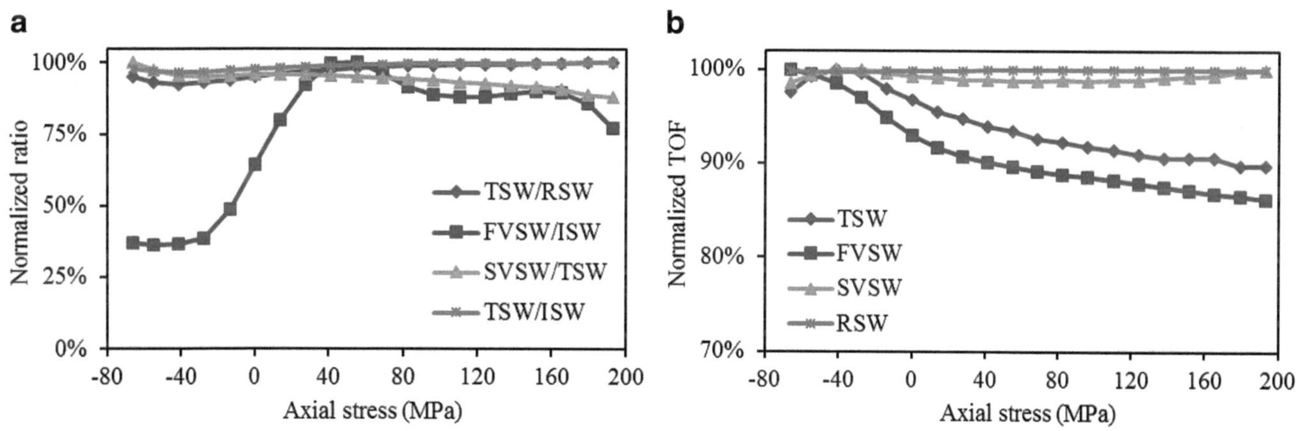


**Fig. 6.7** Two chains configuration. Normalized features extracted from the force profiles measured at sensors S2 and S3 as a function of the axial stress

with the amplitude of the vibration-induced solitary wave propagating in the first chain (feature FVSW/ISW). However, this feature is not helpful at predicting buckling as there is no change before buckling.

The comparison between Figs. 6.5 and 6.7 suggests that the TOF relative to the beam's vibration measured by the single chain provides the largest sensitivity to the presence of axial stress. This is somehow expected as the presence of the second chain restraints the motion of the linear system.

## 6.4 Effect of the Granules' Properties

We studied the effect of the granules' properties on the sensitivity of the proposed method. We examined single chains of particles made of seven particles' diameters namely 5, 10, 15, 20, 25, 30, and 35 mm, and the eleven materials listed in Table 6.3. Thus, a total of 77 different chains were examined. For the analysis, we simulated the same beam. We applied an initial velocity of 0.2566 m/s to the first particle of the chain, and the results presented in this section refers to the solitary pulse measured at the sensing site S2.

Figure 6.8 shows a 3-D diagram where the ratio RSW/ISW is presented as a function of the axial stress and particles' diameter. The cases of Copper and Aluminum are only presented. Overall, we observe that at a given stress the amplitude of the reflected wave decays as the diameter of the granules increases. When the particles are smaller and lighter the acoustic energy carried by the granular system is not sufficient to trigger the beam's vibration, the energy bounces back to the chain, and the ratio RSW/ISW is close to 1. As the particles become bulkier, the amount of energy carried by the reflected pulse decreases at advantage of the beam's vibration. This is visible in Fig. 6.9 which displays the results associated with the ratio VSW/ISW. Here the amount of energy carried by the pulse generated by the beam's vibration increases with the bulkiness of the particles. Moreover, the trend of the amplitudes is not monotonous at any particle diameter or at any material; this trend confirms what we already discussed in Fig. 6.4. The force applied to the beam by the solitary pulse depends on the physical

**Table 6.3** The properties of the considered materials for particles

| Material | Density (kg/m$^3$) | Young's modulus (GPa) | Poisson's ratio |
|---|---|---|---|
| Tungsten | 19,250 | 411 | 0.28 |
| Copper | 8,960 | 120 | 0.34 |
| Nickel | 8,908 | 200 | 0.31 |
| Stainless steel 302 | 8,100 | 196 | 0.33 |
| Stainless steel 316 | 8,000 | 193 | 0.3 |
| Iron | 7,874 | 211 | 0.29 |
| Stainless steel 440C | 7,800 | 200 | 0.28 |
| Zinc | 7,140 | 108 | 0.25 |
| Titanium | 4,506 | 116 | 0.32 |
| Aluminum | 2,700 | 70 | 0.35 |
| Magnesium | 1,738 | 45 | 0.29 |

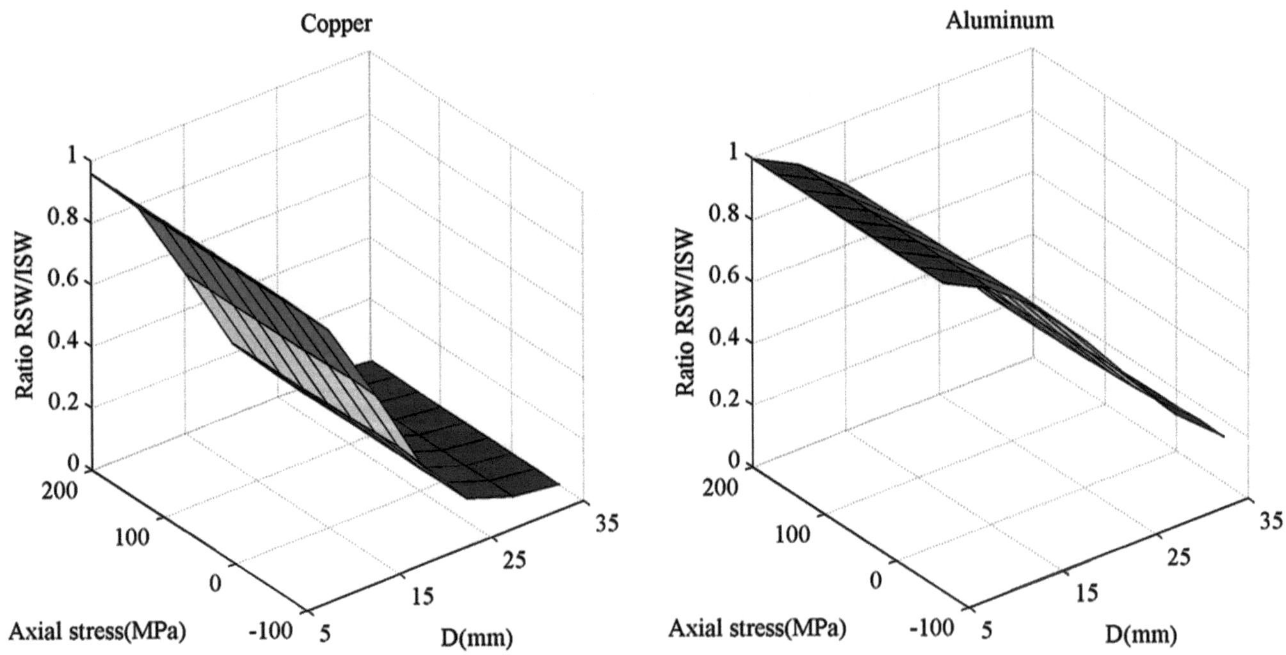


**Fig. 6.8** One chain configuration. Ratio associated with the amplitude of the reflected and incident waves as a function of the axial stress and the particles' diameter

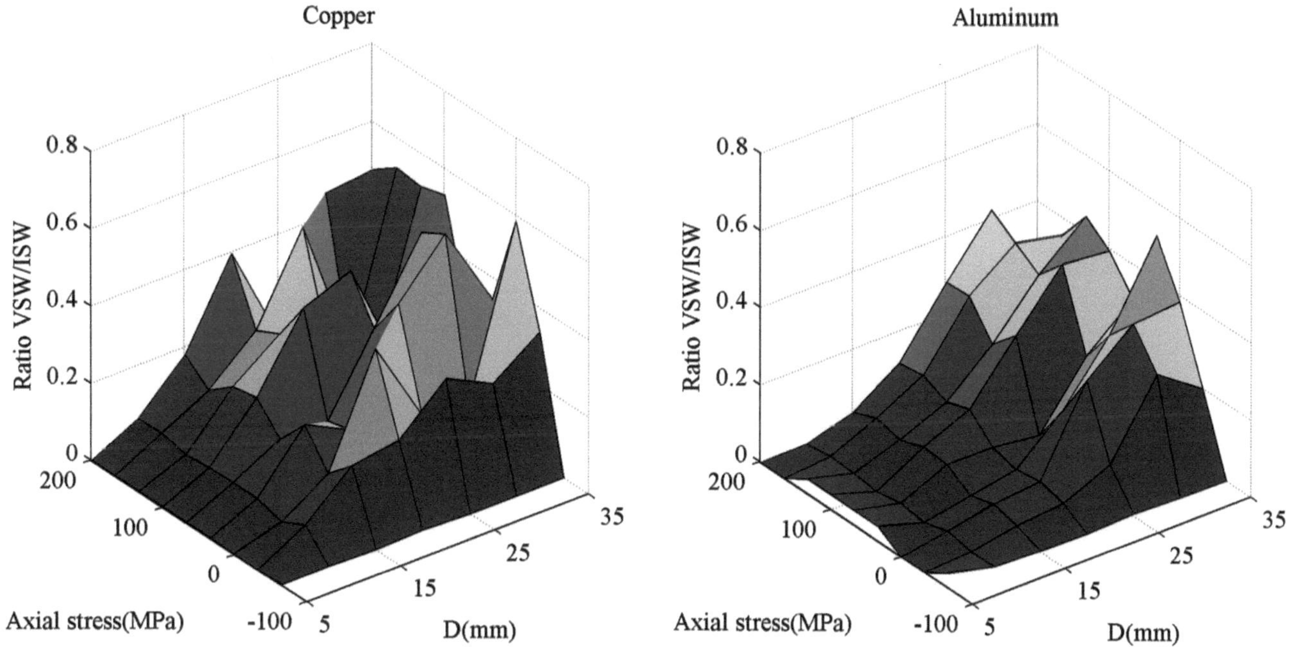


**Fig. 6.9** One chain configuration. Ratio associated with the amplitude of the vibration-induced and incident waves as a function of the axial stress and the particles' diameter

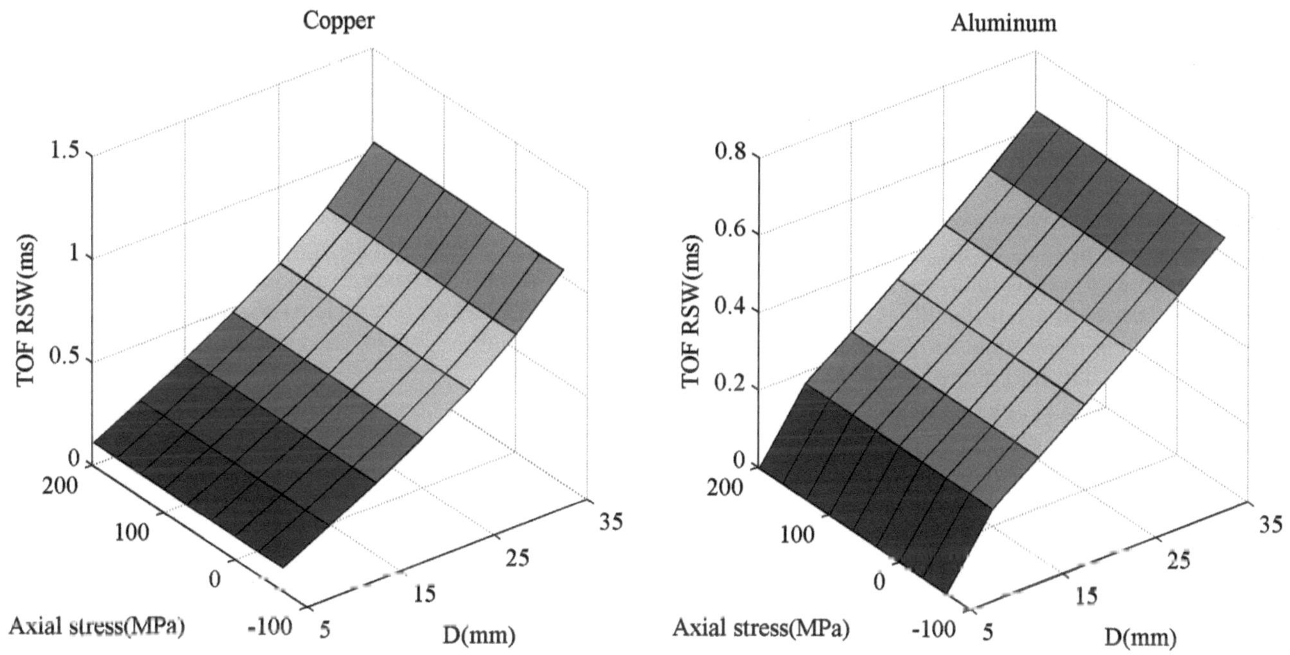


**Fig. 6.10** One chain configuration. Time of flight of the reflected wave as a function of the axial stress and the particles' diameter

and mechanical properties of the particles; in addition, the beam's motion is the combination of many modes of vibration. We conclude that for this complex dynamical system, the feature associated with the amplitude of the vibration-induced solitary wave is not monotonic. At a given diameter the variation of the ratio is similar to what found and discussed in Sect. 6.2.3.

Similar to Figs. 6.8 and 6.9, Figs. 6.10 and 6.11 show the feature associated with the time of flight of the reflected and the vibration-induced solitary wave, respectively. The TOF of the reflected wave (Fig. 6.10) does not change significantly with stress irrespective of the particles diameter or the particles materials. Thus, the same conclusion drawn in Fig. 6.4 can

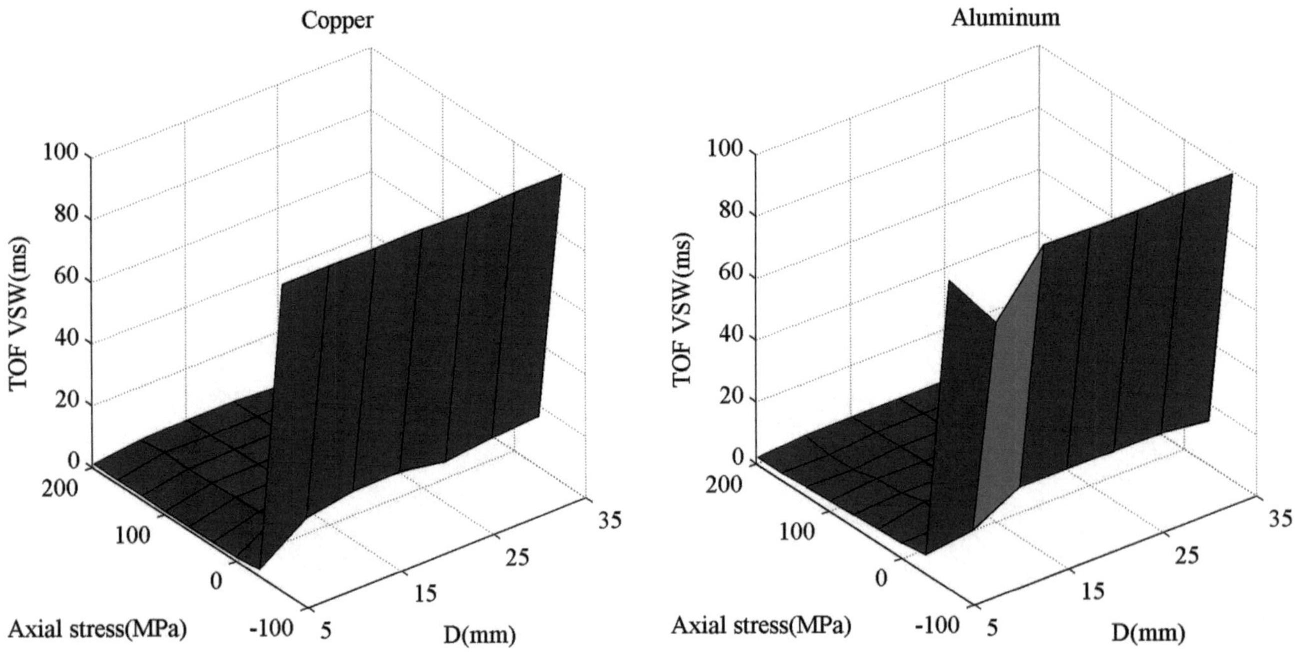


**Fig. 6.11** One chain configuration. Time of flight of the vibration-induced solitary wave as a function of the axial stress and the particles' diameter

be applied here. The result is consistent with the previous findings, i.e. as the axial stress does not affect the beam's stiffness, there is little influence on the generated RSW. At a given stress, the feature's value increases proportionally to the increase of the particle diameter, i.e. proportionally to the relative distance between the sensing sites. In fact, the variation of the travel time due to the change of speed is lower than the variation associated to the increased distance between the sensors. Overall, this trend is identical for all the different materials considered in this study. Figure 6.11 shows the TOF relative to the vibration-induced solitary wave as a function of the axial stress and the particle diameters. The value of this feature rapidly increases as the stress approaches buckling, due to the sharp decrease of the beam's natural frequency of vibration. Based on results of all analysis, the TOF of the wave propagating along the heaviest particles, made of Tungsten, presents the largest gradient with respect to the axial stress. This is likely related to the particles Young's modulus, which contributes to the stiffness constant $\eta_c$ and to the differential equations of particles (Eqs. 6.1 and 6.2). Particles with small stiffness constant generate an amount of dynamic force lower than same diameter particles with larger stiffness, and therefore lower beam's vibration, which results into shorted period of vibration.

In conclusion, with the exception of a few materials and features, we cannot conclude that the largest diameter or the heaviest particle or both provide the highest sensitivity.

## 6.5 Conclusions

In this paper, we presented a numerical study on the mechanical interaction between a granular medium and a slender beam. In particular, a chain of spherical particles is considered as the carrier of a single HNSW that reaches the interface between the granular medium and the beam. The effect is that part of the acoustic energy is reflected back to the chain and part triggers the vibration of the beam. In this study, we evaluated the effect of axial stress applied to the beam on certain characteristics of the solitary pulses. The objective of the study was the development of a noninvasive low-cost system to assess axial stress in beam-like structures. The results were then compared to a similar setup where a second chain of particles adhered at the opposite face of the beam. In this case, the features associated with the solitary waves transmitted across the thickness of the beam were considered.

We found that the time of flight and the amplitude of the solitary waves are affected by the axial stress and their sensitivity is more evident when a single chain is considered. However, the single chain case may be detrimental as it requires the contact between the granular medium and the structure of interest. This contact can be easily guaranteed when two chains are used instead. With respect to ultrasonic methods based on acoustoelasticity, the proposed method may provide unique advantages in terms of larger sensitivity, portability, and cost.

Once the sensitivity of a single-chain configuration was ascertained, we considered particles of different diameter and materials to find the optimal combination that provides the largest sensitivity to the variation of axial stress. We found that the largest and heaviest particles do not necessarily provide the best design and we believe that a generalization cannot be made.

**Acknowledgments** The authors acknowledge the support of the U.S. Federal Railroad Administration under contract DTFR53-12-C-00014. Partial support came from the U.S. National Science Foundation, grant #1200559.

## References

1. Li F, Yu L, Yang J (2013) Solitary wave-based strain measurements in one-dimensional granular crystals. J Phys D Appl Phys 46(15):155106
2. Ni X, Rizzo P, Daraio C (2011) Laser-based excitation of nonlinear solitary waves in a chain of particles. Phys Rev E 84(2):026601
3. Nesterenko VF (1983) Propagation of nonlinear compression pulses in granular media. J Appl Mech Tech Phys 24:733–743
4. Fraternali F, Porter MA, Daraio C (2009) Optimal design of composite granular protectors. Mech Adv Mater Struct 17:1–19
5. Sen S, Manciu M, Wright JD (1998) Solitonlike pulses in perturbed and driven Hertzian chains and their possible applications in detecting buried impurities. Phys Rev E 57:2386–2397
6. Nesterenko VF, Daraio C, Herbold EB, Jin S (2005) Anomalous wave reflection at the interface of two strongly nonlinear granular media. Phys Rev Lett 95:158702
7. Ni X, Rizzo P, Yang J, Katri D, Daraio C (2012) Monitoring the hydration of cement using highly nonlinear solitary waves. NDT E Int 52:76–85
8. Ni X, Rizzo P (2012) Highly nonlinear solitary waves for the inspection of adhesive joints. Exp Mech 52:1493–1501
9. Yang J, Silvestro C, Sangiorgio SN, Borkowski SL, Ebramzadeh E, De Nardo L, Daraio C (2012) Nondestructive evaluation of orthopaedic implant stability in THA using highly nonlinear solitary waves. Smart Mater Struct 21:012002
10. Cai L, Rizzo P, Al-Nazer L (2013) On the coupling mechanism between nonlinear solitary waves and slender beams. Int J Solids Struct 50(25):4173–4183
11. Ni X, Rizzo P (2012) Use of highly nonlinear solitary waves in NDT. Mater Eval 70(5):561–569
12. Tedesco J, Mcdougal W, Ross C (1999) Structural dynamics-theory and applications. Addison Wesley, Menlo Park

# Chapter 7
# Are Today's SHM Procedures Suitable for Tomorrow's BIGDATA?

**Thomas J. Matarazzo, S. Golnaz Shahidi, Minwoo Chang, and Shamim N. Pakzad**

**Abstract** Large SHM datasets often result from special applications such as long-term monitoring, dense sensor arrays, or high sampling rates. Through the development of novel sensing techniques as well as advances in sensing devices and data acquisition technology, it is expected that such large volumes of measurement data become commonplace. In anticipation of datasets magnitudes larger than today's, it is important to evaluate current SHM processing methods at BIGDATA standards and identify potential limitations within computational procedures. This paper will focus on the processing of large SHM datasets and the computational sensitivity of common SHM procedures. Processing concerns encompass efficiency and scalability of SHM software, particularly the computational sensitivity of common system identification and damage detection algorithms with respect to a large amount of sensors and samples.

**Keywords** BIGDATA • Computational cost • Signal processing • System identification • Damage detection

## 7.1 Introduction

The data acquisition process plays a key role in the monitoring of structural systems. Over the years, sensing networks have evolved from basic to complex. Today, structural response quantities are measured with high resolution in time and space by means of wired and wireless contact sensors as well as noncontact digital camera lenses. In order to infer about structural health condition, the measured signals are processed through various SHM algorithms (system identification, model calibration, and damage detection). Therefore, along with improvement in sensing technology, it is critical to evaluate the performance of SHM algorithms to study their scalability potential and overcome possible limitations in processing large SHM datasets. Toward this goal, this paper investigates performance of a subset of the current SHM procedures with growth of network size and data acquisition interval. This investigation is conducted in two parts: the preprocessing of measured data as the first step in any SHM procedure, and the computational cost of certain system identification and damage detection algorithms. Although it is expected that BIGDATA storage requirements will be far greater than limits of local hard disks and RAM, this aspect of the problem is outside of the scope of this paper.

## 7.2 Exactly How Big Is "BIGDATA"?

Today, a dataset earns the title "large" from the sizes of its dimensions and storage space. SHM datasets are typically multivariate times series with two dimensions: number of sensors and number of time samples. However, the mere storage requirements, say in megabytes, of a dataset is not enough to constitute the title "BIGDATA". Inherently, BIGDATA requires "BIGPROCESSING", a subsequent analysis of BIGDATA, of which computational complexity varies among SHM applications and techniques.

T.J. Matarazzo (✉) • S.G. Shahidi • S.N. Pakzad
Department of Civil and Environmental Engineering, Lehigh University, ATLSS Engineering Research Center, 117 ATLSS Drive, Imbt Labs, Bethlehem, PA 18015, USA
e-mail: tjm310@lehigh.edu; sgs310@lehigh.edu; pakzad@lehigh.edu

M. Chang
Department of Civil and Environmental Engineering, Utah State University, Logan, UT 84322, Australia
e-mail: cmw0321@gmail.com

© The Society for Experimental Mechanics, Inc. 2015

C. Niezrecki (ed.), *Structural Health Monitoring and Damage Detection, Volume 7*, Conference Proceedings of the Society for Experimental Mechanics Series, DOI 10.1007/978-3-319-15230-1_7

A dataset may be considered BIGDATA in cases where its dimensions are large relative to the analysis procedure; however, categorizing BIGDATA sizes for each SHM procedure would be cumbersome as well as ineffective. For example, it would be difficult to diligently identify BIGDATA dimension cut-offs for a selection of system identification (SID) algorithms. Additionally, such SID BIGDATA cut-offs would not be consistent with those from preprocessing, damage detection, and (FE) element model updating perspectives. Each threshold would be unique and distant – causing an undesirable discordance among SHM subdisciplines. Furthermore, with the evolution of computational hardware and software, such results would only be accurate and therefore useful for a limited period of time.

As a result, we can conclude the question "Exactly how big is BIGDATA?" is a red herring; it is not our objective to determine an accurate response to this inquiry. Instead, the definition of BIGDATA should remain a subjective projection of datasets magnitudes larger than modern sizes [1], continuously changing with time.

## 7.3 Computational Sensitivity of SHM Procedures

In lieu of precise estimates for BIGDATA dimensions, we can begin to examine BIGPROCESSING capabilities of modern SHM procedures. More specifically, the computational sensitivity of SHM methods with respect to each data dimension provides a useful tool for measuring general BIGDATA sufficiency. The sensitivity analysis approach has two main goals: (I) assess scalability of each SHM procedure and (II) identify operations/subroutines that are susceptible to bottlenecking for certain data volumes.

The following sections present a BIGDATA (by today's standards) sensitivity analysis for selected SHM procedures. The following analyses consider a range of BIGDATA dimensions. As stated previously, the exact dimensions are not critical – the objective is to establish the basic relationship between common SHM processes and computational efforts as these dimensions can increase dramatically. In this paper, computational costs are measured in CPU run time. It is important to note that for a given, stand-alone SHM function, run times vary greatly among CPUs. It is clear that an algorithm's performance fluctuates among implementations, i.e., how it is coded and compiled, but computer hardware and software specifications also contribute significantly to the variability of this measure.

## 7.4 Preprocessing Application

Preprocessing is an integral step present in all SHM procedures. Whether it is included within the data acquisition system or it is a successive routine, preprocessing encompasses the operations in between recording data and the desired SHM function. For the purposes of this paper, SHM preprocessing is comprised of two main steps: data loading and data cleansing. The latter is assumed to consist of detrending, filtering (via FFT for example [2]), and downsampling. For typical data sizes, preprocessing computational efforts are often negligible when compared to those of subsequent SHM functions; however, for completeness it is prudent to include sensitivity analyses of such SHM preprocesses.

This section examines built-in Matlab functions corresponding to the aforementioned SHM preprocessing steps. More specifically, the run times of 'load', 'detrend', 'fft', and 'downsample' [3] are considered for large data volumes. The run times of these functions are measured using Matlab's built-in 'tic' and 'toc' functions for a range of data matrices sized between 10 and 1,000 sensors as well as 10,000 and 100,000 samples. All preprocessing scripts are executed with Matlab R2014a on a machine equipped with Windows 7 enterprise 64-bit, an Intel Xeon processor, and 4 GB RAM.

In this application, data matrices were randomly generated using Matlab at the specified dimensions. Each resulting matrix was saved as a .txt file to exemplify common storage and imported into Matlab using 'load'. Other preprocessing procedures were timed directly after a random dataset of certain dimension was generated. Details of the selected Matlab functions can be found within online documentation [3]. A brief description of each use in this study is as follows, where a vector represents a data acquisition channel:

detrend: remove any constant means from the simulated datasets;
downsample: decrease the sampling frequency by a factor of two;
fft: compute discrete Fourier transform for each vector of the dataset.

The simulations were repeated fifty times at each data size, and the median run times were reported. Figure 7.1 displays the speed of each preprocessing operation in seconds of CPU run time. Foremost, it is necessary to note that none of these operations are scalable with respect to number of sensors or samples, i.e., doubling an input (data dimension) of these

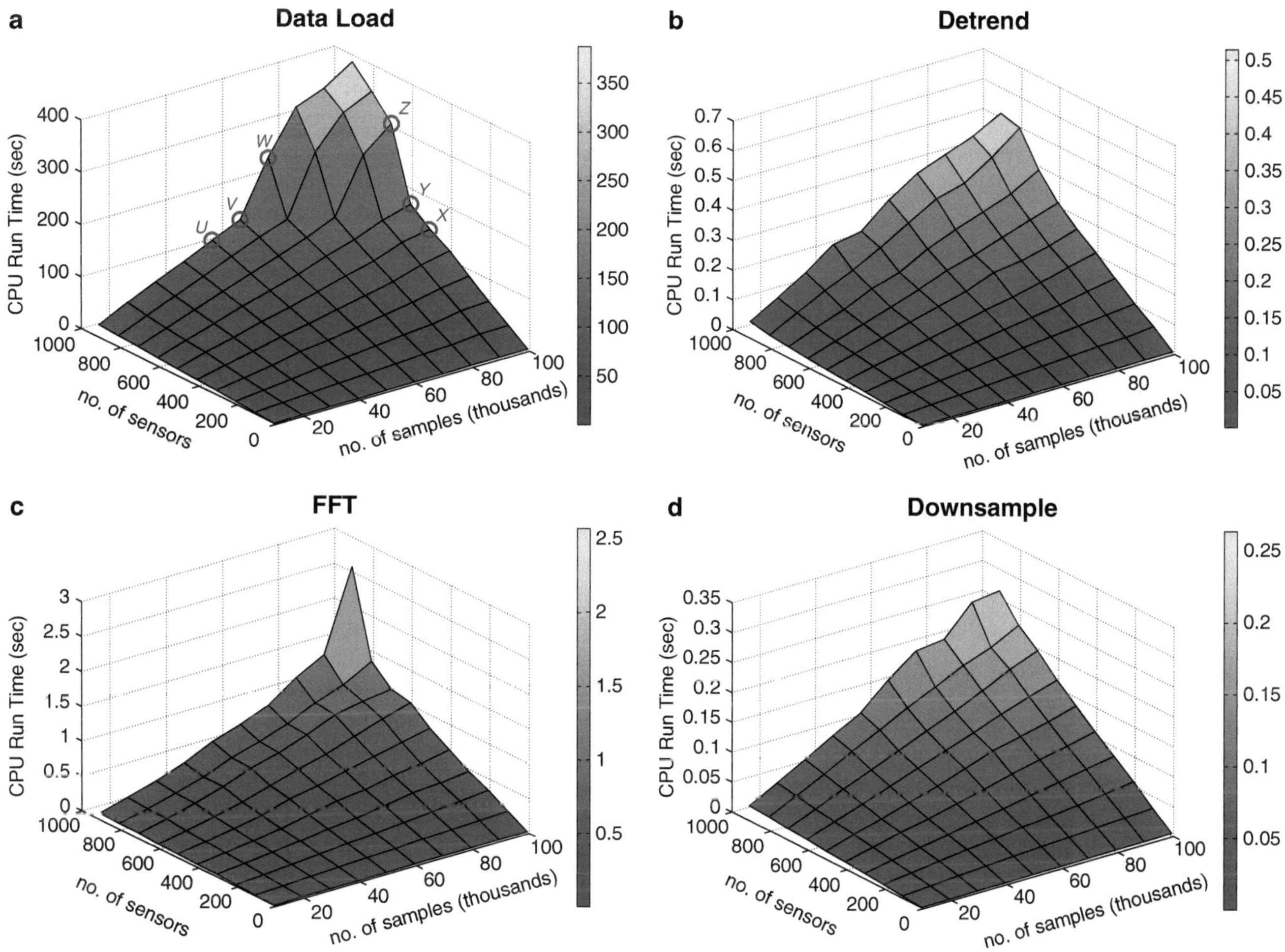


**Fig. 7.1** BIGDATA sensitivity for built-in Matlab functions: (**a**) Load (**b**) Detrend (**c**) FFT (**d**) Downsample; measured CPU run times are in seconds

**Table 7.1** Selected gradient steps from Fig. 7.1a describe bottlenecking behavior

| Gradient | Change in no. of elements (millions) | Change in CPU run time (sec) |
|---|---|---|
| UV | +9.1 | +24 |
| VW | +9.1 | +104 |
| XY | +10.0 | +29 |
| YZ | +10.0 | +136 |

functions does not necessarily double CPU run time. For the specified data ranges, these preprocessing functions behave in a similar manner with respect to each dimension, that is, an increase in sensors is not significantly more expensive than an increase in samples and vice versa – they are both approximately linear in this regard.

The load function in Fig. 7.1a is the most computationally expensive process of the four, with run times roughly two orders of magnitude larger than that of FFT (Fig. 7.1c) for the same size data matrix. Downsampling and detrending functions demonstrated similar, stable behaviors, despite slight variations; these were the least computationally expensive preprocesses of the group which could be expected as they perform basic mathematical operations on the data.

Table 7.1 details selected gradients in Fig. 7.1a which outline an apparent bottlenecking within the load function. More specifically, two steps of each selected gradient are considered: UV, VW, XY, and YZ. In each gradient, the second step (VW and YZ) required significantly more computational effort than the first step (UV and XY) despite a steady increase in data elements (9.1 million and 10 million, respectively). While it may seem tempting to pinpoint a numerical value for a potential BIGDATA threshold in this application, it would only be a point estimation of a random variable, which is of little use. As previously mentioned, it is difficult to identify a precise threshold that would normally trigger this behavior; moreover, it is necessary to recognize its existence.

## 7.5 System Identification Application

System identification (SID) is the process of extracting structural modal properties from measured sensor data. Modern output-only methods such as N4SID [4], AR [5], ERA-NExT [6], ERA-OKID-OO [7], and SSI [8] are well-established in SHM. Typically, these methods require a model order selection procedure, i.e., repeated applications at increasing model complexities to determine the most efficient model order. A stabilization diagram is a graphical tool for presenting the identification results, facilitating model choice and retrieval of estimated modal properties. Model order selection is known to require substantial computational efforts as stabilization diagrams often consider high model orders, e.g., an ARMA model of order forty was necessary to identify ten modes below 0.5 Hz using Golden Gate Bridge accelerometer data [9].

As a result, new identification methods have been developed to identify modal parameters using fewer operations, e.g., ERA-NExT-AVG [10], and/or at lower model orders, e.g., STRIDE [11]. Additionally, software packages such as SMIT [7], ISHMP [12], or SHMTools [13] assist in comparing efficiency and accuracy among methods.

This section investigates BIGDATA computational sensitivity for two output-only SID methods: AR and ERA-OKID-OO. The following analysis considers large datasets between 10 and 1,000 sensors, as well as between ten thousand to one million samples in Matlab using the same specifications (CPU) as the previous section. Due to computational intensity of the algorithms and large data dimensions under consideration, run times were found by combining timed CPU runs with simulations. It was assumed that each algorithm produced a stabilization diagram presenting even model orders from 2 to 100. Therefore, the total run time to process each data matrix size includes fifty identification implementations.

The sensitivity analysis results are summarized in Fig. 7.2 where run times are estimated in hours. Again, it is clear that none of these methods are scalable with respect to number of sensors or samples. Furthermore, for all three algorithms, the run times are substantially more sensitive to an increase in sensor size than to an increase in sample size. More specifically, sensors affect run times cubically while samples have a linear relationship with computational efforts. It is important to note that the slopes of the lines regressing samples on to run times are dependent on the number of sensors in the analysis; a larger number of sensors yields a higher slope. This result is most evident for AR in Fig. 7.2a. The predicted run times for AR in Fig. 7.2a require approximately two orders of magnitudes greater than those of ERA-OKID-OO.

These sensitivity results describe a general behavior of certain SID methods when datasets become very large – much larger than typical sizes by today's standards. It is not yet possible to predict model performance or accuracy of modal estimates for these cases; such an assessment would require large volumes of data that approach BIGDATA magnitudes in at least one dimension.

## 7.6 Damage Detection Application

This section investigates computational cost of a subset of data-driven damage detection methods. Through last decades these methods have been widely used as efficient ways to identify occurrence of any damage induced change in structural properties without using a finite element simulation of the structure [14, 15]. Certain steps are common among damage

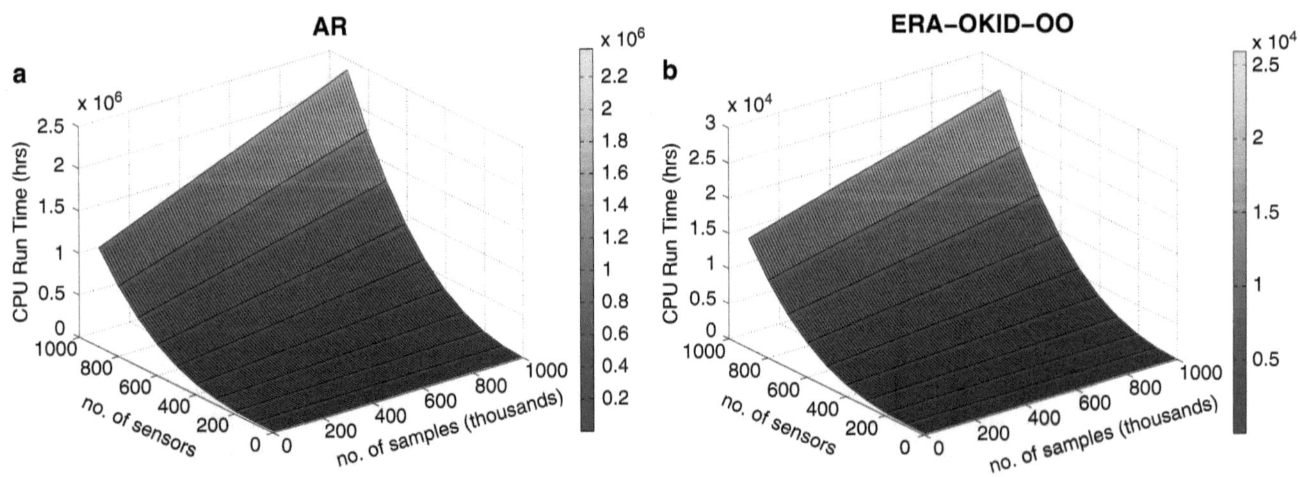


**Fig. 7.2** BIGDATA sensitivity for AR and ERA-OKID-OO output-only identification methods; total CPU run times are estimated in hours

detection algorithms, e.g., training monitoring data using a time series model. The goal of this training step is to extract features from measured signals, features that are sensitive to damage, yet robust to environmental factors. These features are generated from data collected throughout subsequent inspection periods. Statistical tests have been developed to detect changes in such features, which indicate structural damage [16, 17].

Since damage features are often scalar values, univariate statistical tests are utilized to detect possible changes among feature vectors associated with one location. It is expected that the computational cost of such statistical testing is less than those of the feature extraction process. Therefore, in this study, only CPU time for the creation of damage features is measured and considered. In order to estimate total CPU time required to process SHM datasets for damage detection, feature extraction for one sensor (or one sensor pair) is measured while the number of time samples varies from ten thousand to one million. The total CPU time is calculated for all possible sensor groupings in the network.

Three regression models were used to generate damage features. First, Single Variate Regression (SVR) regresses responses from two sensing locations onto each other, in which regression coefficients are interpreted as damage features. Second, autoregressive (AR) models regress the response at one location onto a delayed version of itself, where the largest time lag defines the model order. Finally, autoregressive with exogenous (ARX) input models are considered. In the ARX model formulation, the responses of two locations at different time lags are used for regression. These methods are selected since (1) they do not require prior information about topology of structure and material properties, (2) they are adoptable to various types of measured structural responses, and (3) once the regression analysis is performed, several features can be extracted for damage detection [18–22].

Figure 7.3 shows the CPU time in hours for these damage detection methods. Figure 7.3a shows estimated computation time required for AR modeling (with model order fifty). Fig. 7.3b shows the CPU time for feature extraction from SVR models (that are in fact zero order ARX models). Finally, Fig. 7.3c shows the computational cost of ARX modeling for feature generation. It is observed that AR modeling requires the least computational time, while ARX models require the highest,

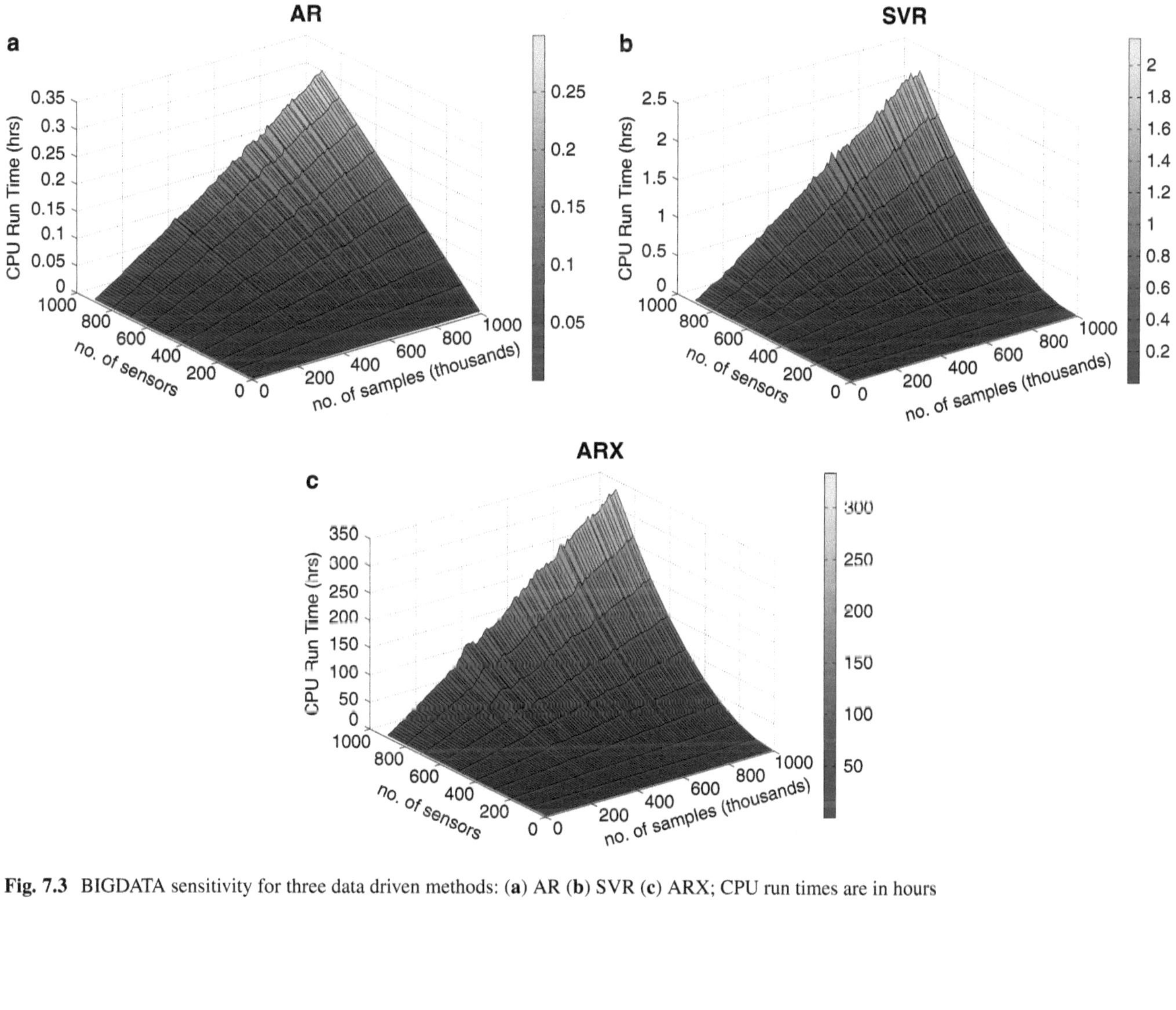


**Fig. 7.3** BIGDATA sensitivity for three data driven methods: (**a**) AR (**b**) SVR (**c**) ARX; CPU run times are in hours

nearly three orders of magnitudes higher than AR modeling for a network composed of one thousand sensors. A comparison of required computational time between SVR and ARX models (Figs. 7.3b, c) indicates a significant increase in effort is necessary for higher model orders. Although, AR modeling is computationally less intensive [21], presents examples of application of SVR and ARX models that outperformed AR models in damage localization. This comparison was performed at a pre-specified model order and did not include the model order selection attempts [23], however it establishes an overall behavior of computation costs for these methods. The results imply these model-free methods are adoptable to analyze large SHM datasets; however, certain improvements are necessary to further reduce this computation time, especially in case of models with higher complexity.

## 7.7 Conclusions

This paper discussed BIGDATA from an SHM perspective. There is a tendency to expand sensor networks, collect more samples, and sample at higher rates with the expectation that an increase in data will improve the quality and/or quantity of extractable structural information. This concept contains an underlying assumption that modern state-of-the-art SHM techniques are capable of processing large datasets; however, this has yet to be verified or analyzed precisely.

This paper presented a preliminary sensitivity analysis for an array of common SHM procedures including preprocessing, system identification, and data-driven damage detection. All procedures were coded and compiled in Matlab R2014a including either estimation or direct measurement of corresponding CPU run times. For SHM preprocessing, the speed of Matlab's built-in functions 'load', 'detrend', 'fft', and 'downsample' were considered for large data volumes. In the examples, it was demonstrated that the load function required the most computational efforts out of these four. Figure 7.1 and Table 7.1 identified and described a bottlenecking behavior indicating that Matlab may have limitations in regards to hauling and internally storing large datasets.

Assuming large data dimensions, the run times for a subset of output-only SHM methods were evaluated. Computation costs associated with time-series analysis based damage detection algorithms using SVR, AR, and ARX models were tested at a single model order. SID methods (AR and ERA-OKID-OO) were estimated for a range of 40 model orders since repeated SID implementations are necessary to produce stabilization plots and to establish adequate model orders. A salient realization from this study was that run times are substantially more sensitive to an increase in sensor size than to an increase in sample size. In general, BIGDATA SID efforts were magnitudes larger than those for preprocessing or damage detection.

In these cases, even basic data operations such as loading may require significantly more computational efforts. Therefore, for more intensive procedures like SID, it is necessary to establish a-priori confidence that such extra efforts guarantee worthwhile information. In other words, does this increase in knowledge justify the cost? Perhaps the cost is so prohibitive that it would be more efficient to redesign the most demanding sub-operations. Physics-based damage detection methods (using FE models) are also known to have computationally demanding operations for intricate FE models. A similar investigation in this field includes merits outside the SHM specialization, introducing BIGDATA to a broader civil engineering community.

**Acknowledgement** Research funding is partially provided by the National Science Foundation through Grant No. CMMI-1351537 by Hazard Mitigation and Structural Engineering program, and by a grant from the Commonwealth of Pennsylvania, Department of Community and Economic Development, through the Pennsylvania Infrastructure Technology Alliance (PITA).

## References

1. Manyika J, Chui M, Brown B, Bughin J, Dobbs R, Roxburgh C, Hung Byers A (2011) Big data: the next frontier for innovation, competition, and productivity, p 143. Retrieved from http://www.mckinsey.com/insights/business_technology/big_data_the_next_frontier_for_innovation
2. Cooley JW, Tukey JW (1965) An algorithm for the machine calculation of complex fourier series. Math Comput 19(90):297–301
3. Mathworks (2014) Functions documentation. Retrieved from http://www.mathworks.com/help/matlab/functionlist.html
4. Van Overschee P, De Moor B (1992) N4SID: subspace algorithms for the identification of combined deterministic-stochastic systems. Automatica 30(1):75–93
5. He X, De Roeck G (1997) System identification of mechanical structures by a high-order multivariate autoregressive model. Comput Struct 64(1–4):341–351
6. James III GH, Carrie TG, Lauffer JP, James GH, Carne TG, Ill GHJ (1993) The natural excitation technique (NExT) for modal parameter extraction from operating wind turbines, Albuquerque, p 46

7. Chang M, Pakzad SN (2013) Observer Kalman filter identification for output-only systems using interactive structural modal identification toolsuite (SMIT). J Bridg Eng 19(5):04014002. doi:10.1061/(ASCE)BE.1943-5592.0000530
8. Peeters B, De Roeck G (1999) Reference-based stochastic subspace identification for output-only modal analysis. Mech Syst Signal Process 13(6):855–878. doi:10.1006/mssp.1999.1249
9. Pakzad SN, Fenves GL (2009) Statistical analysis of vibration modes of a suspension bridge using spatially dense wireless sensor network. J Struct Eng 135(7):863–872. doi:10.1061/(ASCE)ST.1943-541X.0000033
10. Chang M, Pakzad SN (2012) Modified natural excitation technique for stochastic modal identification. J Struct Eng 139(10):1753–1762. doi:10.1061/(ASCE)ST.1943-541X.0000559
11. Matarazzo TJ, Pakzad SN (2015) Structural Identification using Expectation Maximization (STRIDE): an iterative output-only method for modal identification. (ASCE) J Eng Mech. doi:10.1061/(ASCE)EM.1943-7889.0000951
12. Sim S, Spencer Jr BF (2009) Decentralized strategies for monitoring structures using wireless smart sensor networks. Newmark Structural Engineering Laboratory (NSEL) at the Department of Civil and Environmental Engineering of University of Illinois at Urbana-Champaign
13. Flynn EB, Kpotufe S, Harvey D, Figueiredo E, Taylor S, Dondi D, Mollov T, Todd MD, Rosing ST, Park G, Farrar C (2010) SHMTools: a new embeddable software package for SHM applications. In: Tomizuka M (ed) Sensors and smart structures technologies for civil, mechanical, and aerospace systems. Society of Photo-Optical Instrumentation Engineers, San Diego
14. Sohn H, Farrar CR (2001) Damage diagnosis using time series analysis of vibration signals. Smart Mater Struct 10(3):446–451. doi:10.1088/0964-1726/10/3/304
15. Worden K, Manson G, Fieller NRJ (2000) Damage detection using outlier analysis. J Sound Vib 229(3):647–667. doi:10.1006/jsvi.1999.2514
16. Lei Y, Kiremidjian AS, Nair KK, Lynch JP, Law KH, Kenny TW, Carryer ED, Kottapalli A (2003) Statistical damage detection using time series analysis on a structural health monitoring benchmark problem. In: Proceedings of the 9th international conference on applications of statistics and probability in civil engineering. pp. 6–9
17. Nair KK, Kiremidjian AS, Law KH (2006) Time series-based damage detection and localization algorithm with application to the ASCE benchmark structure. J Sound Vib 291(1–2):349–368. doi:10.1016/j.jsv.2005.06.016
18. Nigro MB, Pakzad SN, Dorvash S (2014) Localized structural damage detection: a change point analysis. J Comput Aided Civ Infrastruct Eng 29(6):416–432. doi:10.1111/mice.12059
19. Dorvash S, Pakzad SN, Labuz EL, Ricles JM, Hodgson IC (2014) Localized damage detection algorithm and implementation on a large-scale steel beam-to-column moment connection. Earthq Spectra. doi:10.1193/031613EQS069M
20. Dorvash S, Pakzad SN, Labuz EL (2004) Statistics based localized damage detection using vibration response, (Deutsch 1979)
21. Shahidi SG, Nigro MB, Pakzad SN, Pan Y (2014) Structural damage detection and localisation using multivariate regression models and two-sample control statistics. Struct Infrastruct Eng, pp 1–17. doi:10.1080/15732479.2014.949277
22. Yao R, Pakzad SN (2012) Autoregressive statistical pattern recognition algorithms for damage detection in civil structures. Mech Syst Signal Process 31:355–368. doi:10.1016/j.ymssp.2012.02.014
23. Figueiredo E, Figueiras J, Park G, Farrar CR, Worden K (2011) Influence of the autoregressive model order on damage detection. J Comput Aided Civ Infrastruct Eng 26(3):225–238. doi:10.1111/j.1467-8667.2010.00685.x

# Chapter 8
# Static Deformation Analysis for Structural Health Monitoring of a Large Dam

**Jiann-Shiun Lew**

**Abstract** The structural health monitoring of a dam is important for maintaining the safe operation and the longevity of the dam system. The damage of a structure can be detected from the variation of structural features, such as structural vibration and static deformation. The structural health of a large dam cannot be continuously monitored from structural vibration since vibration is insignificant except when earthquakes happen. The use of the daily measured static deformation can be an alternative way to continuously monitor the long-term structural health of a dam. This paper presents an investigation with the use of the measured static deformation for the structural health monitoring of Fui-Tsui dam, which is located in a very active seismic zone of Taiwan. A modeling technique is applied to characterize the measured static deformation as a function of the measured physical parameters. The proposed technique can also be used for the prediction of the long-term deformation for the structural health monitoring of Fui-Tsui dam.

**Keywords** Structural health monitoring • Large dam • Static deformation • Modeling • Statistical analysis

## 8.1 Introduction

Structural health monitoring (SHM) is important in maintaining the performance, serviceability, and reliability of civil and aerospace structures. The damage of a structure can be detected from the variation of structural features, such as structural vibration and static deformation. Various methods, based on the dynamic and static tests, have been developed to address structural health monitoring and damage identification [1–3]. However, uncertainties in SHM systems, such as changing environmental conditions, operational variations and boundary conditions [4, 5], may significantly affect the measurements and reduce the reliability of damage identification with false indications of damage [6, 7]. Therefore, it is necessary to develop robust and reliable approaches to identify damage status to set an early warning threshold before a disaster occurs.

The SHM of civil infrastructures, such as bridges and dams, is crucial to the safe operation and the longevity of the system [8, 9]. Significant developments of SHM of civil structures have originated from major construction projects, such as large dams and long-span cable-supported bridges. A regular inspection of dams became a legal requirement in the UK after the failure of a 30 m embankment dam claimed the lives of 254 people near Sheffield, UK in 1864 [10]. The traditional procedure for the evaluation of the structural health condition of dams is through visual inspection [11]. Visual inspections have some shortcomings such as significant labor demanding and inaccessible critical locations. The problems of visual inspections can be addressed with the field tests, which are conducted to monitor the dynamic and static characteristics of structures. For a large dam structure, such as the Fei-Tsui dam, it is a challenging task to excite the structure vibration. The vibration measurement data are obtained only when an earthquake happens. The use of the static deformation measurement can be an alternative way to continuously monitor the structural health of a dam [12].

This paper presents an investigation of SHM of Fei-Tsui dam with the use of the daily measured static deformations at various locations. Based on the results of the previous study [1, 13], the water level and temperature are the two most important variables affecting static deformation. The examination in this paper is based on a modeling technique [13], where the static deformation is modeled as a function of the measured water level and temperature. The results demonstrate that the identified model, based on the proposed technique, not only well represents the experimental data but also gives the precise prediction of the long-term deformation for SHM. The goal of this research is to develop a reliable and effective technique to identify any possible abnormality happened to Fei-Tsui dam for safe operation and failure prevention.

J.-S. Lew (✉)
Center of Excellence in Information Systems, Tennessee State University, Nashville, TN 37209, USA
e-mail: lew@coe.tsuniv.edu

© The Society for Experimental Mechanics, Inc. 2015



C. Niezrecki (ed.), *Structural Health Monitoring and Damage Detection, Volume 7*, Conference Proceedings of the Society for Experimental Mechanics Series, DOI 10.1007/978-3-319-15230-1_8

## 8.2 Static Deformation Measurement of Fei-Tsui Dam

Fei-Tsui arch dam, as shown in Fig. 8.1, is 122.5 m high and the dam crest is 510 m in length. It is designed with a double curvature arch along the height of the dam. The capacity of this reservoir is about 400 million m$^3$. Since this dam is located in a very active seismic zone of Taiwan, both dynamic and static monitoring systems of the dam are deployed in this structure [1]. There are several static measurement systems in this Feu-Tsui dam. One of the important static measurement systems is the monitoring of dam deformation. Figure 8.2 shows the three vertical profiles (along vertical lines of NPL1, NPL2 and NPL3) where plumb lines were installed along the dam height. Along each profile, for example in profile NPL2Y, there are five measurement points (at level 172.5, 150, 115, 90 and 57.5 m) to measure the radial deformation (Y direction) of the dam. The daily dam deformations and water level were measured at a fixed time in the morning. These static deformation measurements were collected automatically with the use of the laser beams starting from January 1, 1987, and they were recorded to the unit of 0.1 mm with the initial deformation set as zero. The daily maximum and minimum temperatures at the dam site were also measured. The data were not recorded during the time of maintenance, so the data were unavailable for these days.

**Fig. 8.1** Areal photo of Fei-Tsui arch dam

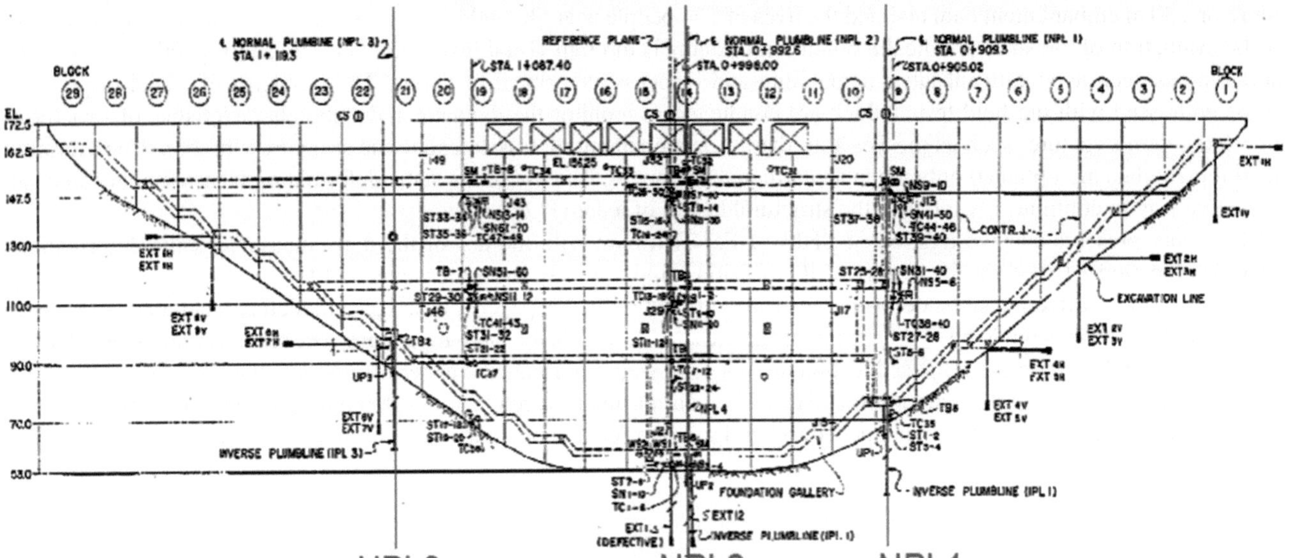


**Fig. 8.2** Locations of static deformation measurement of Fei-Tsui dam: three vertical plumb lines (*NPL1*, *NPL2* and *NPL3*) along the dam height

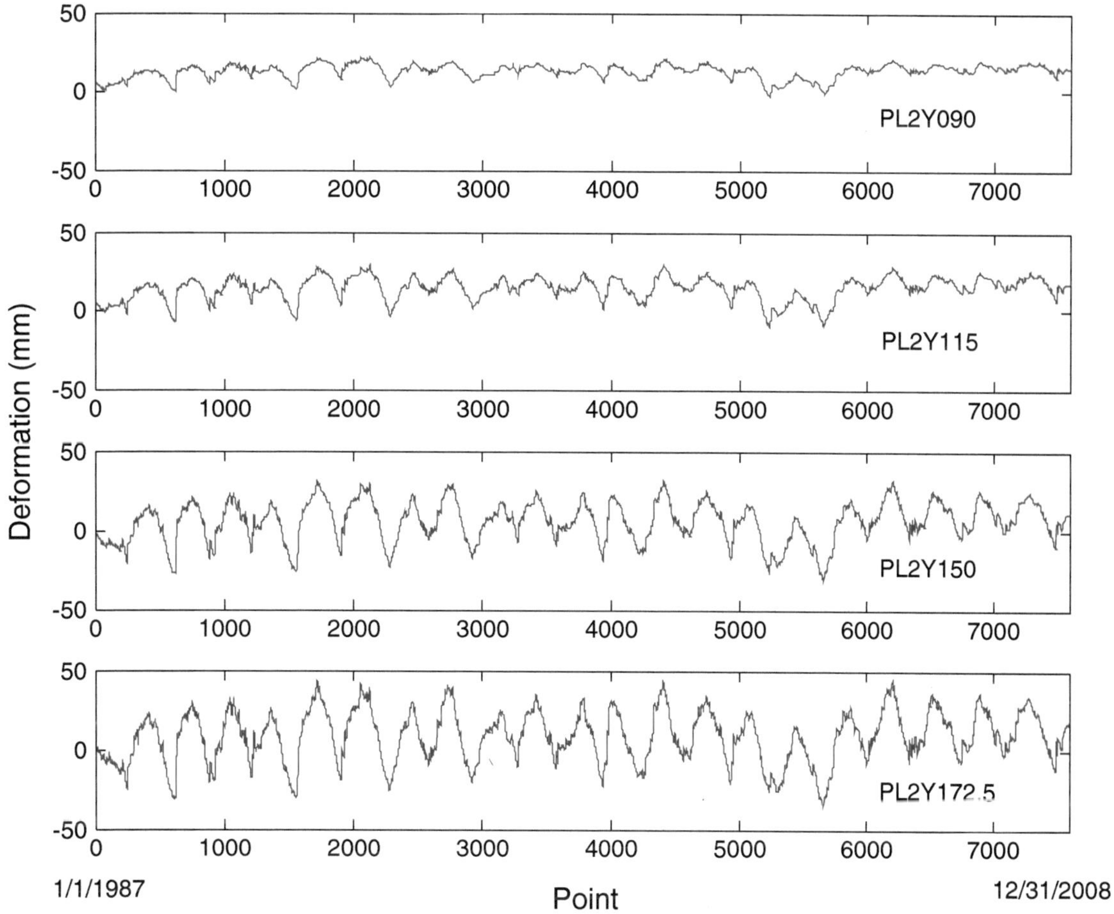


**Fig. 8.3** Recorded static radial deformations of Fei-Tsui arch dam along plumb NPL2 line at different heights

Figure 8.3 plots the measured static deformation at four locations with different heights of Feu-Tsui dam along profile NPL2. The abscissa of Fig. 8.3 indicates the number of the point (from 1/1/1987 to 12/31/2008, a total of 7,600 data points). The magnitude of the deformation increases as the height of the measurement point increases, and it is relatively insignificant at the height of 90 m. The signatures of the deformations at different heights are very similar to each other. The investigation in this paper is demonstrated by the results of the analysis of the static deformations at these four locations with different heights.

## 8.3 Modeling Technique

The previous study [1] shows that the deformation is mainly affected by the water level and temperature. The proposed approach has two stages with iterations [13]. In the first stage, a least-squares technique is used to curve fit the measured data of static radial deformation as a linear function of the measured water level and temperature. A cost function is formed as

$$J_1 = \sum_{i=n_0}^{n_1} \left[a_1 h\left(d_i\right) + a_2 t\left(d_i\right) + a_3 - y\left(d_i\right)\right]^2, \tag{8.1}$$

where $h(d_i)$, $t(d_i)$, and $y(d_i)$ are the measured water level, low temperature, and deformation at the $i$th measurement date $d_i$. The unique solution of $a_i$ of this linear model can be obtained by minimizing this quadratic cost function. The error between the measured data and the identified least-squares model can be computed as

$$y_r\left(d_i\right) = y\left(d_i\right) - y_a\left(d_i\right), \tag{8.2}$$

where $y_a$ is the identified linear model. This identified model represents the global effects of the water level and temperature on the static deformation during this time period. The previous study shows that the model error $y_r$ behaves like a sine-wave shifting upwards or downwards and the period of this sine-wave is one year [13]. To include this sine-wave error in the model, a cost function is formed to curve fit the model error $y_r$ as

$$J_2 = \sum_{i=n_0}^{n_1} \left[c_1 \sin\left(2\pi c_2 d_i + c_3\right) + c_4 + c_5 d_i - y_r\left(d_i\right)\right]^2, \tag{8.3}$$

where $c_1$ is the amplitude of the sine-wave signal with the period $1/c_2$ (day) and the initial phase $c_3$, coefficient $c_4$ is the initial mean value of the sine-wave and this mean value changes linearly with time at a rate of $c_5$. This is a nonlinear least-squares problem. Here the solution is obtained with the use of the *Matlab* program "*Fminsearch*", based on the Nelder-Mead simplex (direct search) method.

Next, a technique with iterations, based on the integration of the preceding linear least-squares technique and the nonlinear least-squares technique, is presented to get the identified model. This technique has the following steps:

1. Choose $n_0$ and $n_1$ for the data points used in the model identification.
2. Form the cost function $J_1$ in Eq. (8.1) and get a solution of the linear least-squares model.
3. Compute the model error $y_r\left(d_i\right) = y\left(d_i\right) - y_a\left(d_i\right)$ of the linear least-squares model.
4. Form the cost function $J_2$ in Eq. (8.3) and minimize this cost function to get the solution of variables $c_i$ and compute this sine-wave correction term as

$$y_s\left(d_i\right) = c_1 \sin\left(2\pi c_2 d_i + c_3\right) + c_4 + c_5 d_i. \tag{8.4}$$

5. Compute the model error as

$$y_e\left(d_i\right) = y\left(d_i\right) - y_a\left(d_i\right) - y_s\left(d_i\right), \quad i = n_0, \cdots, n_1. \tag{8.5}$$

6. Generate data

$$y_1 = y - y_s. \tag{8.6}$$

7. Use data $y_1$ instead of $y$ to update the linear least-squares solution of parameters $a_i$.
8. Check if the changes of the updated solution of $a_i$ are smaller than the specified errors: Yes, then stop; No, then go to step 3.

This approach has two stages with iteration. The iteration is stopped when the changes of the identified variables $a_i$ are smaller than the specified errors. Then the solution of the identified variables $a_i$ and $c_j$ of the identified model is obtained.

## 8.4 Results of Static Deformation Monitoring of Fei-Tsui Dam

Next we will investigate the results of the application of the proposed approach to the static deformation of Feu-Tsui dam. In this example, the measured deformations between 5,801 and 6,800 data points (from 1/23/2004 to 10/23/2006) at various locations along NPL2 line are used for the study. Figure 8.4 shows the results of the analysis of the deformation at location NPL2Y90. The dashed line in Fig. 8.4a represents the data generated from the identified model, which are computed as

$$y_m\left(d_i\right) = y_a\left(d_i\right) + y_s\left(d_i\right) = a_1 h\left(d_i\right) + a_2 t\left(d_i\right) + a_3 + c_1 \sin\left(2\pi c_2 d_i + c_3\right) + c_4 + c_5 d_i. \tag{8.7}$$

Figure 8.4a demonstrates that the identified model well represents the experimental data. Figure 8.4b shows the model error $y_e$ between the experimental data and the predicted data based on the identified model, the range between 2 standard deviations (dotted lines), and the range between 3 standard deviations (dashed lines). The standard deviation $\sigma$ of the model error $y_e$ in Fig. 8.4b is 0.5986 mm, and the mean value $\mu$ of $y_e$ is $2.0 \times 10^{-6}$ mm, very close to 0. Figure 8.5 shows the results of the predicted deformation from 10/24/2006 to 11/28/2007 (6,801–7,200 data points), which is estimated based on Eq. (8.7) with the measured water level and temperature. SHM is based on the difference between the measured deformation

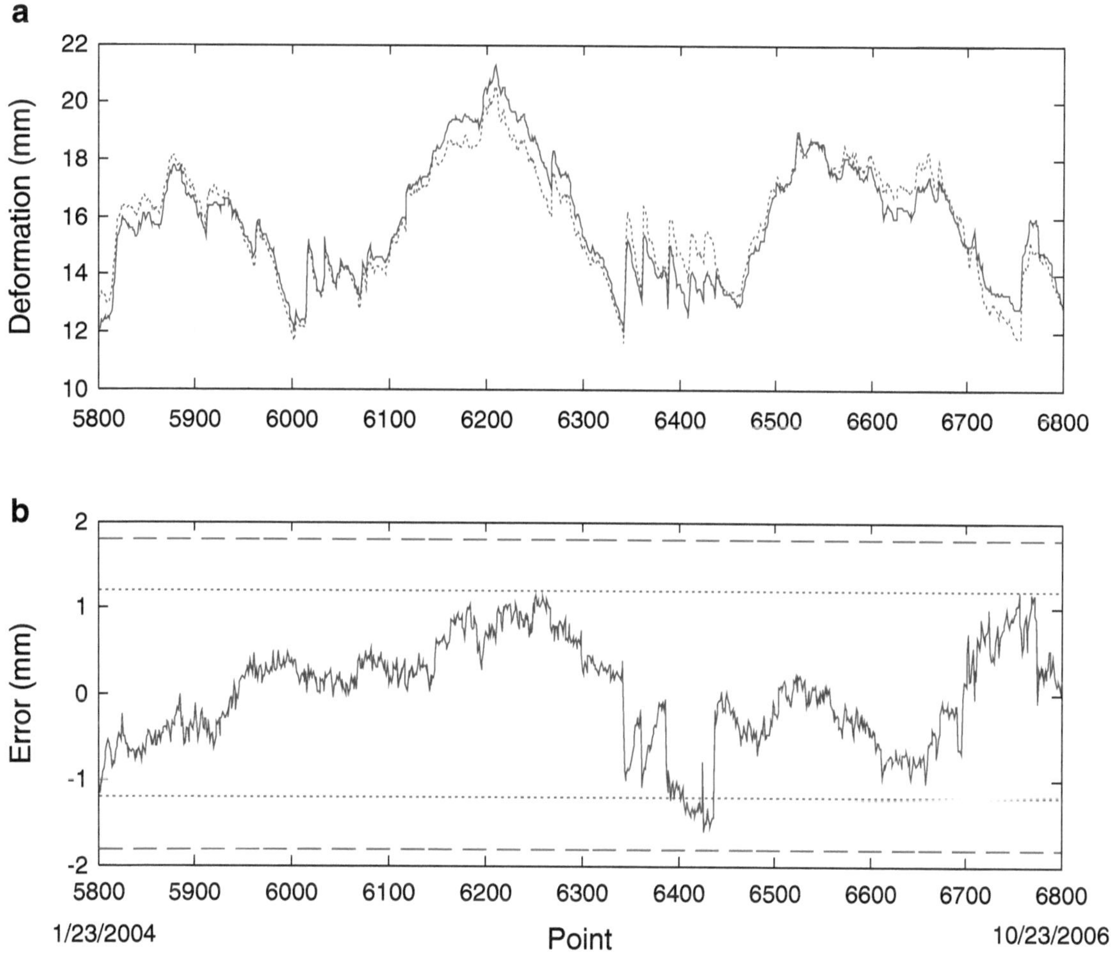


**Fig. 8.4** Results of deformation analysis of data at NPL2Y90: (**a**) — $y$, ··· $y_m$; (**b**) — $y_e$, ··· $(\mu \pm 2\sigma)$, — $(\mu \pm 3\sigma)$

and the predicted deformation. Figure 8.5a shows that the predicted deformation is very close to the measured deformation between 10/24/2006 and 11/28/2007. Figure 8.5b shows the model error $y_e$ between the experimental data and the predicted data based on the identified model, the range between 2 standard deviations (dotted lines), and the range between 3 standard deviations (dashed lines). Figure 8.6 shows the standard deviation of the model error $y_e$, and some of the identified parameters as functions of the iteration number for the analysis of the deformation at location NPL2Y90. The measured water level and low temperature from 5,801 to 7,200 data points (from 1/23/2004 to 11/28/2007) are shown in Figs. 8.7 and 8.8.

From the results in Figs. 8.4–8.9, the following observations are noted:

1. The standard deviation of the model error $y_e$ decreases with iteration and converges to 0.5986 mm, and it has a negligible change after 10 iterations. The coefficient $a_1$ corresponds to the effect of the water level, and it converges to 0.4146. The increase of 1 m water level averagely produces 0.4146 mm deformation at NPL2Y90.
2. The absolute value of $a_2$ decreases as the iteration number increases, as shown in Fig. 8.6, while the amplitude $c_1$ of the sine-wave increases as the iteration number increases. The decrease of the absolute value of $a_2$ reduces the effect of temperature on the deformation and also reduces the fluctuation due to the temperature variation, as shown in Fig. 8.8. The increase of the sine-wave amplitude $c_1$ augments the effect of the seasonal temperature (average temperature) variation on the deformation. The optimal solution process is to find a balance between these two factors, the daily temperature effect and the seasonal temperature effect. The negative value of the coefficient $a_2$ indicates that the increase of temperature reduces the deformation. It requires additional iterations (more than 10) for these two variables ($a_2$ and $c_1$) to converge. As expected, the period $1/c_2$ (days) of the sine-wave converges to a value close to 365.

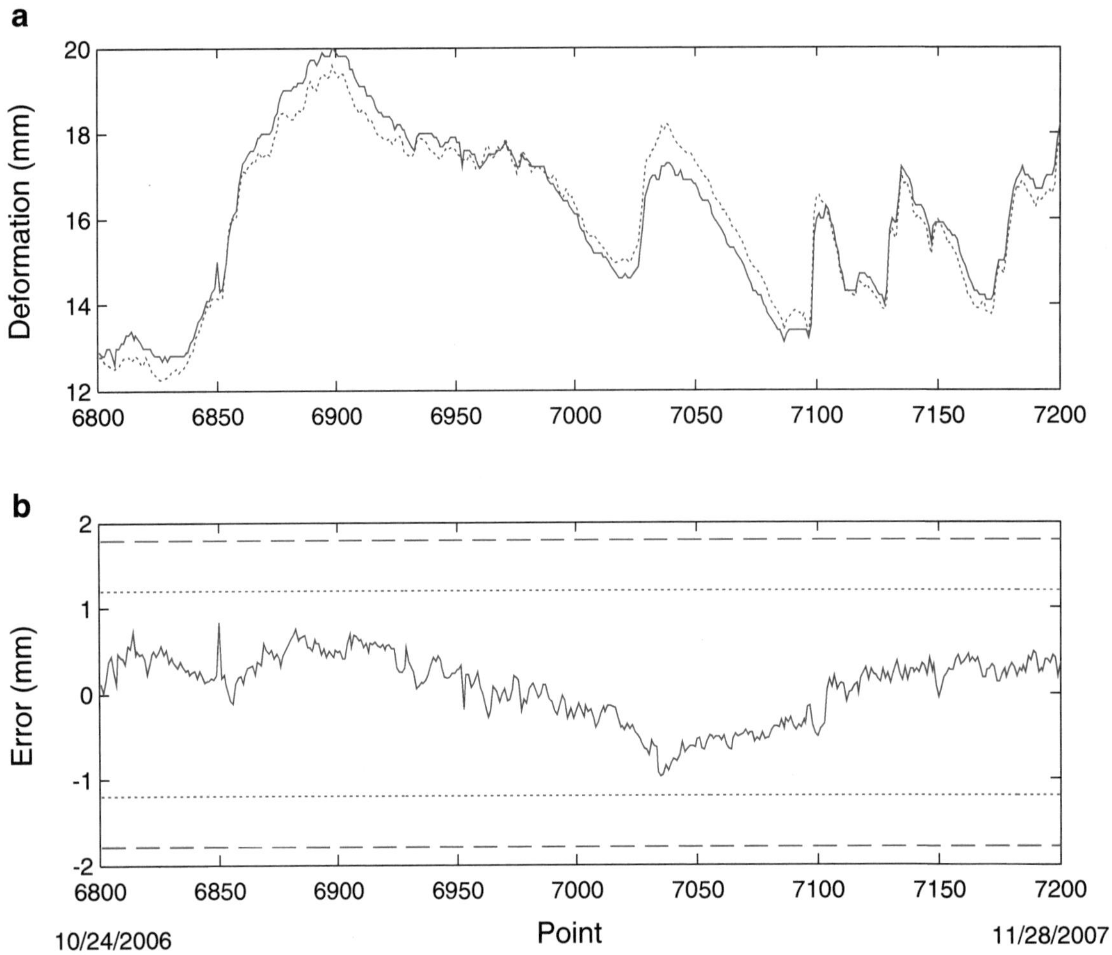


**Fig. 8.5** Results of prediction deformation of data at NPL2Y90: (**a**) — $y$, ··· $y_m$; (**b**) — $y_e$, ··· $(\mu \pm 2\sigma)$, — $(\mu \pm 3\sigma)$

3. The previous study indicates that the deformation increase has a time delay corresponding to the quick increase of the water level. The quick water increase happens from 6,342 to 6,344 data points (from 7/17/2005 to 7/19/2005), where water increases from 153.84 to 163.8 m, and this cause the model error $y_e$ changes from 0.3904 to −0.8848 mm, which corresponds to the variation of 2.1302 $\sigma$.
4. Figure 8.7 show that there are five peaks of water levels between 6,341 and 6,440 data points with quick water level increases and decreases, which is from 7/16/2005 to 10/23/2005 in the typhoon season. This causes the significant changes of model error. The model error $y_e$ has a big increase from −1.2558 to −0.1826 mm from 6,437 to 6,438 data points (10/20/2005 to 10/21/2005), which corresponds to the change of 1.7986 $\sigma$. But the measured water level and temperature change insignificantly from 10/20/2005 to 10/21/2005. This abnormal condition requires further investigation.

Figures 8.9–8.14 show the results of the analysis of the deformations at locations NPL2Y115, NPL2Y150, and NPL2Y172. To demonstrate the results, two metrics are defined to represent the magnitude of deformation at various measured locations as

$$\overline{y} = \max\{y(d_i)\} - \min\{y(d_i)\}, \quad \sigma_y = std\{y(d_i)\}. \tag{8.8}$$

In this example, these two metrics are computed from the measured deformations from 5,801 to 6,800 data points. Table 8.1 summarizes the results for the deformation analysis at various measured.

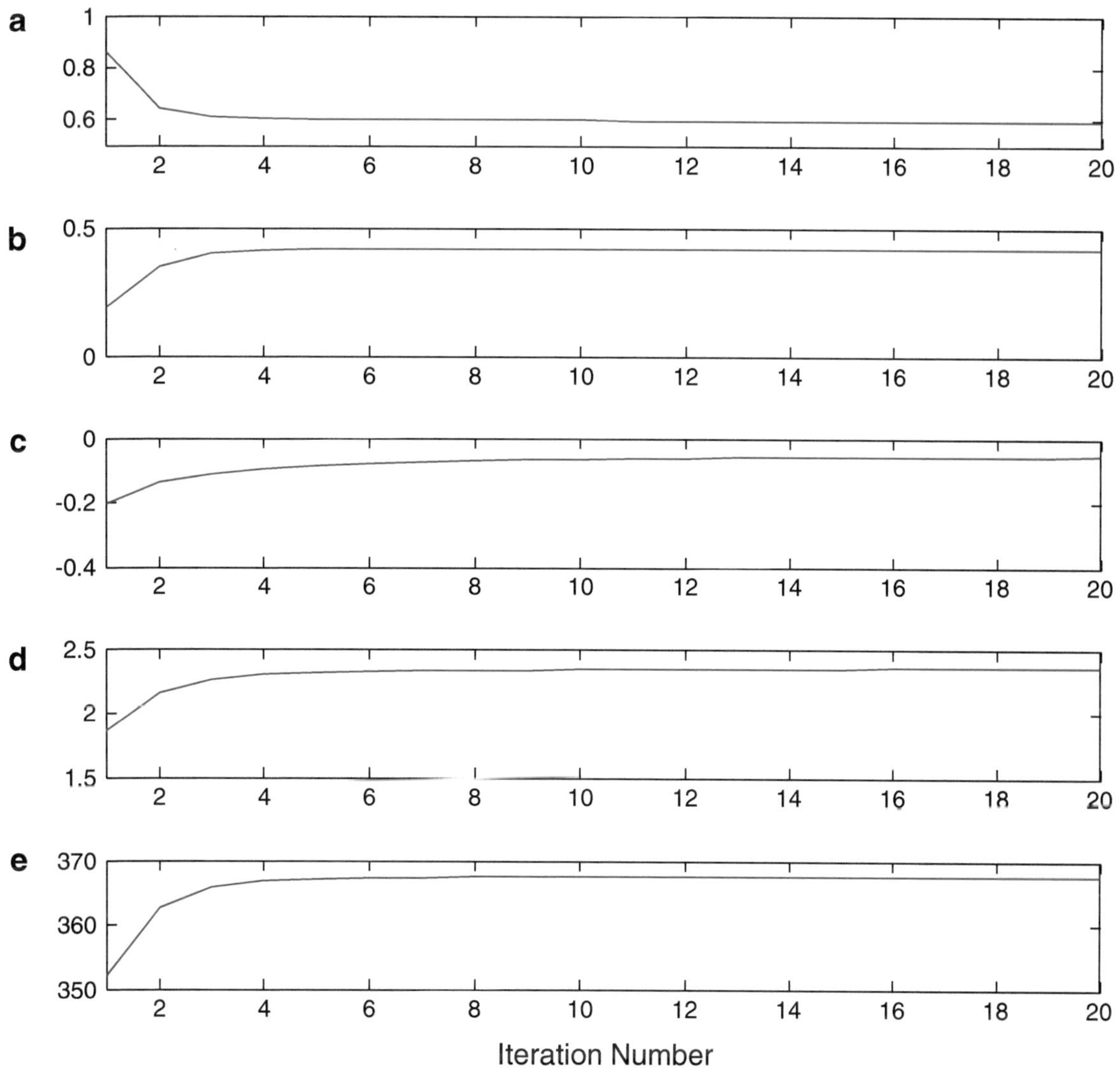


**Fig. 8.6** Standard deviation of model error $y_e$ and identified variables as functions of the iteration number: (**a**) $\sigma$ (mm); (**b**) $a_1$; (**c**) $a_2$; (**d**) $c_1$; (**e**) $1/c_2$

From the results, the following observations are noted:

1. From Figs. 8.4–8.14, the measured deformations at these four locations have similar signatures. The measured locations of NPL2Y90 and NPL2Y115 are far below water surface, and the model errors at these two locations have similar signatures. The measured location of NPL2Y150 is below and close to water surface and the measured location of NPL2Y172 is above water surface. The model errors at these two locations have similar signatures, also the model errors of the prediction deformation at these two locations are similar.
2. Table 8.1 shows that the magnitudes $\overline{y}$ of the deformation at locations NPL2Y172, NPL2Y150, and NPL2Y115 are around 6, 4, and 2 times that at location NPL2Y90, respectively. The effects of temperature on deformation ($a_2$ and $c_1$) at locations NPL2Y172, NPL2Y150, and NPL2Y115 are around 6, 4, and 2 times that at location NPL2Y90, respectively. But the effects of water level on deformation ($a_1$) at these locations are in different ratios since the location of the measured deformation, which is below or above water surface, affects the influence of water level on deformation. Further research will be conducted to address this issue.

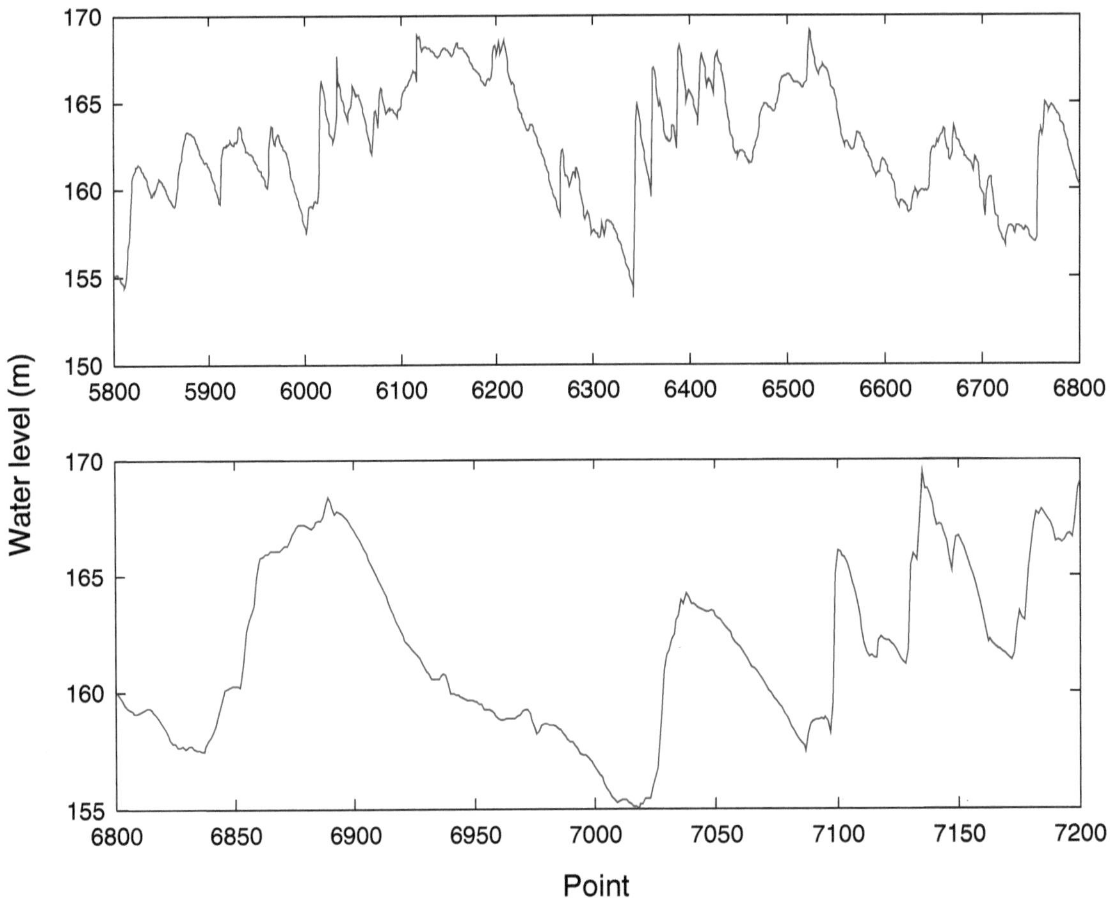


**Fig. 8.7** Measured water level from 5,801 to 7,200 data points

3. The significant change of water level, which can be caused by flooding or drought, may results in the large change of the model error and the bias of the identified model. Especially, this may result in the bias of the model error of the prediction deformation, such as the bias and the significant change of the model error from 7,150 to 7,200 data points at locations NPL2Y172 and NPL2Y150. This may require the further investigation of the dam health, and the identified model may need to be updated.
4. The identified parameters $a_i$ and $c_j$ may change with the measured water level, temperature, and other unmeasured and unknown factors, such as dam foundation shift and environmental uncertainty. In the modeling process, these identified parameters are assumed as constant. Further investigation will be conducted to address this issue.

The results demonstrate that the identified model, which is used to characterize the static deformation of Fei-Tsui dam, changes slowly with time. This identified model, used to predict the static deformation for SHM, needs to be updated every few years or when any of the following conditions appear.

- Significant water lever changes caused by flooding or drought may result in the significant change of the model and model error.
- Anomalies occur and further investigation for damage status is required.
- The prediction error has clear bias.
- Natural events, such as significant earthquakes and floods, happen and it may induce change in or damage to the structure and its surrounding.

The proposed modeling approach has been also applied to the measured deformation data in different periods of time. Results from the investigation of other examples provide the same conclusion as the demonstrated example.

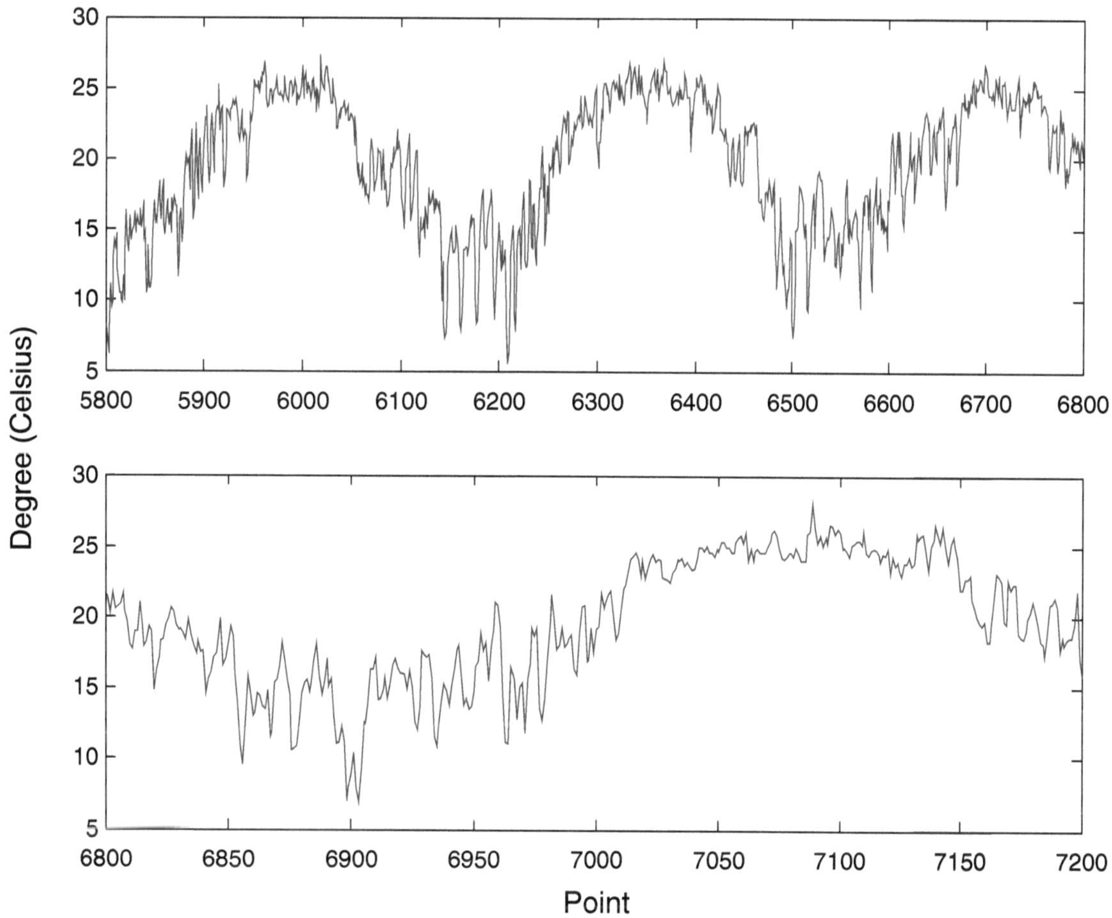


**Fig. 8.8** Measured low temperature from 5,801 to 7,200 data points

## 8.5 Conclusions

This paper presents an investigation of structural health monitoring of Fei-Tsui dam with the use of the daily measured static deformations at various locations. The results demonstrate that the identified model, based on the proposed technique, well represents the experimental data and provides accurate estimation of the prediction deformation. The results show that the quick increase of the water level results in the delay of deformation increase and so causes a significant change of model error during the transition time. The significant change of water level, which can be caused by flooding or drought, may results in the significant change of the model error and the bias of the identified model. This may require the further investigation of the health condition of dam and the update of the identified model. The proposed modeling technique also gives a precise prediction of the long-term future static deformation from the measured water level and temperature, which provides an efficient and effective tool for structural health monitoring of Fei Tsui dam.

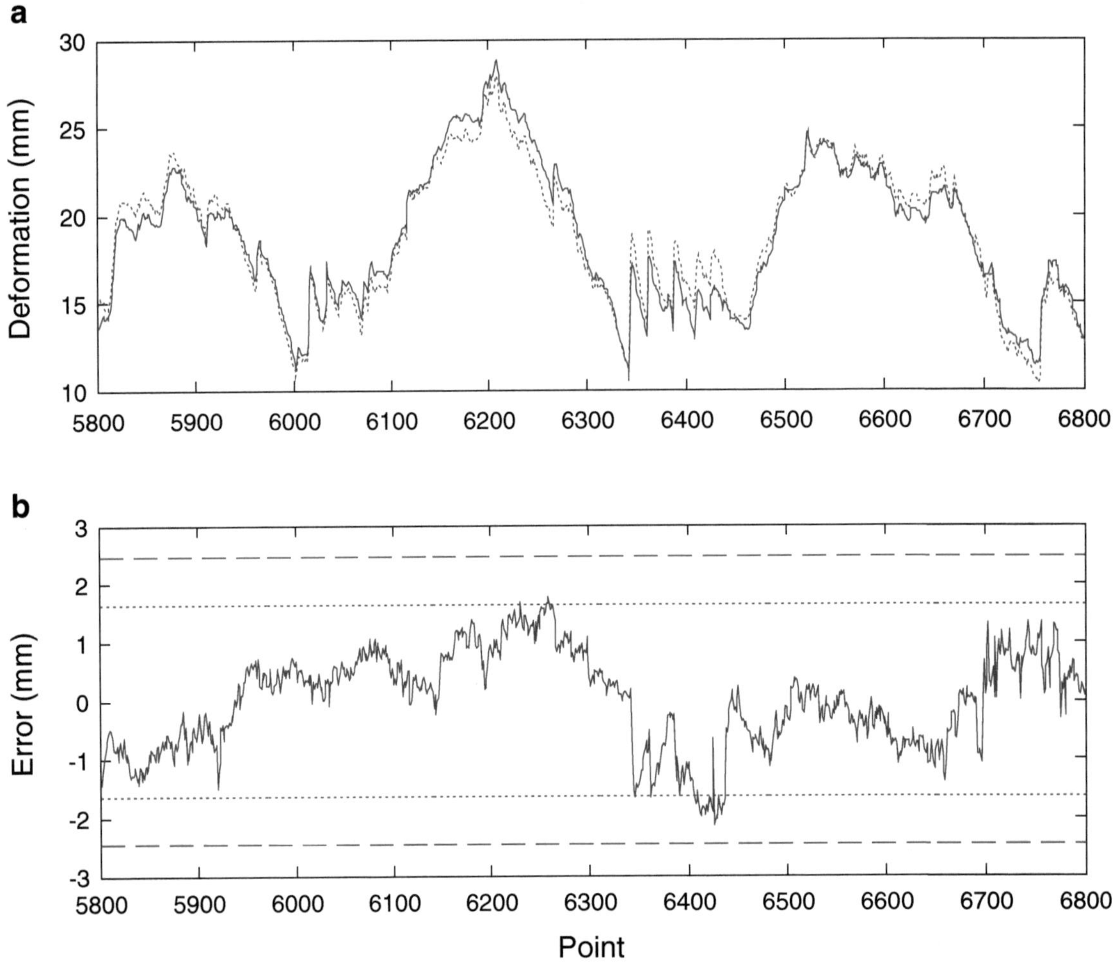


**Fig. 8.9** Results of deformation analysis of data at NPL2Y115: (**a**) — $y$, $\cdots$ $y_m$; (**b**) — $y_e$, $\cdots$ $(\mu \pm 2\sigma)$, — $(\mu \pm 3\sigma)$

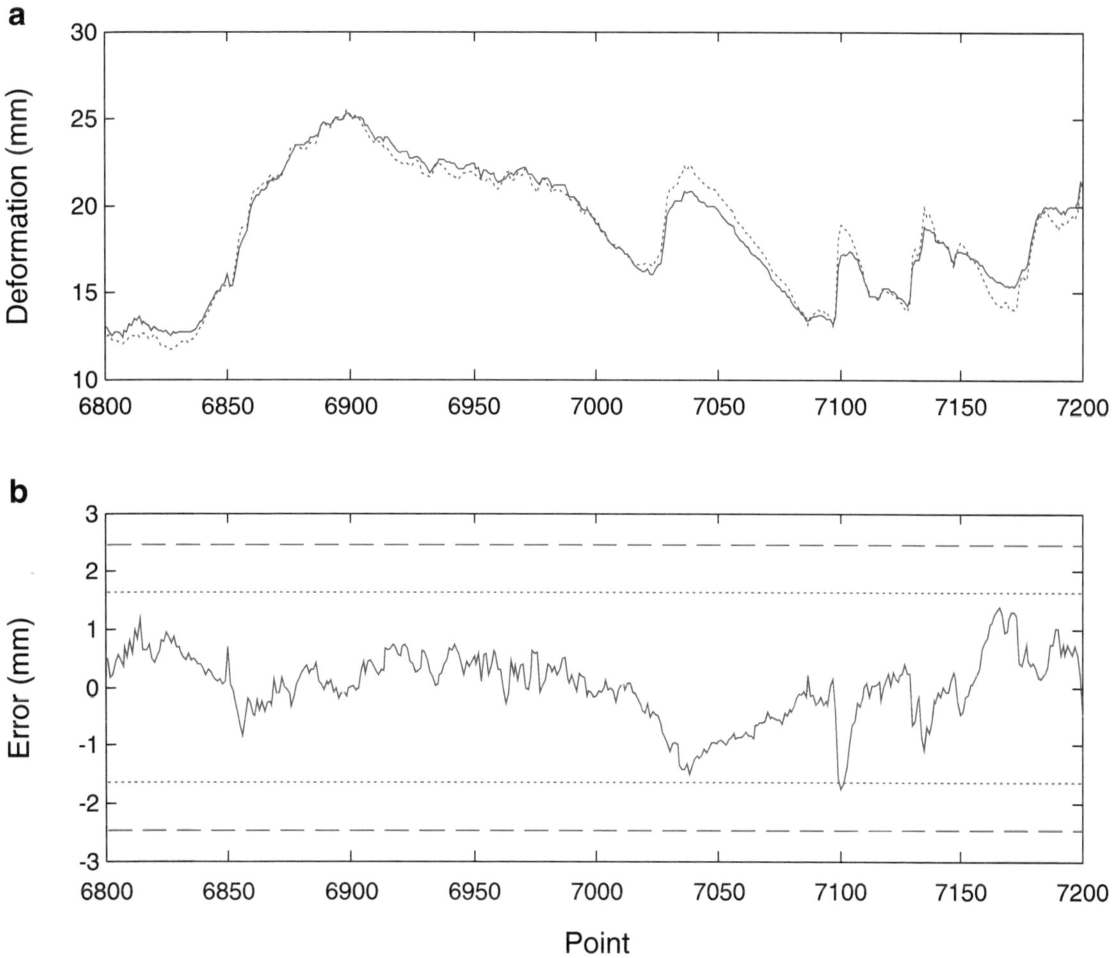


**Fig. 8.10** Results of prediction deformation of data at NPL2Y115: (**a**) — $y$, ··· $y_m$; (**b**) — $y_e$, ··· $(\mu \pm 2\sigma)$, — $(\mu \pm 3\sigma)$

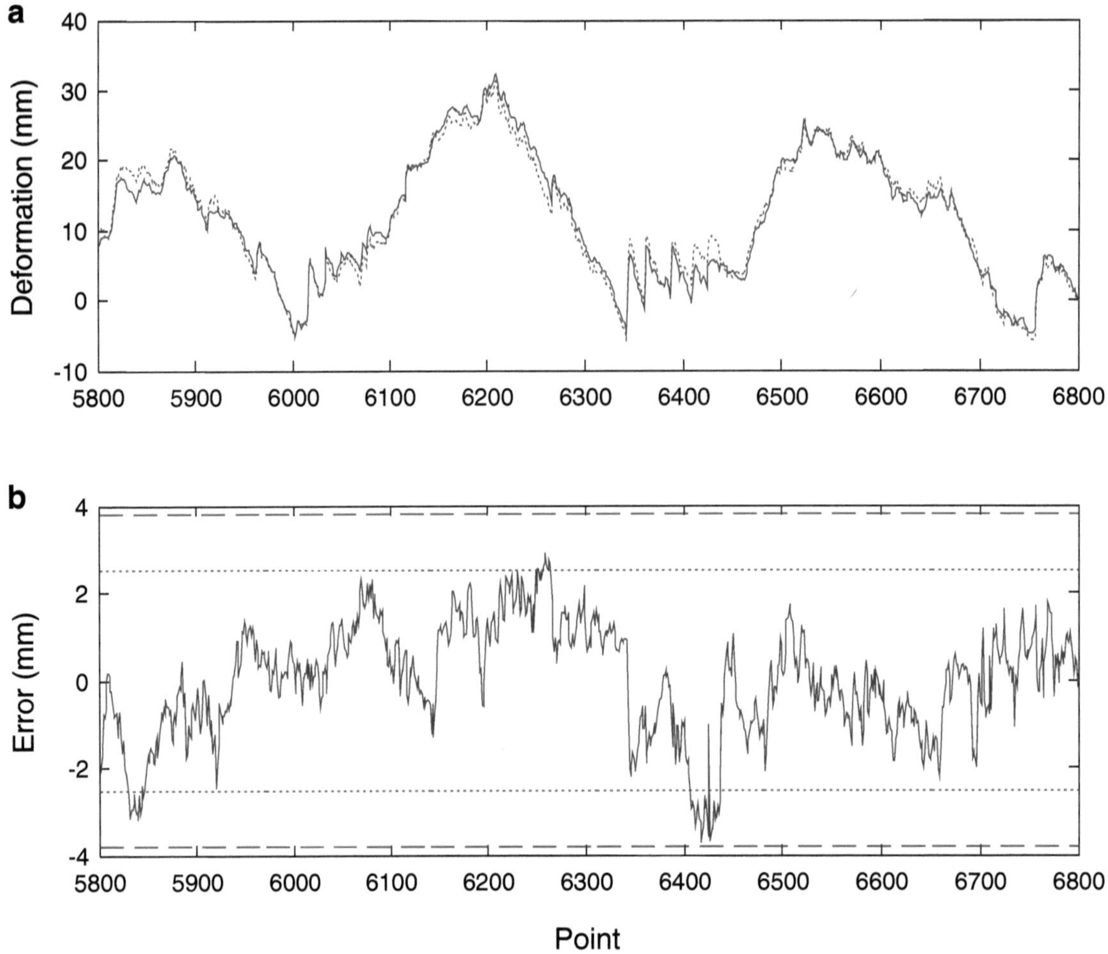


**Fig. 8.11** Results of deformation analysis of data at NPL2Y150: (**a**) — $y$, ··· $y_m$; (**b**) — $y_e$, ··· $(\mu \pm 2\sigma)$, — $(\mu \pm 3\sigma)$

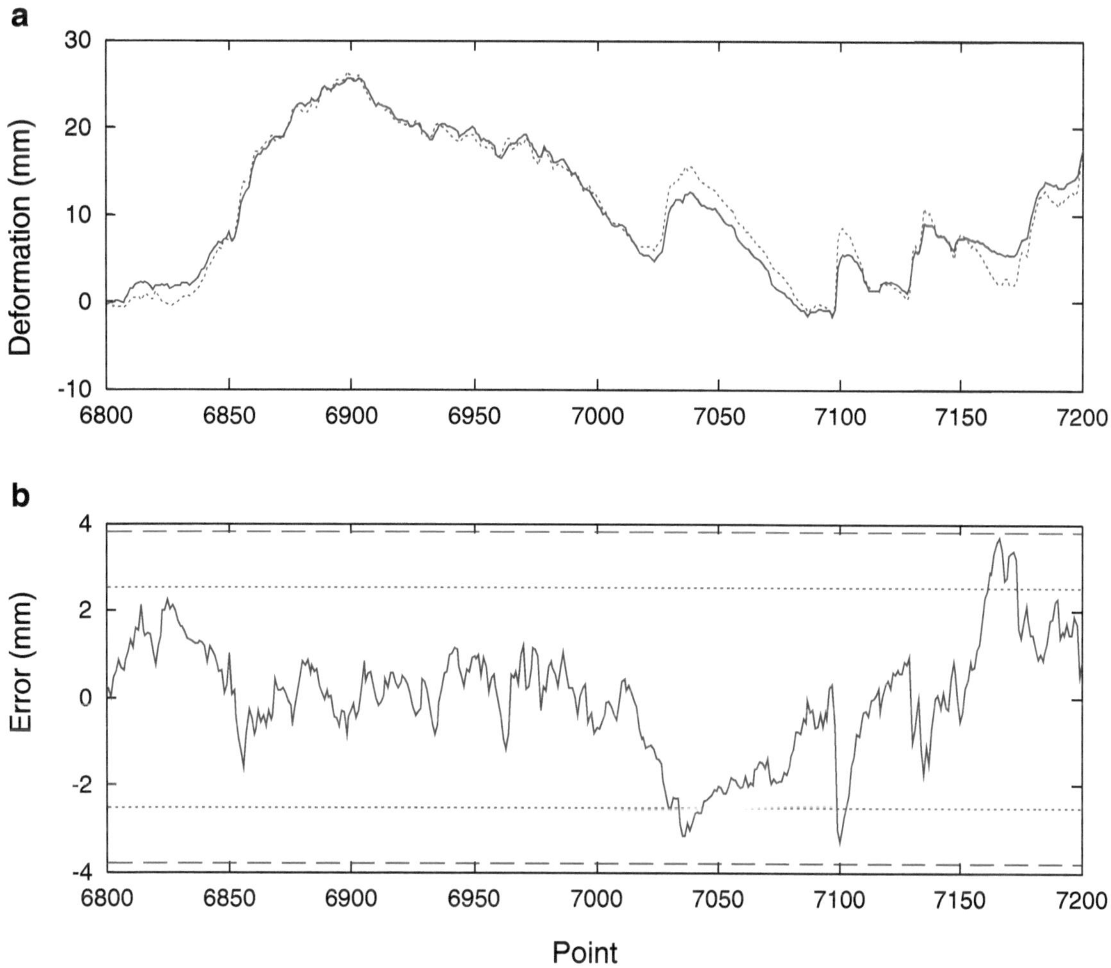


**Fig. 8.12** Results of prediction deformation of data at NPL2Y150: (**a**) — $y$, ··· $y_m$; (**b**) — $y_e$, ··· $(\mu \pm 2\sigma)$, — $(\mu \pm 3\sigma)$

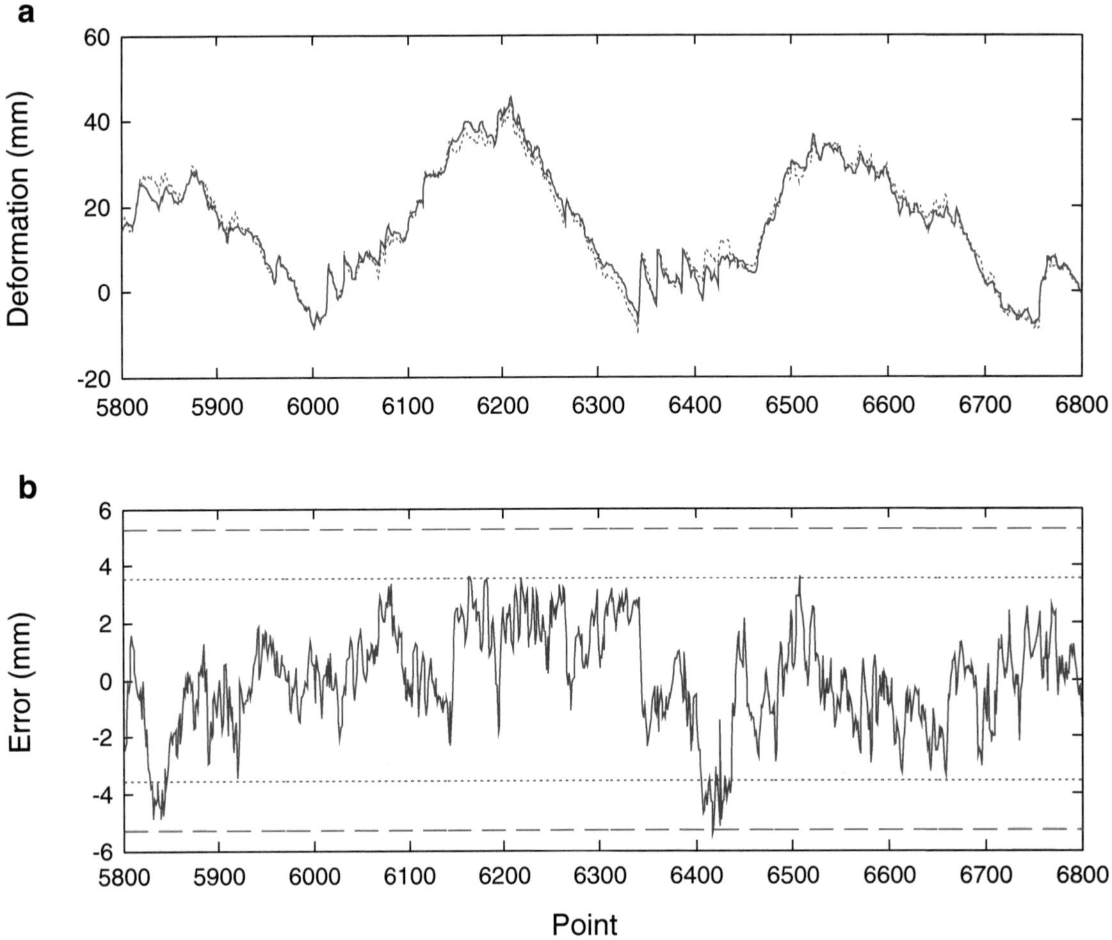


**Fig. 8.13** Results of deformation analysis of data at NPL2Y172: (**a**) — $y$, ··· $y_m$; (**b**) — $y_e$, ··· $(\mu \pm 2\sigma)$, — $(\mu \pm 3\sigma)$

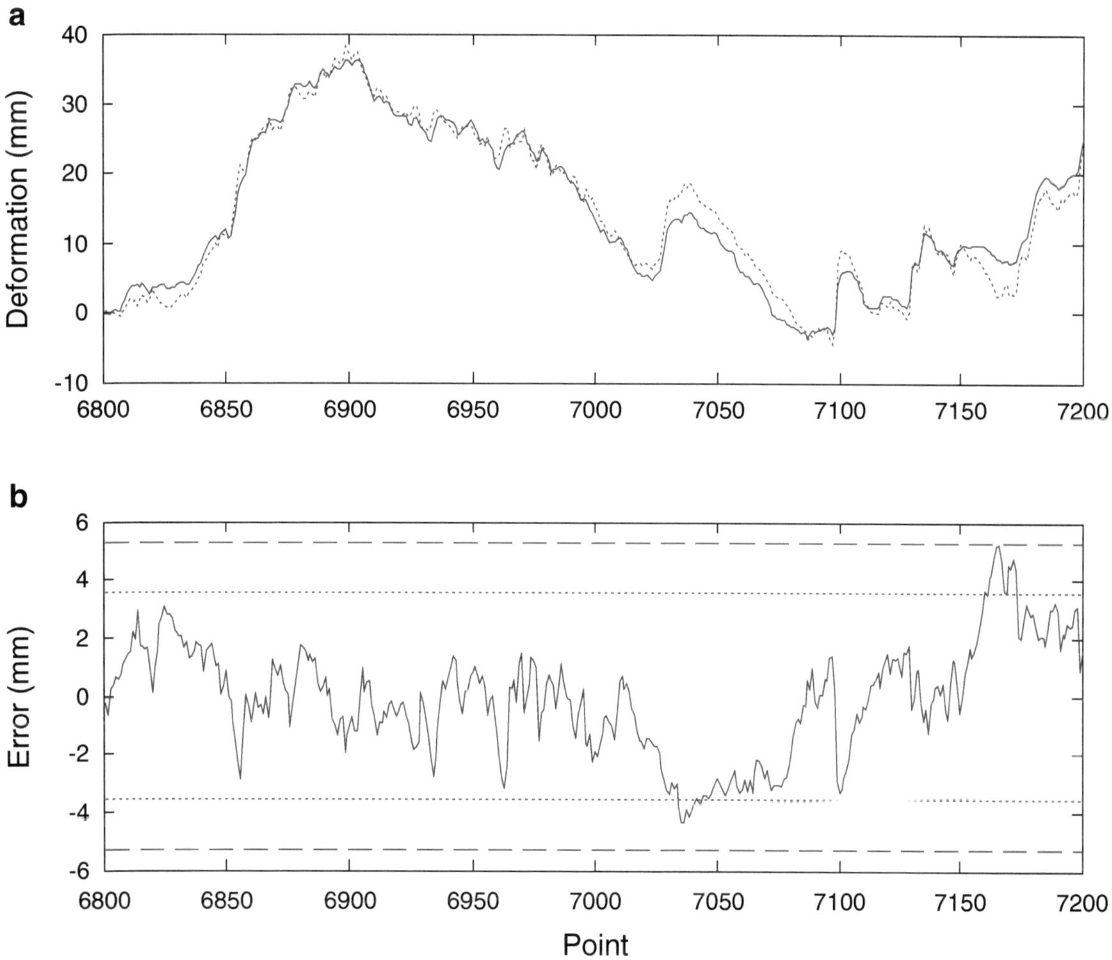


**Fig. 8.14** Results of prediction deformation of data at NPL2Y172: (**a**) — $y$, ··· $y_m$; (**b**) — $y_e$, ··· $(\mu \pm 2\sigma)$, — $(\mu \pm 3\sigma)$

**Table 8.1** Results of deformation analysis at various locations

| | NPL2Y90 | NPL2Y115 | NPL2Y150 | NPL2Y172 |
|---|---|---|---|---|
| $\overline{y}$ | 9.30 | 18.10 | 37.50 | 54.20 |
| $\sigma_y$ | 2.05 | 3.98 | 8.82 | 12.80 |
| $\sigma$ | 0.5986 | 0.8121 | 1.2667 | 1.7647 |
| $a_1$ | 0.4161 | 0.7610 | 1.3374 | 1.7233 |
| $a_2$ | −0.0545 | −0.0901 | −0.2250 | −0.4665 |
| $c_1$ | 2.3555 | 4.6351 | 9.5904 | 12.9338 |

## References


1. Loh CH, Hsu TY, Chen CH (2011) Application of advanced statistical methods for extracting long-term trends in static monitoring data from an arch dam. Struct Health Monit 10(6):587–601
2. Sohn H, Farrar CR, Hemez FM, Czarneki JJ, Shunk DD, Stinemates DW, Nadler BR (2002) A review of structural health monitoring literature: 1996–2001. Los Alamos National Laboratory report, LA-13976-MS
3. Lew J-S (1995) Using transfer function parameter changes for damage detection of structures. AIAA J 33(11):2189–2193
4. Mace BR, Worden K, Mason G (2005) Uncertainty in structural dynamics. J Sound Vib 288:423–429
5. Lew J-S (2008) Reduction of uncertainty effect on damage identification using feedback control. J Sound Vib 318:903–910
6. Peeters B, DeRoeck G (2001) One-year monitoring of the Z24-bridge: environmental effects versus damage events. Earthq Eng Struct Dyn 30:149–171
7. Sohn H, Dzwonczyk M, Straser EG, Kiremidjian AS, Law KH, Meng T (1999) An experimental study of temperature effect on modal parameters of the Alamosa Canyon bridge. Earthq Eng Struct Dyn 28:879–897
8. Brownjohn JMW (2007) Structural health monitoring of civil infrastructure. Phil Trans R Soc A 365:589–622

9. Su JZ, Xia Y, Chen L, Zhao X, Zhang QL, Xu YL, Ding JM, Xiong HB, Ma RJ, Lv XL, Chen AR (2013) Long-term structural performance monitoring system for the Shanghai tower. J Civ Struct Health Monit 3:49–61
10. Fanelli MA (1992) In: Blockley D (ed) The safety of large dams, engineering safety. McGraw-Hill, Berkshire, pp 205–223
11. Bukenya P, Moyo P, Beushausen H, Oosthuizen C (2014) Health monitoring of concrete dams: a literature review. J Civ Struct Health Monit 4:235–244
12. Hsu TY, Loh CH (2010) Damage detection accommodating nonlinear environmental effects by nonlinear principal component analysis. Struct Control Health Monit 17:338–354
13. Lew J-S, Loh C-H (2014) Structural health monitoring of an arch dam from static deformation. J Civ Struct Health Monit 4:245–253

# Chapter 9
# Operational Vibration-Based Response Estimation for Offshore Wind Lattice Structures

**P. van der Male and E. Lourens**

**Abstract** The design for fatigue for offshore wind turbine structures is characterized by uncertainty, resulting from both loading specifications and numerical modelling. At the same time, fatigue is a main design driver for this type of structures. This study presents a strategy to monitor the accumulated fatigue damage in real-time, employing a joint input-state estimation algorithm. Measuring the operational vibrations at well-chosen locations enables the estimation of strain responses at unmeasured locations. The estimation algorithm is applied to a wind turbine on a lattice support structure, for which the response estimates of the lattice members are based on measurements on the turbine tower only. This restriction follows from the difficulty to reliably and robustly measure at locations on the lattice structure. Artificial measurement data is generated with a full-order finite element model, while the strains are estimated with an erroneous reduced-order design model, after inclusion of measurement noise. The strain estimates show that the main frequency content can be captured relatively accurately, except for a small bias and some high frequency disturbance, corresponding to a weakly observable higher mode. This second aspect shows the importance of a trade-off between the accuracy of the reduced-order finite element model and the ill-conditioning of the observability matrix.

**Keywords** Offshore wind • Response estimation • Operational vibrations • Lattice structure • Sensor network

## 9.1 Introduction

For offshore wind turbine support structures, fatigue is a main design driver [1]. Fatigue-based design requires an adequate prediction of the environmental conditions combined with an accurate description of the structural properties, valid for the design life-time of, for example, 20 years. Estimates of the accumulated fatigue damage are, however, characterized by a large degree of uncertainty. This uncertainty stems from (a) the loading specifications and (b) the numerical models used to predict the response.

Given the random nature of both aerodynamic and hydrodynamic conditions, and the dependency on the structural response of the corresponding forces, a large number of scenarios needs to be evaluated, each of which relating the operational state of the turbine to the state of the environment. The aerodynamics within an offshore wind farm are not yet fully understood and moreover, the calculation of the loads on a rotating blade attached to a vibrating tower is generally based on inaccurate aerofoil data [2]. Considering hydrodynamics, the choice of an appropriate wave climate and a corresponding wave load calculation procedure is not straightforward [3]. Additional aspects, such as scour development, marine growth and secondary steel complicate the hydrodynamic analysis even further [4, 5]. Since the aerodynamic and hydrodynamic actions on an offshore wind turbine are coupled through the structural response, the evaluation of the environmental conditions cannot be separated [6]. Similar difficulties are encountered when regarding the dynamic action due to sea ice [7].

Vorpahl et al. [8] conducted an extensive comparative study for a large number of relevant design calculation codes. Already for the estimation of the first natural frequency of a monopile-founded turbine, considerable scatter was observed. Furthermore, measurements of natural frequencies of installed offshore wind turbines showed higher values than designed for [9]. Unsufficient understanding of the interaction between the structure and the soil in terms of damping [10] and stiffness [9] could be an important source of this deviation. Zaaijer [11] considered the uncertainty regarding the foundation design, which potentially leads to large discrepancies between modelled and observed dynamic behaviour.

The uncertainty in both loading and modelling necessitates the use of safety factors, which most likely lead to overly conservative designs or underestimated service lives. Real-time monitoring of the strains in existing wind turbine support

P. van der Male (✉) • E. Lourens
Department of Hydraulic Engineering, Delft University of Technology, Stevinweg 1 2628, Delft, CN, The Netherlands
e-mail: p.vandermale@tudelft.nl

© The Society for Experimental Mechanics, Inc. 2015

C. Niezrecki (ed.), *Structural Health Monitoring and Damage Detection, Volume 7*, Conference Proceedings of the Society for Experimental Mechanics Series, DOI 10.1007/978-3-319-15230-1_9

structures would provide us with knowledge of the actual fatigue damage accumulation, enabling an estimation of the remaining service life-time. One approach could be to continually monitor the strain at a number of critical locations. Knowing these locations beforehand, however, could be problematic – consider for instance a lattice structure. Moreover, the failure of a single sensor could frustrate the monitoring process considerably. A need thus exists for a robust integrated fatigue monitoring strategy capable of decreasing the current uncertainty regarding damage development in support structures offshore.

With the exception of [12], very little has been done with respect to integrated structural health monitoring of offshore wind turbines. Their focus, however, was on damage detection, sensor-fault detection and load identification. In this contribution, the strain response in a lattice support structure – allowing for the estimation of the accumulated fatigue damage – is estimated on the basis of operational vibrations using a joint input-state estimation algorithm proposed by Lourens et al. [13], in which it is assumed that no prior knowledge of the dynamic characteristics of the input forces is available. Particular attention is paid to the placement of the sensors, which should be within reach for maintenance. The choice for a lattice support structure follows from the expectation that in moderate water depths its application will be more common. Previous studies to the dynamics of lattice support structures can be found in [14] and [15].

To generate artificial measurement data, a reference finite element model, consisting of a simplified wind turbine on a lattice foundation, is constructed. The response data results from combined aerodynamic and hydrodynamic loading. After inclusion of measurement noise, the generated data and an erroneous design model are used to estimate the input forces, the states, and subsequently the strains required to predict fatigue. The erroneous model deviates significantly from the true finite element model, illustrating that despite a relatively weak model representation, the accumulated fatigue damage can be estimated accurately. Apart from the deliberate inclusion of modelling errors, Papadimitriou et al. [16] presented a similar, successful application of the joint input-state estimator for fatigue prediction.

## 9.2 Method

### *9.2.1 Joint Input-State Estimation*

The joint input-state estimator used for the current analysis was presented by Lourens et al. [13]. The algorithm allows for the estimation of states, in terms of displacements and velocities, and the input forces on the basis of a limited number of measurement signals, displacements and accelerations, given that the location of the input forces is known. The starting point of the estimation algorithm is the modally reduced formulation of the system under consideration:

$$\ddot{\mathbf{q}}(t) + \boldsymbol{\Gamma}\dot{\mathbf{q}}(t) + \boldsymbol{\Omega}^2\mathbf{q}(t) = \boldsymbol{\Phi}^{\mathrm{T}}\mathbf{S}_{\mathbf{p}}\mathbf{p}(t) \tag{9.1}$$

Here, $\mathbf{q}(t) \in \mathbb{R}^{n_m}$ represents the vector of generalized coordinates and $\mathbf{p}(t) \in \mathbb{R}^{n_p}$ the input force vector, with $n_m$ the number of modes and $n_p$ the number of input forces. The matrix $\boldsymbol{\Gamma} \in \mathbb{R}^{n_m \times n_m}$ is the modal damping matrix and $\boldsymbol{\Omega} \in \mathbb{R}^{n_m \times n_m}$ a diagonal matrix, containing the natural frequencies related to the $n_m$ modes on its diagonal. The corresponding mass normalized mode shapes are collected in the matrix $\boldsymbol{\Phi} \in \mathbb{R}^{n_{dof} \times n_m}$, with $n_{dof}$ the number of degrees of freedom of the unreduced space-discretized model, and the mode vectors $\varphi_j$, for $j = 1, \ldots, n_m$ as its columns. The force selection matrix $\mathbf{S}_{\mathbf{p}} \in \mathbb{R}^{n_{dof} \times n_p}$ specifies the force locations. A dot indicates a derivative with respect to time and the superscript T implies a transpose.

The measured quantities are combined in the output vector $\mathbf{d}(t) \in \mathbb{R}^{n_d}$, with $n_d$ the number of measured locations:

$$\mathbf{d}(t) = \mathbf{S_a}\boldsymbol{\Phi}\ddot{\mathbf{q}}(t) + \mathbf{S_v}\boldsymbol{\Phi}\dot{\mathbf{q}}(t) + \mathbf{S_d}\boldsymbol{\Phi}\mathbf{q}(t), \tag{9.2}$$

The selection matrices $\mathbf{S_a}$, $\mathbf{S_v}$ and $\mathbf{S_d} \in \mathbb{R}^{n_{dof} \times n_d}$ specify the locations of the acceleration, velocity and displacement/strain measurements, respectively. After adopting the state-space formulation for both Eqs. (9.1) and (9.2), where $\mathbf{x}(t) = \left[\mathbf{q}(t)\ \dot{\mathbf{q}}(t)\right]^{\mathrm{T}}$, and discretizing the continuous-time components, the system can be rewritten in terms of the following discrete-time combined deterministic-stochastic state-space model [13]:

$$\mathbf{x}_{k+1} = \mathbf{A}\mathbf{x}_k + \mathbf{B}\mathbf{p}_k + \mathbf{w}_k \tag{9.3}$$

$$\mathbf{d}_k = \mathbf{G}\mathbf{x}_k + \mathbf{J}\mathbf{p}_k + \mathbf{v}_k \tag{9.4}$$

with the discretizations $\mathbf{x}_k = \mathbf{x}(k\Delta t)$, $\mathbf{p}_k = \mathbf{p}(k\Delta t)$ and $\mathbf{d}_k = \mathbf{d}(k\Delta t)$, for $k = 1, \ldots, N$, where $\Delta t$ is the sampling time step and $N$ is the number of samples. The matrices $\mathbf{A} \in \mathbb{R}^{2n_m \times 2n_m}$ and $\mathbf{B} \in \mathbb{R}^{2n_m \times n_p}$ represent the discretized system matrices, that can be related to their time-continuous counterparts $\mathbf{A_c}$ and $\mathbf{B_c}$ in the following manner:

$$\mathbf{A} = \mathrm{e}^{\mathbf{A_c}\Delta t} \tag{9.5}$$

$$\mathbf{B} = [\mathbf{A} - \mathbf{I}]\, \mathbf{A_c}^{-1}\mathbf{B_c} \tag{9.6}$$

with the identity matrix $\mathbf{I} \in \mathbb{R}^{2n_m \times 2n_m}$. The output influence matrix $\mathbf{G} \in \mathbb{R}^{n_d \times 2n_m}$ and the direct transmission matrix $\mathbf{J} \in \mathbb{R}^{n_d \times n_p}$ are defined as

$$\mathbf{G} = \left[\, \mathbf{S_d}\mathbf{\Phi} - \mathbf{S_a}\mathbf{\Phi}\mathbf{\Omega}^2 \;\; \mathbf{S_v}\mathbf{\Phi} - \mathbf{S_a}\mathbf{\Phi}\mathbf{\Gamma} \,\right] \tag{9.7}$$

$$\mathbf{J} = \left[\mathbf{S_a}\mathbf{\Phi}\mathbf{\Phi}^{\mathrm{T}}\mathbf{S_p}\right] \tag{9.8}$$

Process and measurement noise, resulting, respectively, from unmodelled inputs or modelling errors and sensor inaccuracies, are represented by the stochastic components $\mathbf{w}_k \in \mathbb{R}^{2n_m}$ and $\mathbf{v}_k \in \mathbb{R}^{n_d}$. These noise processes are assumed to be stationary, zero-mean and white. Furthermore, the noise processes $\mathbf{w}_k$ and $\mathbf{v}_k$ are assumed to be uncorrelated. The joint input-state estimation algorithm requires the covariance matrices for the separate noise processes, represented by $\mathbf{Q} = \mathrm{E}\left\{\mathbf{w}_k\mathbf{w}_l^{\mathrm{T}}\right\} \in \mathbb{R}^{2n_m \times 2n_m}$ and $\mathbf{R} = \mathrm{E}\left\{\mathbf{v}_k\mathbf{v}_l^{\mathrm{T}}\right\} \in \mathbb{R}^{n_d \times n_d}$, for $k, l = 1, \ldots, N$, to be known. Additionally, the algorithm requires an initial unbiased state estimate $\widehat{\mathbf{x}}_0$, where the hat indicates an estimated quantity, and its error covariance matrix $\mathbf{P}_0 \in \mathbb{R}^{2n_m \times 2n_m}$ to be available.

The estimation algorithm represents an extension of the filter presented by Gillijns and De Moor [17], developed for linear systems with direct transmission. This filter was proved to be optimally in estimated states and inputs in a minimum-variance unbiased sense. The extension stems from the numerical instabilities which arise when $n_d$ exceeds $n_m$. This instability issue can be circumvented by limiting the order of the potentially rank-deficient matrix multiplications to the number of modes accounted for [13].

A contribution by Maes et al. [18] generalizes the joint input-state estimator by allowing for correlation between the processes $\mathbf{w}_k$ and $\mathbf{v}_k$ via the correlation matrix $\mathbf{S} \in \mathbb{R}^{2n_m \times n_d}$:

$$\mathbf{S} = \mathrm{E}\left\{\mathbf{w}_k\mathbf{v}_l^{\mathrm{T}}\right\} \tag{9.9}$$

Under the assumption of white noise stochastic input processes, the matrix entries of $\mathbf{S}$ can be estimated from operational data. Since the stochastic input processes in the specific case under consideration, an offshore wind turbine, are not close to being white, this generalization is not accounted for.

### 9.2.2 *Response Estimation*

With the estimated force time histories $\widehat{\mathbf{p}}_k \in \mathbb{R}^{n_p}$ and state sequences $\widehat{\mathbf{x}}_k \in \mathbb{R}^{2n_m}$, response predictions $\widehat{\mathbf{d}}_k \in \mathbb{R}^{n_r}$ at $n_r$ unmeasured locations can be constructed. Hereto, the observation equation (Eq. (9.4)) is employed [16]:

$$\widehat{\mathbf{d}}_k = \mathbf{G}\widehat{\mathbf{x}}_k + \mathbf{J}\widehat{\mathbf{p}}_k \tag{9.10}$$

It should be noted that the output influence matrix $\mathbf{G}$ and the direct transmission matrix $\mathbf{J}$ are now constructed to correspond to the response prediction, for which the selection matrices $\mathbf{S_a}$, $\mathbf{S_v}$ and $\mathbf{S_d}$ specify the locations, as well as the type of response to be predicted.

The response estimation does not require the locations of the input forces to be accurately known. If the location of the input forces is unknown, the joint input-state estimator can be applied to estimate equivalent forces at arbitrarily chosen locations, causing the same measured response. Furthermore, if an erroneous model is applied for the response estimation, the force estimations could potentially serve to reduce the effects of the modelling errors on the response predictions, as will be shown in Sect. 9.3.2.

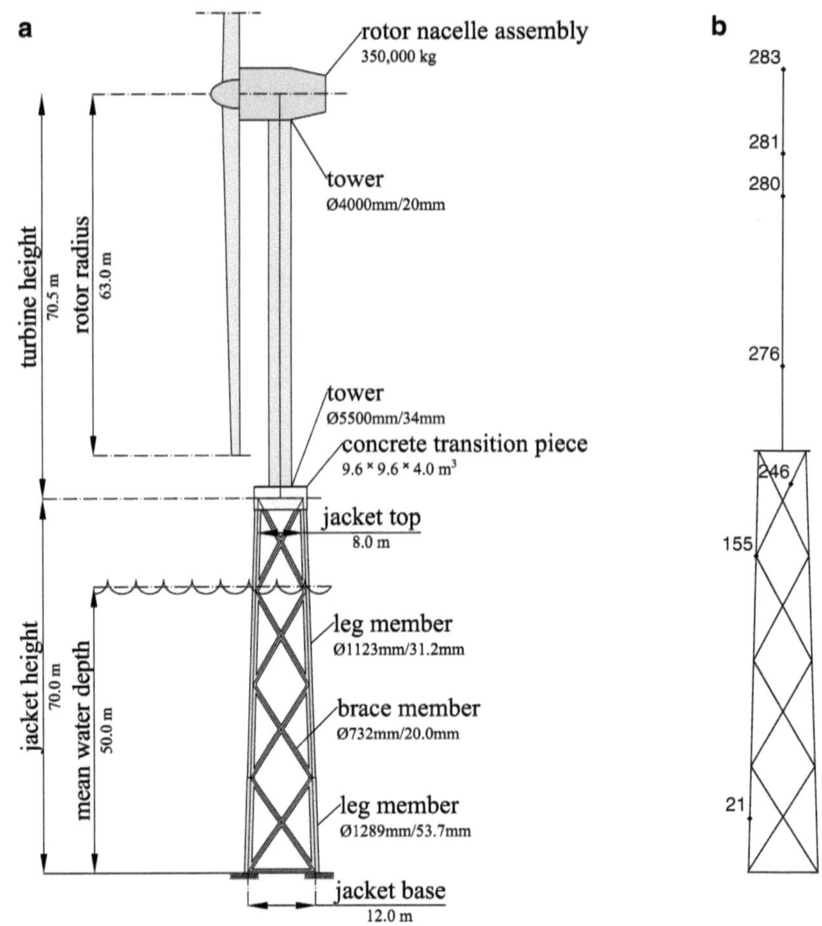


**Fig. 9.1** (**a**) Combined wind turbine and lattice support structure (**b**) Finite element representation of the wind turbine structure with indicated nodes for force positioning, sensor placement and response estimations

### *9.2.3 Wind Turbine Model*

The analysis is based on a lattice structure, as described by De Vries et al. [15], supporting a 5 MW reference turbine, for which the main characteristics are presented by Jonkman et al. [19]. Figure 9.1 illustrates the geometry of the combined turbine and support structure, including the main geometric and material characteristics. The combined turbine and lattice structure is modelled in 2D by means of the finite element method. Euler-Bernoulli beam elements, possessing six degrees of freedom, are employed to represent the steel members and the turbine tower. The rotor-nacelle assembly is reduced to a lumped mass and at the jacket base the structure is connected rigidly fixed to the soil. Compression due to the self-weight of the structure, which reduces the effective stiffness, is not accounted for. By varying the stiffness characteristics of the concrete transition piece, the model is updated such, that the first two natural frequencies show good agreement with those presented in [15].

As a basis for the joint input-state estimation, a modal representation of the wind turbine model is required, see Eq. (9.1). Figure 9.2 presents the first ten mode shapes of the modelled turbine structure. The natural frequencies of the first ten modes, including a brief description of the mode, are presented in Table 9.1. Structural damping is accounted for in terms of classical modal damping, i.e. a diagonal damping matrix is adopted. For each mode, a damping value of 1.0 % of the critical damping is assumed.

### *9.2.4 Stochastic Wind Force*

The time-dependent wind force is determined on the basis of the actuator disc concept, elaborated on in [20], where a 1D free field turbulence is simulated on the basis of the spectral properties of a Kaimal power density spectrum $S_{uu}$ as a function of frequency $f$:

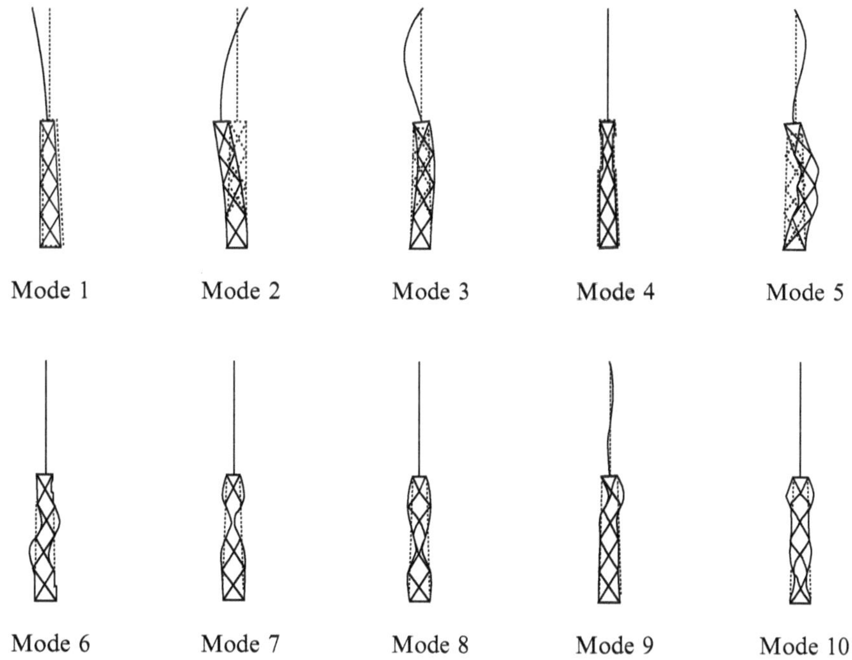


**Fig. 9.2** Results of the finite element modal analysis for the first ten modes

**Table 9.1** Natural frequencies corresponding to the first ten modes, as derived from the finite element modal analysis

| No. | Natural frequency [Hz] | Description |
|---|---|---|
| 1 | 0.364 | First global lateral mode |
| 2 | 1.07 | Second global lateral mode |
| 3 | 4.89 | Third global lateral mode |
| 4 | 6.20 | First global vertical mode |
| 5 | 6.77 | Fourth global lateral mode |
| 6 | 10.4 | Lateral jacket mode (second and third frame from top – in-phase |
| 7 | 10.4 | Local lateral jacket mode (second frame from top – anti-phase) |
| 8 | 10.9 | Local lateral jacket mode (third frame from top – anti-phase) |
| 9 | 12.6 | Local lateral jacket mode (first frame from top – in-phase) |
| 10 | 13.8 | Local lateral jacket mode (first and fourth frame from top – anti-phase) |

$$S_{uu}(f) = \sigma_u^2 \frac{4L_{1u}/\overline{U}}{\left(1 + 6fL_{1u}/\overline{U}\right)^{5/3}}, \text{ for } f > 0 \ Hz \tag{9.11}$$

$\bar{U}$ represents the mean wind velocity, $L_{1u}$ is a turbulence length scale and $\sigma_u$ is the standard deviation of the turbulence velocity. To generate a wind force signal, an optimally functioning turbine is assumed, implying that the turbine operates at the Lanchester-Betz limit. Furthermore, it is assumed that the induced velocity through the rotor disc follows the instantaneous turbulent wind velocity. This assumption implies that the entire wake changes instantaneously, such that equilibrium in the wake is maintained at all times.

Figure 9.3a shows the Kaimal spectrum adopted for the generation of the wind force signal. The spectrum reveals a main energy contribution from the frequencies below 1 Hz. Assuming a random phase distribution, and a cut-off frequency of 3 Hz, a wind force signal of 546 s is generated, in accordance with [21]. Figure 9.3b depicts a 100 s window of this time signal. Since the finite element model does not include a detailed rotor representation, the total wind force is assumed to act concentratedly at the rotor nacelle assembly at the tower top, as can be seen in Fig. 9.3c. It should be noted that, despite the turbulence frequency cut-off, the wind force signal contains higher frequency contributions. This results from the nonlinear dependency of the wind force on the turbulent wind velocity.

The apparent damping, resulting from a rotating rotor, is assumed at 4.0 % of the critical damping for the first structural mode. For higher modes, the contribution of this aerodynamic damping is scaled on the basis of the modal deflection at the tower top.

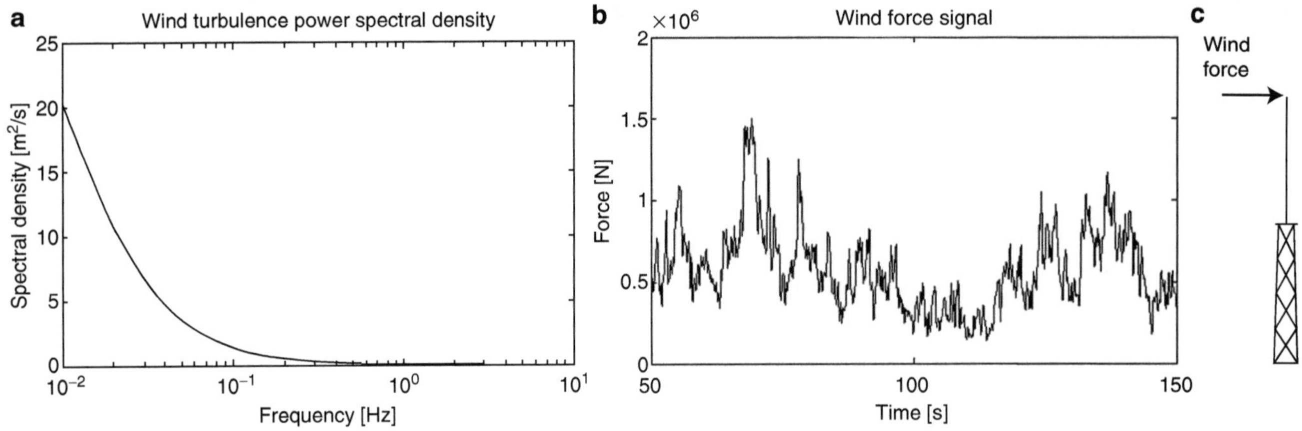


**Fig. 9.3** (**a**) Kaimal spectrum derived for mean wind velocity of 10 m/s, a turbulence intensity of 10 % and a turbulence length scale of 150 m (**b**) Turbulent-wind force signal for NREL5 offshore wind turbine, and (**c**) positioning of the wind force on the FE model

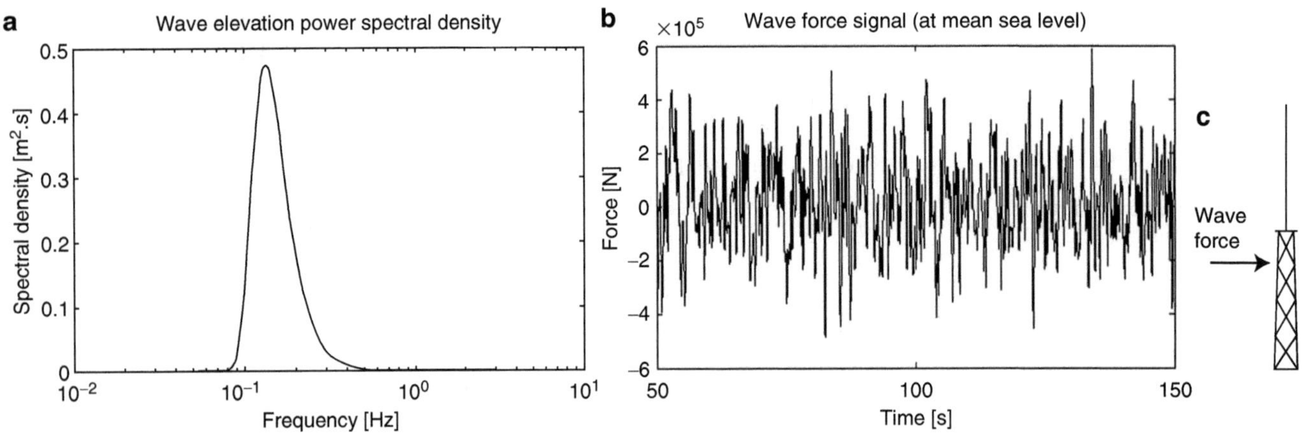


**Fig. 9.4** (**a**) Pierson-Moskowitz spectrum derived for a mean wind velocity of 10 m/s (**b**) Wave force signal for on lattice structure, and (**c**) positioning of the wave force on the finite element model

### *9.2.5 Stochastic Wave Force*

The Morison equation is commonly adopted to estimate hydrodynamic actions on slender vertical members. The equations derive the total hydrodynamic force by superposing an inertia and a drag force contribution, depending on the wave particle acceleration and velocity, respectively. A procedure for defining a hydrodynamic force on the basis of the Morison equation, using the linear wave theory for defining the wave particle actions, is elaborated on in [22]. The original Morison equation was derived for vertically-oriented cylindrical piles and for very small pile diameter to wave length ratios. To determine the wave forces on the current lattice structure, some engineering adjustments have to be implemented. As an equivalent structural diameter, the combined width of the members exposed to the hydrodynamic action is taken. Equally, an equivalent cross-sectional area results from the sum from the different lattice elements. For the hydrodynamic inertia and drag coefficients, common values of 2.0 and 0.7, respectively, are adopted.

The wave particle actions for a specific sea state can be derived from a relevant wave elevation spectrum. In this particular case, use is made of a Pierson-Moskovitz spectrum, depending on the mean wind velocity $\bar{U}$, see Fig. 9.4a:

$$S_{hh}(f) = \alpha \frac{g^2}{(2\pi f)^5} e^{-\beta \left( \frac{g}{2\pi f \bar{U}} \right)^4}, \tag{9.12}$$

where $\alpha = 0.0081$ and $\beta = 0.74$ are coefficients and $g$ is the gravitational acceleration. Compared to the wind turbulence spectrum of Fig. 9.3a, the wave elevation spectrum contains its energy at somewhat higher frequencies. The peak energy is much smaller. On the basis of the wave elevation distribution, the associated wave kinematics can be determined.

The magnitude of the distributed wave force is calculated at mean sea level. This force is assumed to act within a wave impact zone of ±5.0 m with respect to this level. After integration over this wave impact zone, a concentrated wave force signal at mean sea level is derived. In deriving this force signal, of which a 100 s window is presented in Fig. 9.4b, a 1.0 m/s current is added to the wave particle velocity. In correspondence to the wind force generation, the signal has a length of 546 s, while a cut-off frequency of 3 Hz has been adopted. The force location is chosen at the K-joint, connecting the first and second jacket frame, see Fig. 9.4c. Since the Morison equation relates the drag force nonlinearly to the wave particle velocity, the wave force contains frequency content above the cut-off frequency of the wave elevation energy spectrum.

The added damping resulting from the hydrodynamic action is neglected. This hydrodynamic damping would result from the response velocity of the jacket members and its contribution can be assumed to be small.

### 9.2.6 Sensor Network

Given the hostile environmental offshore conditions, the measurement of the structural motion requires a robust network of sensors. This robustness is first pronounced in the number of required sensors: if only a limited number of sensors is needed, the costs to build in sufficient redundancy remain low. Second, the positioning of the sensors affects the robustness of the system. This implies that no sensors should be placed under the water level, because these locations are not easily reached when maintenance is necessary. Moreover, to prevent sensors from early failure, sensors within the wave splash zone should be avoided. Therefore, the estimation of the dynamic response will be based on a network consisting of sensors attached to the turbine tower only.

The design of a sensor network for optimal estimation of inputs and states is considered by Maes et al. [23]. In this contribution, the criteria for the invertibility of a system, and therefore the possible application of the joint input-state estimator, are derived in terms of stability and identifiability. For the joint input-state estimator to be stable, two requirements need to be fulfilled. First, the number of acceleration and/or velocity sensors, $n_{d;a}$ and $n_{d;v}$, respectively, needs to be equal to or larger than the number of input forces:

$$n_{d;a} + n_{d;v} \geq n_p \tag{9.13}$$

Second, the number of displacement/strain sensors $n_{d;d}$ needs to be equal to or larger than the number of input forces:

$$n_{d;d} \geq n_p \tag{9.14}$$

In this particular case, the separate estimation of the wind and wave force requires a network consisting of two acceleration and/or velocity sensors and two displacement/strain sensors.

Identifiability relates to the controllability of the input forces, the observability of the states and the direct invertibility of the measurement outputs towards states and input. The controllability of the input forces can by assessed by determining the rank of the controllability matrix $\mathbf{C} \in \mathbb{R}^{2n_m \times (n_p \times 2n_m)}$:

$$\mathbf{C} \equiv \left[ \mathbf{B}\ \mathbf{AB} \ldots \mathbf{A}^{2n_m-1}\mathbf{B} \right] \tag{9.15}$$

If $\mathbf{C}$ is of full rank, i.e. rank $(\mathbf{C}) = 2n_m$, the system is controllable, implying that the assumed forces are positioned such, that all modes can contribute to the response. Alternatively, the controllability can be assessed by determining the rank of the modal projections of the force selection matrix $\mathbf{S_p}\varphi_j$, for which it should apply that rank $\left(\mathbf{S_p}\psi_j\right) = 1$ for $j = 1, \ldots, n_m$.

In a similar manner, the observability of the system can be assessed by determining the rank of the observability matrix $\mathbf{H} \in \mathbb{R}^{(2n_m \times n_d) \times 2n_m}$:

$$\mathbf{H} \equiv \begin{bmatrix} \mathbf{G} \\ \mathbf{GA} \\ \vdots \\ \mathbf{GA}^{2n_m-1} \end{bmatrix} \tag{9.16}$$

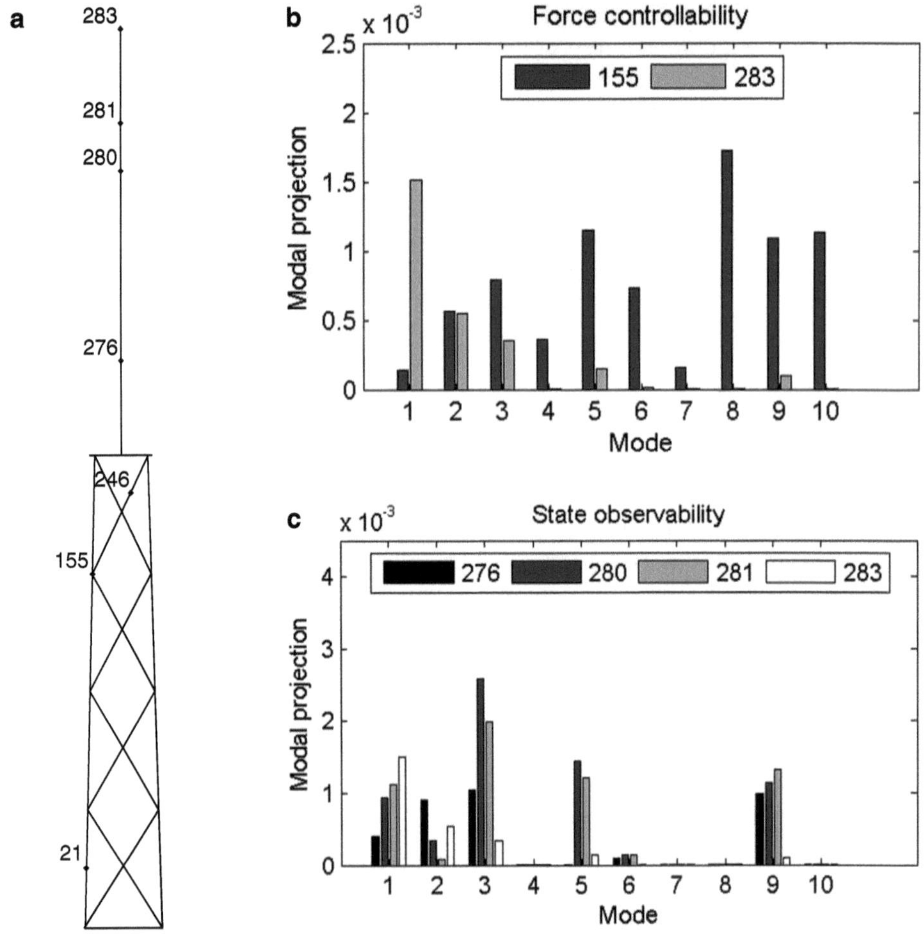


**Fig. 9.5** (**a**) Selected nodes for force positioning, sensor placement and response estimations (**b**) Modal projections to the wave force location (node 155) and the wind force location (node 283) (**c**) Modal projections to the sensor locations, node 276, node 280, node 281 and node 283

If **H** is of full rank, i.e. rank$(\mathbf{H}) = 2n_m$, the system is observable, implying that the sensors are positioned such, that all states can be identified uniquely, irrespective of the initial conditions. Alternatively, the observability can be assessed by determining the rank of the modal projections of the sensor selection matrix $\mathbf{S_d}\varphi_j$, for which it should apply that rank$\left(\mathbf{S_d}\boldsymbol{\varphi}_j\right) = 1$ for $j = 1, \ldots, n_m$.

Direct invertibility requires the number of structural modes accounted for with the reduced-order model to be equal to or larger than the number of forces to be identified, i.e. $n_m \geq n_p$. The implication of direct invertibility is that input forces can be estimated from the system output, without delay in time.

Figure 9.5a presents the finite element model of the structure, where a number of nodes is specifically indicated. The known locations of the wind and wave force, respectively, are node 283 and node 155. The four sensor locations are node 276, 280, 281 and 283. The remaining nodes, 21 and 246, will serve as response estimation locations in Sects. 9.3.1 and 9.3.2. These locations are chosen to illustrate the difference in quality of response estimations over the height of the lattice structure. Particularly for node 21, the estimate is expected to be inaccurate, due to the large distance from the sensor locations [24].

The modal projections $\mathbf{S_p}\varphi_j$ for the first ten modes are depicted in Fig. 9.5b. It can be clearly seen that the wind force location projects mainly on the first mode. Mode 2 and 3 could serve the wind force identification too. The modal projection of the wave force location is highest for the higher modes: mode 5, 8, 9 and 10. Given the frequency content of the wave force, however, not much excitation of these higher modes is to be expected.

Figure 9.5c shows the modal projections $\mathbf{S_d}\varphi_j$ of the four sensor locations for the first ten structural modes. The sensors only measure lateral motions. Based on this figure, it can be concluded that positioning four sensors at the chosen tower locations enables the identification of states related to mode 1, 2, 3, 5 and 9. It is clear that for some modes the modal projection is very small. Mode 4, for instance, represents a vertical global mode, for which the states are difficult to capture with sensors that only function laterally. Mode 6, 7, 8 and 10 represent local jacket modes, with very small tower amplitudes.

Combining Figs. 9.5b, c, it can be concluded that the network consisting of sensors on the tower only will not allow for the dynamic response estimation resulting from higher mode excitation. The frequency content of both turbulence and wave elevation is cut off at 3 Hz, implying that the frequency content of the wind and wave force above 3 Hz will be limited – despite the nonlinear force formulation. Still, the inclusion of mode 4, 7, 8 and 10 in the estimation analysis results in a numerically rank-deficient observability matrix, implying that the system is ill-conditioned. To a lesser extent, this also applies for mode 6. To prevent the system from being ill-conditioned, the modes 4, 7, 8 and 10 are excluded from further analysis, leaving a reduced-order model consisting of six modes. A convergence analysis with respect to the full-order model has shown that the reduced-order model enables the generation of accurate measurement signals. If mode 6 is excluded from the analysis too, the ill-conditioning of the observability matrix would be reduced. This further reduction of the modal basis would, however, have too large an effect on the accuracy of the model.

The adopted sensor network consists of accelerometers at node 281 and 283 and strain gauges at node 276 and 280. The reasoning behind is that the accelerations are expected to be largest near the tower top. Strains, on the other hand, are more pronounced in the lower section of the tower.

## 9.3 Results

### *9.3.1 Response Estimation in the Absence of Modelling Errors*

First, the response prediction is tested for the robustness to the inclusion of measurement noise only. Application of the force time signals to the finite element model results in simulated measurement data at the chosen sensor locations. To the measurement signals $\mathbf{d}_k$, consisting of $N$ time samples each, some Gaussian white noise is added, resulting in the polluted output vector $\tilde{\mathbf{d}}_k \in \mathbb{R}^{n_d}$ per time step $k$. Given the stationary random nature of the measurement signals, the noise is chosen to be related to the standard deviation of the separate measurements, $\boldsymbol{\sigma}_{\mathbf{d}} \in \mathbb{R}^{n_d \times n_d}$:

$$\tilde{\mathbf{d}}_k = \mathbf{d}_k + \gamma \boldsymbol{\sigma}_{\mathbf{d}} \mathbf{r}_k, \tag{9.17}$$

where $\gamma$ represents the noise level and $\boldsymbol{\sigma}_{\mathbf{d}}$ is a diagonal matrix with the standard deviations of the time signals as its diagonal entries. $\mathbf{r}_k \in \mathbb{R}^{n_d}$ is a vector composed of random values taken from a normal distribution with zero mean and a standard deviation of one.

For $\gamma$ a value of 0.05 is applied, corresponding to 5 % measurement noise. Equation (9.17) allows for the exact calculation of the entries of the measurement covariance matrix $\mathbf{R}$:

$$\mathbf{R} = \gamma \boldsymbol{\sigma}_{\mathbf{d}}^2 \tag{9.18}$$

The initial states are assumed to be zero. Since no process noise is present, the entries of the error covariance matrices $\mathbf{P}_0$ and $\mathbf{Q}$ are chosen very small, namely $1 \cdot 10^{-20}$.

The prediction of the response in the lattice structure requires force and state estimations, obtained by means of the joint input-state estimator from the noise-contaminated measurement signals. These estimated forces and states, $\widehat{\mathbf{p}}_k$ and $\widehat{\mathbf{x}}_k$, respectively, are subsequently used to estimate the response as described in Sect 9.2.2. Since the modal basis consists of six modes, a total of twelve states is estimated by the estimation algorithm. Given the main frequency content of the force signals – below 3 Hz – and the natural frequencies of the system, the estimation of the first and second modal states are most relevant. Figure 9.6 presents the state estimation corresponding to these modes. The inclusion of the measurement noise still allows an accurate estimation of the first modal states. The estimation of the second modal states is shown to be obviously more noisy.

Using Eq. (9.10), the strain estimates are derived at node 21 and 246 (see Fig. 9.5a) and presented in Fig. 9.7. These particular locations are chosen to illustrate to what extent the response of lattice members can be estimated by means of tower measurements only. Figure 9.7 shows that the low-frequency strain response is captured relatively well, despite the noise-contamination of the measurement signals. Still, some high-frequency disturbance in the estimations can be observed. The disturbance is most pronounced at 10.4 Hz, which corresponds to the sixth natural frequency of the structure, a mode that is weakly observed by the sensor network. The response estimation can be improved by excluding mode 6 from the modal basis. This, however, would also weaken the accuracy of the reduced-order finite element model – see Sect 9.3.2 for a more detailed discussion.

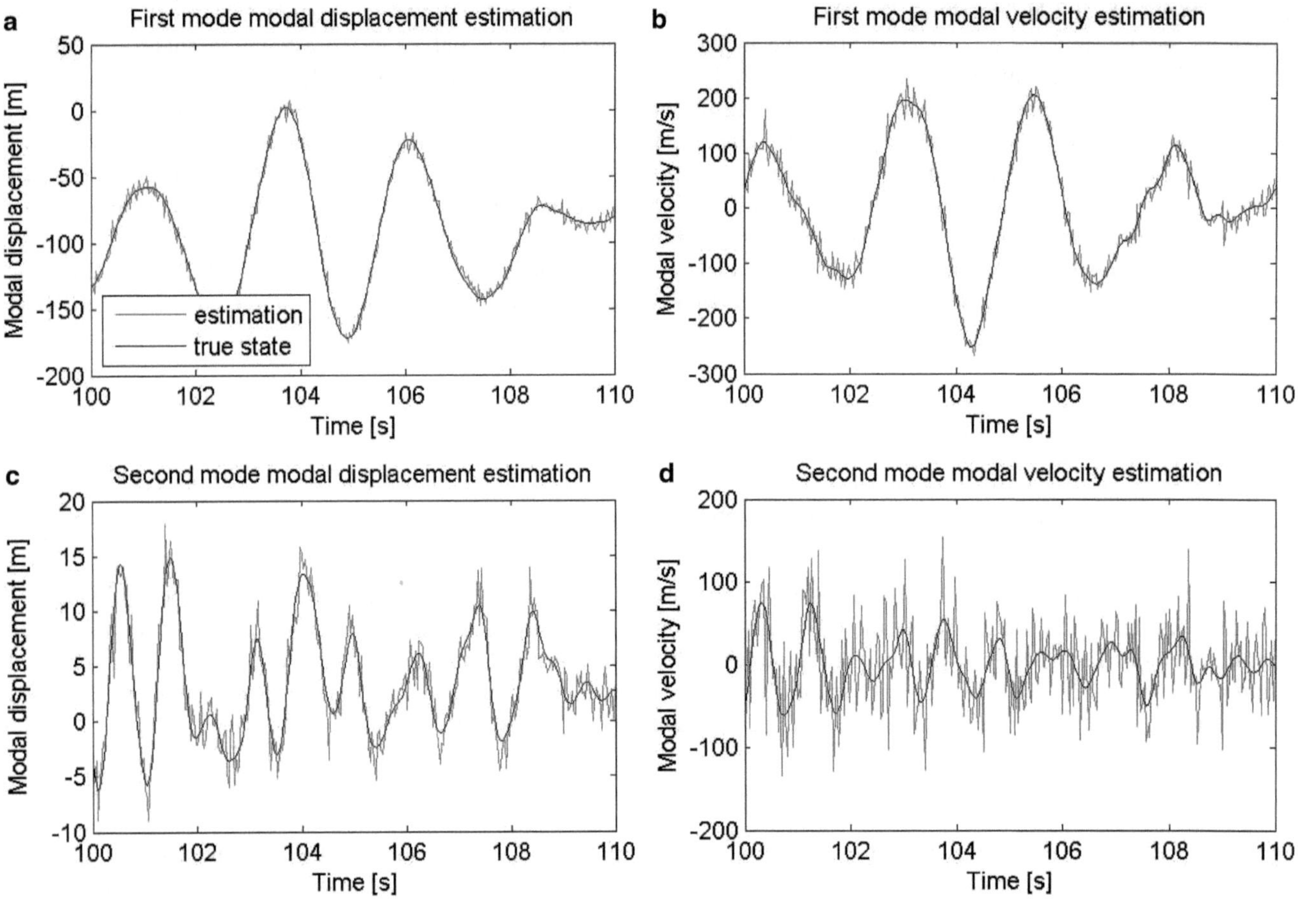


**Fig. 9.6** State estimation for (**a**) the modal displacement of mode 1, (**b**) the modal velocity of mode 1, (**c**) the modal displacement of mode 2 and (**d**) the modal velocity of mode 2

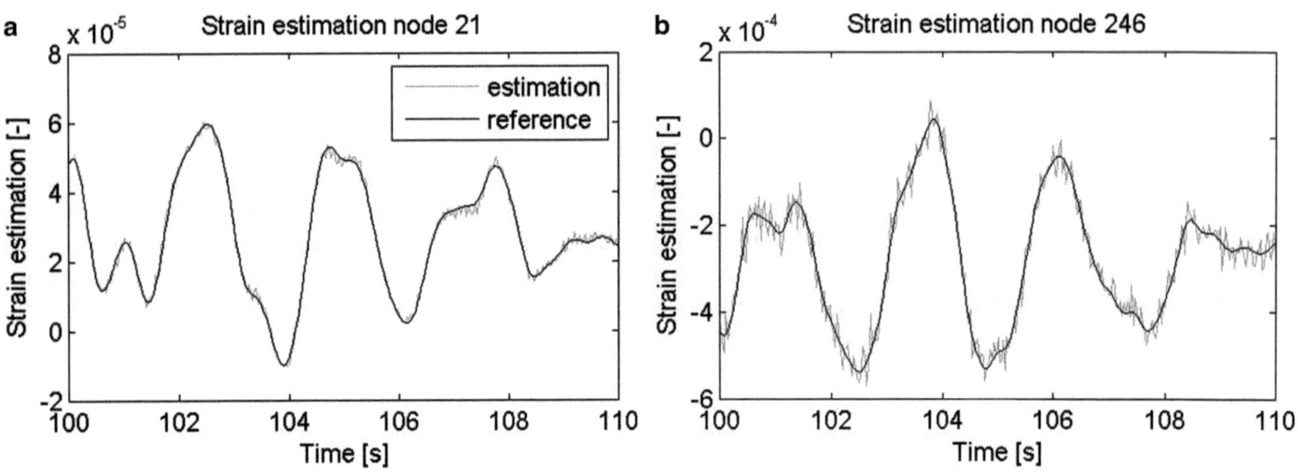


**Fig. 9.7** Time signal representation of the strain response estimation at two lattice members: (**a**) node 21 and (**b**) node 246

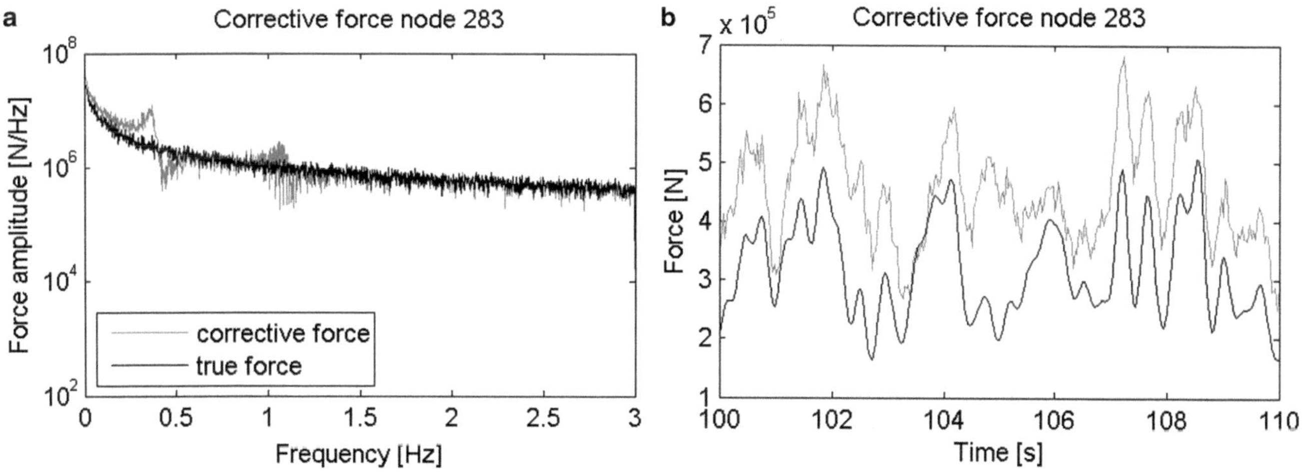


**Fig. 9.8** Corrective force estimation at node 283 with erroneous design model, based on measurements with and without measurements noise, presented in (**a**) the frequency-domain and (**b**) the time-domain

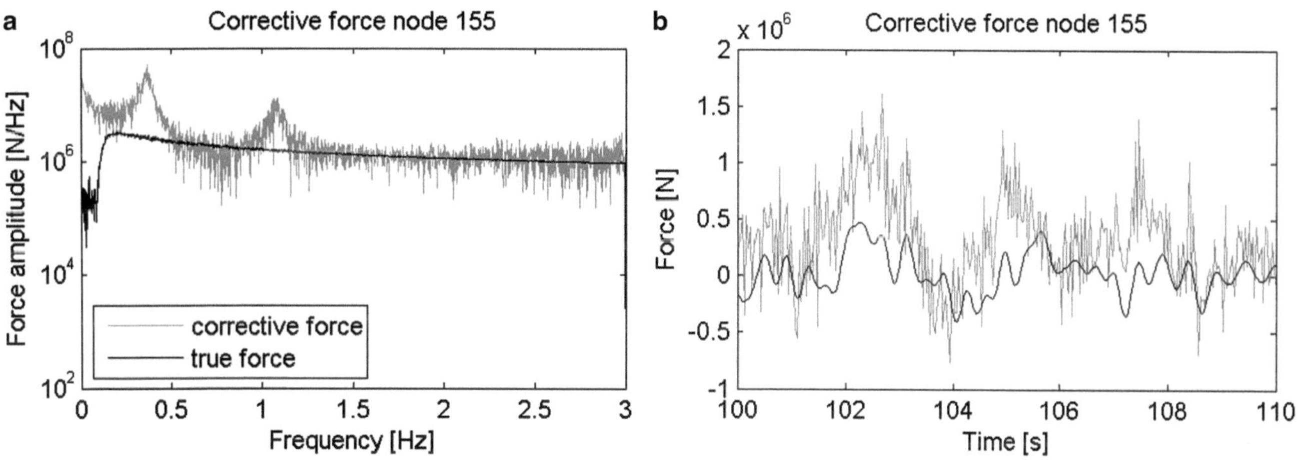


**Fig. 9.9** Corrective force estimation at node 155 with erroneous design model, based on measurements with measurements noise, presented in (**a**) the frequency-domain and (**b**) the time-domain

## *9.3.2 Response Estimation with an Erroneous Model*

A second response estimation is performed, this time with a design model that does not exactly represent the true structure. Process noise is deliberately introduced in the design model, by increasing the first and second natural frequency with 20 %. The first and second natural frequency of the design model are 0.437 and 1.21 Hz, instead of 0.364 and 1.07 Hz of the model with which the data is generated. Again, the measurement data is polluted with 5 % measurement noise.

The inclusion of process noise complicates the estimation of the true forces and states. Instead, the joint input-state estimator enables the estimation of equivalent forces and states, which combined enable the response estimation at unmeasured structure locations. To lesser extent, this was already observed in the previous section, where only measurement noise was accounted for. To optimize the functioning of the estimation algorithm, the entries of the **Q** matrix are adjusted. The covariance of the process noise cannot be as easily estimated as the covariance of the measurement noise. By choosing the square roots of the process noise covariance entries of the same order of magnitude as a small percentage of the states, the covariance matrix entries can be made to correspond with what they represent [24].

When considering the frequency content of the estimated corrective forces, see Figs. 9.8 and 9.9, the force signals clearly compensate for the unmodelled dynamics. The corrective forces contain frequency peaks at the true natural frequencies The corresponding time signals illustrate the deviation of the estimated force from the true force. The effect of the measurement

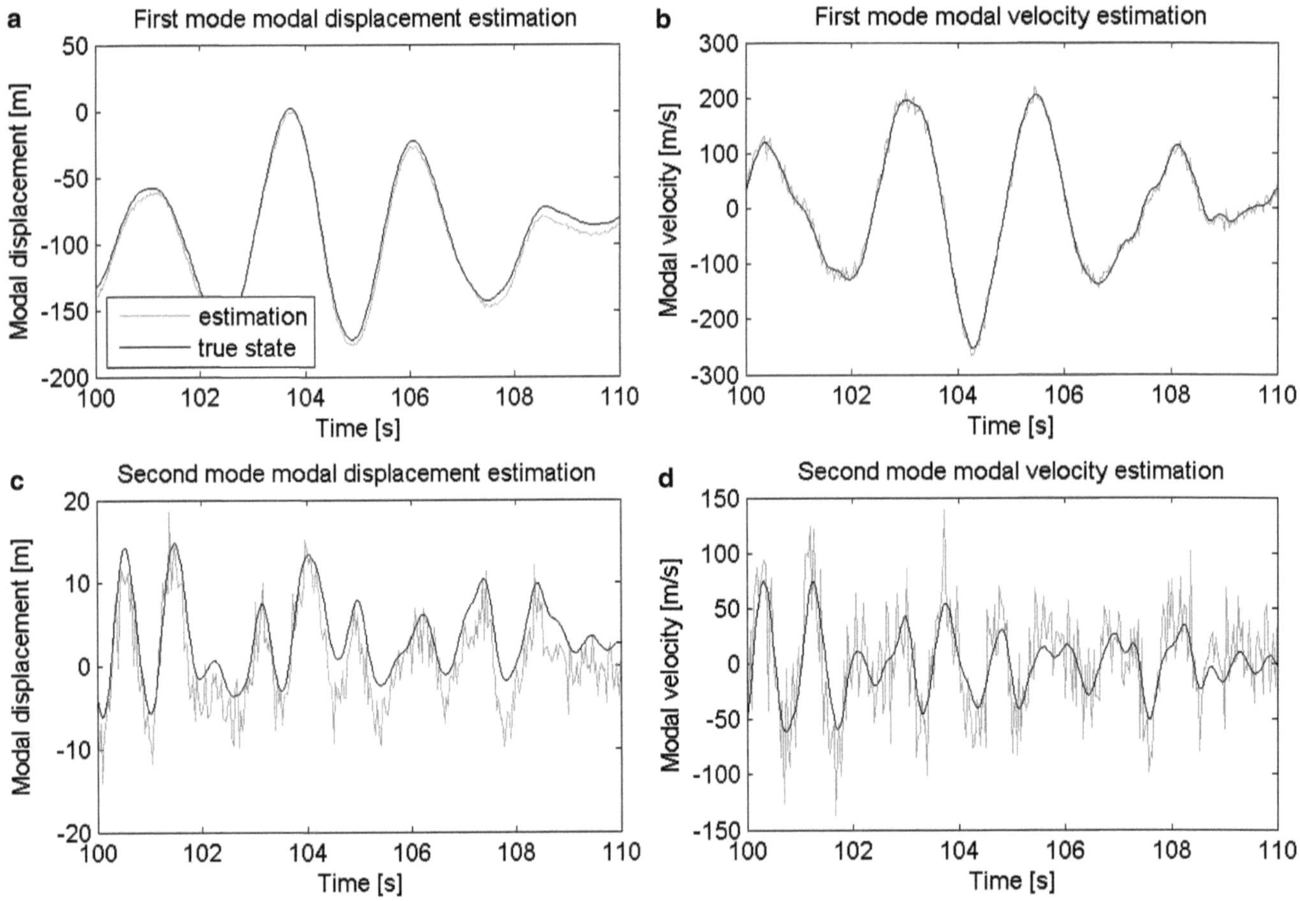


**Fig. 9.10** State estimation with erroneous design model for (**a**) the modal displacement of mode 1, (**b**) the modal velocity of mode 1, (**c**) the modal displacement of mode 2 and (**d**) the modal velocity of mode 2

noise is mainly expressed as the high frequent disturbance of the corrective force at node 155. The compensation of the frequency content at the true natural frequencies enables a relatively accurate estimation of the first and second modal states, see Fig. 9.10. Despite a small bias, the estimated states show good correspondence with the true states.

The small bias in the state estimates can be recognized too in the estimated strain response at node 21 and 246. Figure 9.11 presents short samples of the estimated time signals. This bias is a direct result of the invalidity of the zero-mean assumption used for the process noise $\mathbf{w}_k$. The magnitude of the bias, however, remains unaffected over the length of the time signals, implying that the bias does not corrupt possible fatigue damage estimates. Particularly at node 21, a noisy component in the response estimation is observed. To further asses this noisiness, the frequency content of the estimates is depicted in Fig. 9.12. It becomes clear that the sixth mode contributes excessively to the identified response. This result was already expected when the observability of the states was discussed (see Fig. 9.5c), where it was expected that the inclusion of mode 6 increases the ill-conditioning of the observability matrix.

## 9.4 Discussion and Conclusions

Recognizing that the design for fatigue for offshore wind turbines is characterized by uncertainty, while being a main design driver for this type of structures, this study presents a strategy for the real-time monitoring of the accumulated fatigue damage, employing a joint input-state estimation algorithm. In particular, a wind turbine on a lattice support structure is considered, for which the response estimates of the lattice members are based on measurements at the turbine tower only. This restriction follows from the difficulty to reliably and robustly measure at locations on the lattice structure.

The study is based on a 2D finite element model representing the true offshore wind turbine. Measurement signals are generated from the response to known wind and wave forces, which are defined from conventional load models. First, inputs

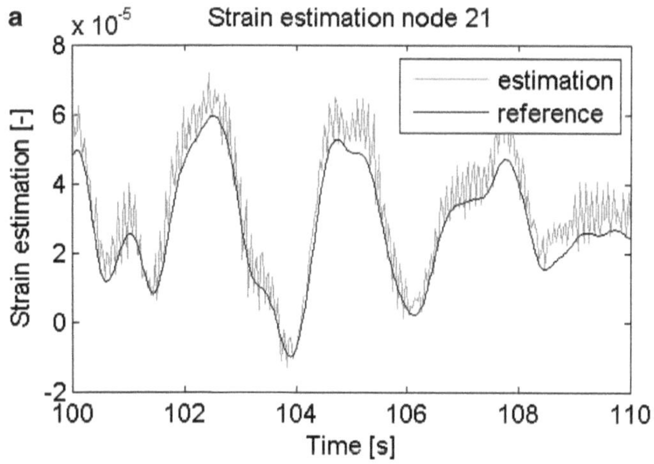


**Fig. 9.11** Time signal representation of the strain response estimation with erroneous design model at two lattice members: (**a**) node 21 and (**b**) node 246

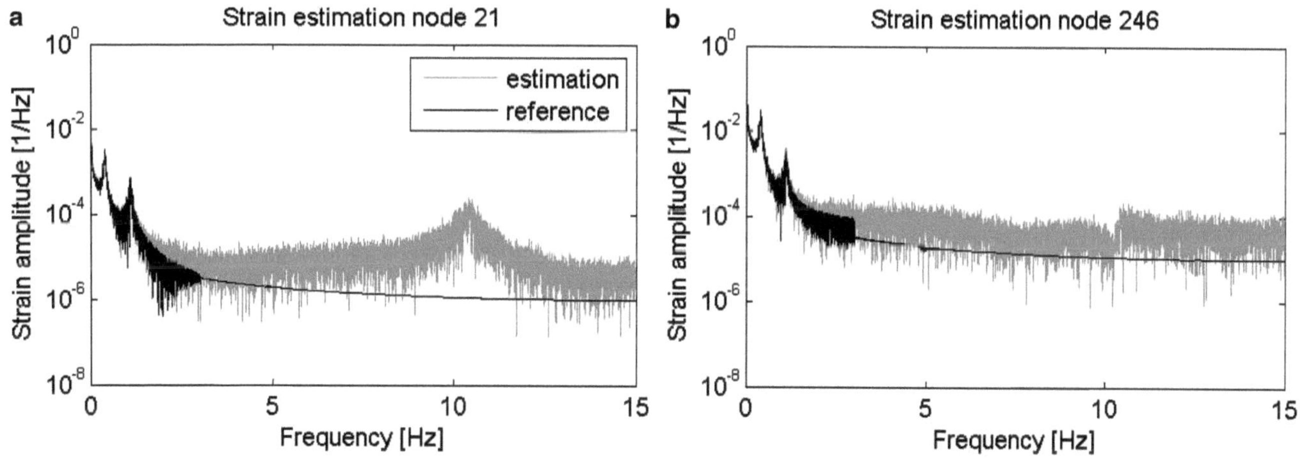


**Fig. 9.12** Frequency domain representation of the strain response estimation with erroneous design model at two lattice members: (**a**) node 21 and (**b**) node 246

and states are estimated from measurements with a noise level of 5 % with a finite element model in the absence of modelling errors. From this, the strain response at chosen locations on the lattice structure is estimated, which could eventually serve the estimation of the accumulated fatigue damage. Subsequently, a design finite element model is constructed by adjusting the main natural frequencies. This design model is applied to estimate the strain response in the lattice structure on the basis of the true response measurements.

The response estimates show that the main frequency content can be captured relatively accurately. The estimations with the design model, with a 20 % error on the first and second natural frequency, do show a small bias with respect to the real response, resulting from the invalidity of the zero-mean assumption used for the process noise. This bias, however, will not harm the quality of the accumulated fatigue damage estimation, since for this only the magnitude of the strain cycles is of interest. Nevertheless, the strain response estimates do contain some high frequency disturbance in which the sixth natural frequency of the system is most pronounced. The occurrence of this disturbance can be related to the ill-conditioning of the observability matrix.

The analysis illustrated the trade-off between the accuracy of the reduced-order finite element model and the ill-conditioning of the observability matrix. The modal basis of the finite element model should account for a sufficient number of modes to describe the dynamic response sufficiently accurate. Certain modes, however, can be hardly observable, implying ill-conditioning of the observability matrix. In this particular case, the observability of mode 6 is small, resulting in the noisy disturbance of the estimated response. Excluding mode 6 from the modal basis would decrease the ill-conditioning of the observability matrix and, of course, result in strain estimates without the disturbance from the sixth natural frequency. On the other hand, further reduction of the modal basis would increase the modelling error in the design model.

It should be noted that the input forces are related to environmental conditions, turbulence and wave elevation, with a limited frequency content. Despite the nonlinear dependency of the actual forces on these conditions, the energy content of the higher frequencies is small. As a result, the structure mainly responds at its first and second natural frequency, and measurements at the tower only enable the local response estimation of the lattice structure. For normal environmental conditions, this restriction can be expected to be valid. For extreme conditions, for instance slamming waves during a storm, higher frequencies are excited, resulting in the dynamic response of local modes in the lattice structure. Since these modes are not observable by the adopted sensor network, the fatigue damage accumulated during these conditions cannot be estimated accurately.

As a final remark it is mentioned that theoretically the sensor network could be reduced to one acceleration/velocity sensor and one displacement sensor. This network would not allow for the estimation of a separate equivalent wind and wave force, but it could be sufficient to estimate the response at unmeasured locations. The sensor locations, however, should be chosen such that the states relevant for the response are observed sufficiently.

## References

1. Muskulus M, Schafhirt S (2014) Design optimization of wind turbine support structures – a review. J Ocean Wind Energ 1(1):12–22
2. Hansen MOL, Sørensen JN, Voutsinas S, Sørensen N, Madsen HA (2006) State of the art in wind turbine aerodynamics and aeroelasticity. Prog Aerosp Sci 42(4):285–330
3. Henderson AR, Zaaijer MB (2014) Hydrodynamic loading on offshore wind turbine support structures. Eng Integr 25:24–31
4. Whitehouse RJS, Harris JM, Sutherland J, Rees J (2011) The nature of scour development and scour protection at offshore windfarm foundations. Mar Pollut Bull 62(1):73–88
5. Segeren MLA (2011) Influence of a boatlanding and J-tubes on wave loads and wall thickness of the monopile support structure design. In: Proceedings of the EWEA offshore 2011 conference, Amsterdam, 29 Nov – 1 Dec 2011
6. Kühn MJ (2001) Dynamics and design optimisation of offshore wind energy conversion systems. PhD thesis, Delft University Wind Energy Research Institute
7. Hendrikse H, Renting FW, Metrikine AV (2014) Analysis of the fatigue life of offshore wind turbine generators under combined ice- and aerodynamic loading. In: Proceedings of the ASME 2014 33rd international conference on ocean, offshore and arctic engineering, San Francisco, 2014
8. Vorpahl F, Strobel M, Jonkman JM, Larsen TJ, Passon P, Nichols J (2014) Verification of aero-elastic offshore wind turbine design codes under IEA wind task XXIII. Wind Energ 17(4):519–547
9. Versteijlen WG, van Dalen KN, Metrikine AV, Hamre L (2014) Assessing the small-strain soil stiffness for offshore wind turbines based on in situ seismic measurements. J Phys Conf Ser 524(1):012088
10. Versteijlen WG, Metrikine AV, Hoving JS, Smid E, de Vries WE (2011) Estimation of the vibration decrement of an offshore wind turbine support structure caused by its interaction with soil. In: Proceedings of the EWEA offshore 2011 conference, Amsterdam, 29 Nov – 1 Dec 2011
11. Zaaijer MB (2006) Foundation modelling to assess dynamic behaviour of offshore wind turbines. Appl Ocean Res 28:45–57
12. Fritzen CP, Kraemer P, Klinkov M (2010) An integrated SHM approach for offshore wind energy plants. In: Proulx T (ed) Proceedings of the IMAC-XXVIII, Jacksonville, pp 727–740, 1–4 Feb 2010
13. Lourens E, Papadimitriou C, Gillijns S, Reynders S, de Roeck G, Lombaert G (2012) Joint input-state estimation for structural systems based on reduced-order models and vibration data from a limited number of sensors. Mech Syst Signal Process 29:310–327
14. Böker C (2009) Load simulation and local dynamics of support structures for offshore wind turbines. PhD thesis, Institute for Steel Construction, Leibniz Universität Hannover
15. de Vries WE, Vemula NK, Passon P, Fischer T, Kaufer D, Matha D, Schmidt B, Vorpahl F (2011) Upwind wp4 d4.2.8 final report wp4.2: support structure concepts for deep water. Technical report, UpWind
16. Papadimitriou C, Lourens E, Lombaert G, de Roeck G, Liu K (2012) Prediction of fatigue damage accumulation in metallic structures by the estimation of strains from operational vibrations. In: Life-cycle and sustainability of civil infrastructure systems: proceedings of the third international symposium on life-cycle civil engineering (IALCCE'12), Vienna, 3–6 Oct 2012, CRC Press
17. Gillijns S, de Moor B (2007) Unbiased minimum-variance input and state estimation for linear discrete-time systems with direct feedthrough. Automatica 43(5):934–937
18. Maes K, Lourens E, van Nimmen K, van den Broeck P, Guillaume P, de Roeck G, Lombaert G (2013) Verification of joint input-state estimation by in situ measurements on a footbridge. In: Chang F-K (ed) 9th international workshop on structural health monitoring, 1, Stanford, pp 343–350, Sept 2013
19. Jonkman J, Butterfield S, Musial W, Scott G (2009) Definition of a 5-MW reference wind turbine for offshore system development. Technical report NREL/TP-500-38060, National Renewable Energy Laboratory, Golden, Feb 2009
20. Burton T, Jenkins N, Sharpe D, Bossanyi E (2011) Wind energy handbook, 2nd edn. Wiley, West Sussex
21. Shinozuka M, Deodatis G (1991) Simulation of stochastic processes by spectral representation. Appl Mech Rev 44(4):191–204
22. Vugts JH (2013) Handbook of bottom founded offshore structures. Part 1. General features of offshore structures and theoretical background. Eburon, Delft
23. Maes K, Lourens E, van Nimmen K, Reynders E, de Roeck G, Lombaert G (2014) Design of sensor networks for instantaneous inversion of modally reduced order models in structural mechanics. Mech Syst Signal Process, http://dx.doi.org/10.1016/j.ymssp.2014.07.018i
24. Lourens E (2012) Force identification in structural dynamics. PhD thesis, Katholieke Universiteit Leuven

# Chapter 10
# Autoregressive Model Applied to the Meazza Stadium for Damage Detection

**G. Busca, A. Cigada, and A. Datteo**

**Abstract** Aerospace, civil and mechanical structures are naturally exposed to damages, which could depend on several sources, such as environment degradations, design faults or unexpected natural events. Statistical pattern recognition has recently emerged as an effective technique for structural health monitoring. Its success depends on the possibility to detect unusual operational scenarios just through a statistical data processing of the structural vibration measurements and without the need of a physical model.

In this paper, we present the application of one of these techniques to the Meazza stadium grandstands, in order to detect different operational and environmental conditions of this structure, by extracting sensitive features from vibration time series. We trained an autoregressive model (AR) on the vibrations data acquired for empty stadium conditions, which were considered the "undamaged" status. Then we tested how this statistical model is able to describe the behaviour of the same structure under different environment conditions, for instance at different temperature values. In the end, we used statistical pattern recognition to detect the "damaged" scenarios represented by the events planned in the stadium, football matches and concerts, when the stands was occupied by public.

**Keywords** Structural health monitoring • Damage detection • Statistical pattern recognition • Autoregressive model • Mahalanobis distance

## 10.1 Introduction

Structural Health Monitoring (SHM) consists in monitoring structures by the use of many sensors and devices to detect the presence of any eventual damage or unusual behaviour. An appropriate SHM system can help to reduce the possibility of catastrophic failure, optimizing maintenance costs, and the downtime for structural rehabilitation.

The authors pose the SHM process in the context of a statistical pattern recognition paradigm [1]. The idea is to fix the normal dynamic operating condition of the structure, recording and extracting features from the vibration data during the daily use, and considering this database as the initial state, describing its standard behaviour; then the data obtained by subsequent acquisitions can be examined to check if these features exhibit meaningful drifts from the standard state. The discerning of the standard state from the damage condition has to be done by fixing a threshold between these two cases. It should be noted that these statistic-based techniques do not use finite element models or any others physical models, which are very difficult to be designed for real complex structures, because they often require hard work on intensive tuning and they are often affected by significant uncertainties caused by user interaction and modelling errors.

In this paper, we show a SHM application only based on signal statistical analysis from the measured vibration data; this approach is very attractive for the development of an automated health monitoring system when the structures are too complex to be modelled.

As a first step, we will process the vibration data acquired from one stand of the Meazza stadium in Milan, by an autoregressive model. For applications on existing structures, data from damaged conditions are often unavailable, at least for structures still used for their original purpose. A database concerning the empty stadium condition is used to set the standard behaviour and to evaluate a threshold of the damage feature for the out-of-normal behaviour recognition. The feature used in this paper is the well-known Mahalanobis distance [2] calculated on the parameters of the autoregressive model.

G. Busca • A. Cigada • A. Datteo (✉)
Department of Mechanical Engineering, Politecnico di Milano, Via G. La Masa, 1, 20156 Milano, Italy
e-mail: alessio.datteo@polimi.it

© The Society for Experimental Mechanics, Inc. 2015



C. Niezrecki (ed.), *Structural Health Monitoring and Damage Detection, Volume 7*, Conference Proceedings of the Society for Experimental Mechanics Series, DOI 10.1007/978-3-319-15230-1_10

The layout of the next paper steps is the following: Sect. 10.2 provides a brief description of the theory adopted in this paper, through the adopted autoregressive model (10.2.1) and the estimation of the optimal model order (10.2.2); in the end a summary about the Mahalanobis distance method will be given (10.2.3). Section 10.3 describes the monitored structure, the Meazza stadium based in Milan, with a structural description of the stand under examination (10.3) and of its dynamical behaviour (10.3.1). Section 10.4 then deals with data analysis, through the basic statistics on the available vibration data (10.4.1), the implementation and results of the autoregressive model (10.4.2). Finally, Sect. 10.5 gives the final comments.

## 10.2 Theory and Description of the Adopted Models

In this section, the AR modelling, the choice of optimal model order and Mahalanobis distance are briefly explained.

### *10.2.1 Autoregressive Models: AR*

In time series analysis, one of the most useful representation to express a time series process is the autoregressive (AR) model. It is a stochastic finite linear model that will be briefly explained in the following. For a complete theory of this topic, please refer to [3]. If Z is a generic stationary process, we can estimate the value of Z at time t just basing the evaluation on its past values plus white noise. Let us fix the value of a process at equally spaced times $t$, $t-1$, $t-2$, ... by $z_t$, $z_{t-1}$, $z_{t-2}$, ... . Also let $\tilde{z}_t$, $\tilde{z}_{t-1}$, $\tilde{z}_{t-2}$, ... be the deviation from the process mean value $\mu$ (assumed stationary); for example, $\tilde{z}_t = z_t - \mu$. Then the value $\tilde{z}_t$ can be written as:

$$\tilde{z}_t = \Phi_1 \tilde{z}_{t-1} + \Phi_2 \tilde{z}_{t-2} + \cdots + \Phi_p \tilde{z}_{t-p} + a_t \tag{10.1}$$

where $p$ is the order of the model, $\Phi_i$ is the constant coefficient of the autoregressive model and $a_t$ is white noise. $a_t$ is a sequence of uncorrelated random values from a fixed distribution with constant mean $E(a_t)$, (usually assumed to be 0) and constant variance $Var(a_t) = \sigma_a^2$).

If we define an autoregressive operator of order $p$ by the following equation:

$$\Phi(B) = 1 - \Phi_1 B - \Phi_2 B^2 - \cdots - \Phi_p B^2 \tag{10.2}$$

Then the autoregressive model may be written as:

$$\Phi(B)\tilde{z}_t = a_t \tag{10.3}$$

The model contains p + 2 unknown parameters $\mu$, $\Phi_1$, $\Phi_2$, ..., $\Phi_p$, $\sigma_a^2$, which in practice have to be estimated from the data.

### *10.2.2 Optimal Order Estimation*

In time series analysis or, more generally, in any data analysis, there may be several adequate models that can be used to represent a given data set. Sometimes, the best choice is easy, other times the choice can be very difficult. Thus, plenty of criteria for model order estimation have been introduced and are described in literature for a proper model selection. Model identification techniques such as Partial Autocorrelation Function (PACF), Akaike's Information Criterion (AIC) and Bayesian Information Criterion (BIC) [4] are used to identify the most adequate model order. These three models are those most used in literature. A way to assess an adequate model is to check if residuals are white noise. For a given data set, when there are multiple adequate models, the selection is normally the model with the minimum order. In the following, we briefly introduce the formulations of some model selection criteria.

#### 10.2.2.1 Partial Autocorrelation Function (PACF)

Given a time series Z, the partial autocorrelation of lag k, denoted α(k), is the autocorrelation between $z_t$ and $z_{t+k}$. We may want to investigate the correlation between $z_t$ and $z_{t+k}$ after their mutual linear dependency on the intervening values $z_{t+1}$, $z_{t+2}, \ldots, z_{t+k-1}$ has been removed.

$$\alpha(1) = Cor\,(z_t,\ z_{t+1}) \tag{10.4}$$

$$\alpha(k) = Cor\left(z_t,\ z_{t+k}\,\middle|\,z_{t+1},\ z_{t+2},\ \ldots,\ z_{t+k-1}\right),\ \text{for k} \geq 2 \tag{10.5}$$

where $|z_{t+1}, z_{t+2}, \ldots, z_{t+k-1}$, means these values are excluded.

The partial autocorrelation of an AR(p) process is zero at lag p + 1 and the consecutive lags. If the sample autocorrelation plot indicates that an AR model may be appropriate, then the sample partial autocorrelation plot is examined to identify the order. The order is selected by the point on the plot where the partial autocorrelations for all higher lags are zero, given a fixed tolerance. The sample PACF has a sampling uncertainty that makes the trend of the function more difficult to fix. This uncertainty has to be accounted for in recognizing the lag for which the function goes to zero. In general the uncertainty contribution is to reduce the order of the model [4].

#### 10.2.2.2 Akaike's Information Criterion (AIC)

Assume that a statistical model of k parameters ($\Phi_i$) is fitted to the data to be evaluated. To assess the quality of the model fitting, Akaike introduced an information criterion which is defined as:

$$\text{AIC (k)} = -2\ln\left[\text{maximum likelihood}\right] + 2\text{ k} \tag{10.6}$$

where $k$ is the number of parameters in model. For more information about maximum likelihood, please refer to [3]. The optimal order of the model is chosen by the value of $k$ so that AIC(k) is minimum. Hence, AIC not only rewards the goodness of fit, but also includes a penalty that is an increasing function of the number of estimated parameters ($\Phi_i$). Increasing the order of the model improves the goodness of the fit. The penalty is against overfitting [4].

#### 10.2.2.3 Bayesian Information Criterion (BIC)

In statistics, the Bayesian information criterion (BIC) is a criterion for model selection among a finite set of models. It is partially based, on the likelihood function [3] and it is closely related to the Akaike information criterion (AIC). When fitting models, it is possible to increase the likelihood by adding parameters ($\Phi_i$), but doing so may come up in overfitting. Both BIC and AIC overcome this problem by introducing a penalty term for the number of parameters in the model, the penalty term is larger in BIC than in AIC.

The BIC is an asymptotic result derived under the assumptions that the data distribution is in the exponential family.

The formula for the BIC is

$$-2\ln p\,(Z|\,M) \approx BIC = -2\ln\widehat{L} + k\ \left(\ln(n) - \ln\left(2\pi\right)\right) \tag{10.7}$$

where $Z$ is the observed data, n is the number of observations, k is the number of parameters of the model and $p(Z|M)$ is the marginal likelihood of the observed data given the model M. The last term corresponds to the integral of the likelihood function $p(Z|\theta, M)$ times the prior probability distribution $p(\theta|M)$ over the parameters $p(Z|\theta, M)$ of the model M for fixed observed data Z. Finally $\hat{L}$ is the maximized value of the likelihood function of the model M, i.e. $\theta$ where $\hat{\theta}$ is the parameter values that maximize the likelihood function.

For large n, (10.7) can be approximated by:

$$BIC = -2\ln\widehat{L} + k\ \ln(n) \tag{10.8}$$

Given any two estimated models, the model with the lower value of BIC is the one to be preferred. The BIC generally depends on the size of n and relative magnitude of n and k. It is important to keep in mind that the BIC can be used to compare models based on different probability distributions [4].

### 10.2.3 *Mahalanobis Distance*

The Mahalanobis distance is the feature that is used in this paper to estimate the supposed unusual behaviour of the tested structure [2]. The Mahalanobis distance is a measurement of the distance between a point P and a distribution D. It is a multi-dimensional generalization of the idea of measuring how many standard deviations P is away from the mean of D.

It has many advantages such as: it takes into account the fact that the variances in each direction are different, it considers the covariance between variables, and it becomes the Euclidean distance for uncorrelated variables with unit variance.

This distance is zero if P is at the mean of the D, and it grows as P moves away from the mean. Along each principal component axis, it measures the number of standard deviations between P and the mean of D.

In the following $^{-}$ denotes a vector, otherwise = denotes a matrix. The Mahalanobis distance of the observations $\bar{\bar{x}} = (\bar{x}_1, \bar{x}_2, \bar{x}_3, \ldots, \bar{x}_n)$ from a group of observation with mean $\bar{\mu} = (\mu_1, \mu_2, \mu_3, \ldots, \mu_n)$ and covariance matrix $\bar{\bar{S}}$, is defined as:

$$Dist_{Mahalanobis}\left(\bar{\bar{x}}\right) = \sqrt{\left(\bar{\bar{x}} - \bar{\mu}\right)^T \bar{\bar{S}}^{-1} \left(\bar{\bar{x}} - \bar{\mu}\right)} \tag{10.9}$$

In this paper the observations ($\tilde{x}$) are the parameters ($\Phi_i$) of the model for many different days, each one having its mean, grouped in the same vector ($\tilde{\mu}$. Moreover, we compute the covariance matrix ($\bar{\bar{S}}$) of all these parameter ($\Phi_i$).

## 10.3 G. Meazza Statium

The Meazza stadium consists of a number of different substructures, subject to continuous modifications, enlargements and reinforcements, see Figs. 10.1 and 10.2. The main substructures are the rings and the roofing structures. The first ring is the oldest substructure, it was built in 1927, and permitted to reach the capacity of 35,000 seats. The second ring was built in

**Fig. 10.1** Inside view of the stadium

Fig. 10.2 Outside view of the stadium

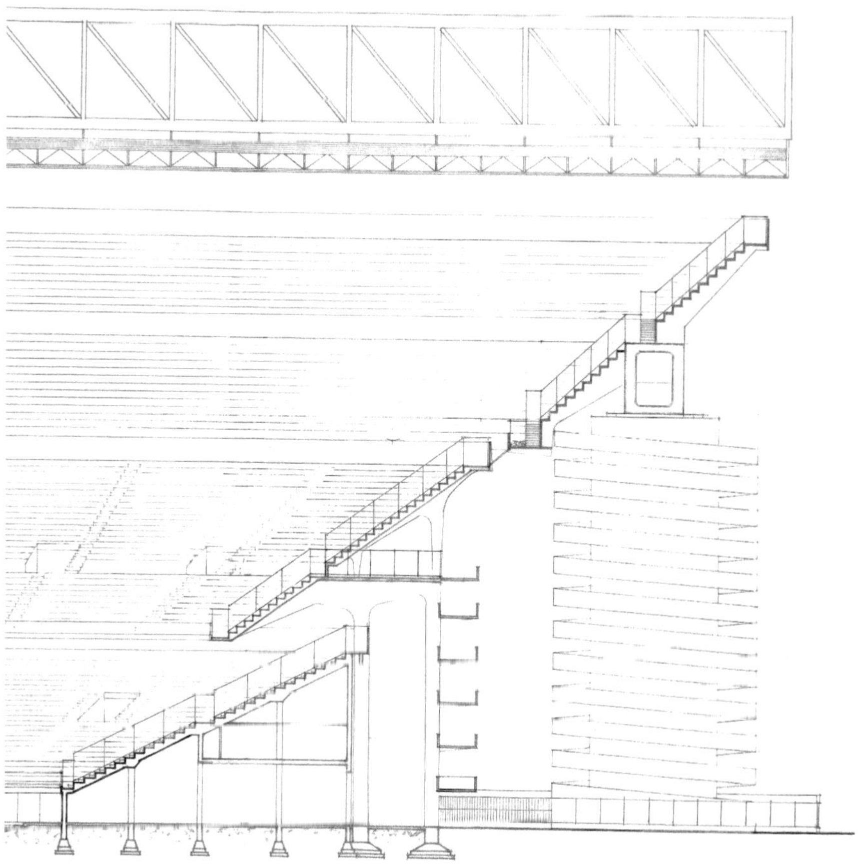

Fig. 10.3 Section of the structure of the stands

1955 to achieve the 55,000 seats. In the end, for the 1990 World Cup the third ring and the roofing structure were added to the existing structure, with a final capacity of about 90,000 (see Fig. 10.3).

The third ring has no links with the other structures ring, this gap is covered by steel plates and there is no way of passing between the two. The stands of the third ring are linked to 10 solid web girders, in prestressed concrete, sustained by 11 towers measuring 30 m in height, based at the external area of the original structure. The stands are composed by three main substructures: the box girders, the cantilevers to support the terraces and terraces itself (see Fig. 10.5). In this paper, we are taking into account the stand between the towers 5 and 6 (see Fig. 10.4).

Fig. 10.4 Map of the third ring

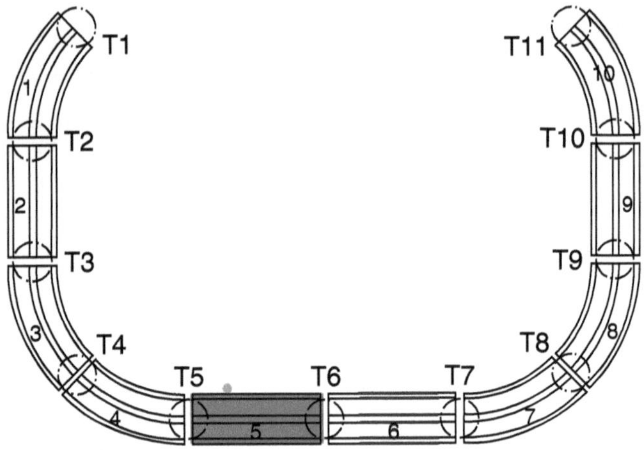


Fig. 10.5 Section of the third ring

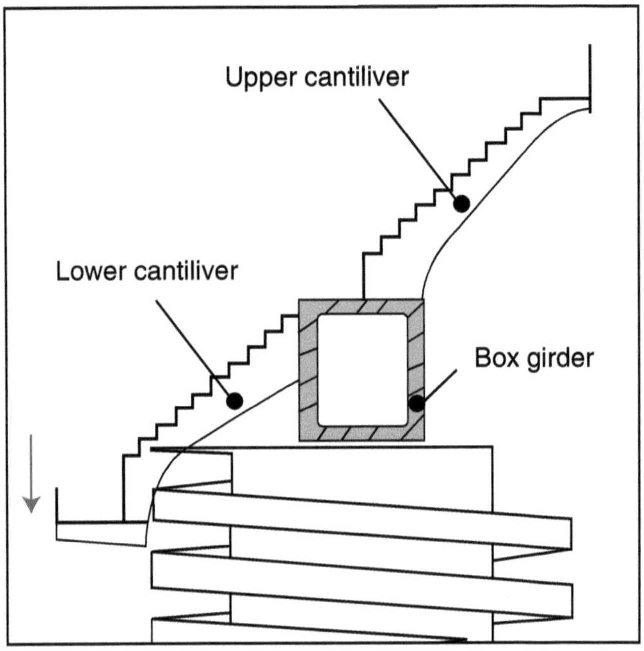


### 10.3.1 *Dynamic Behaviour of the Stand*

The first frequency of the stands is located around 1 Hz, but its amplitude is low and it is linked to a horizontal mode, hardly excited, with the exception of earthquakes. As our main interest is the vertical motion, which is richer in information, then we focus our attention on higher frequencies. In Fig. 10.6, we show the amplitude of the frequency response function (FRF) in the vertical direction as from an experimental modal analysis, carried out in 2010 [5]. The test consist in providing a step sine excitation by means of a hydraulic exciter and accelerations are recorded at a number of points, among them one is near the actuator and another one among the remaining has been randomly chosen just to evaluate the feasibility of the approach. Considering these two records, FRFs were computed, using the accelerometer near the actuator as landmark.

The two main modes are approximately at 3 Hz and 5 Hz. Therefore, attention has been focussed on these two modes to characterize the vertical behaviour of the stand (identified by the green arrows in Fig. 10.6).

In Fig. 10.6 "Force 1" means 900 N, whereas "Force 2" is 1,800 N. The two forcing levels allowed to estimate very close FRFs, confirming the robustness of the tests and the linear behaviour of the structure, in the testing range. The modal analysis, using the polymax procedure, identified the following modes:

Mode 2 and 2(b) is a twin mode, which can be ascribed to the interaction of close stands [5].

All these data are reported to describe the behaviour of the stand, using a robust procedure, which can be considered as a reference for further investigations. For the purpose of this paper, in the following, we take into account the environmental vibration of the stand. We consider an accelerometer out of the permanent monitoring system equipping the stadium and we analyse the records acquired over years. The considered accelerometer is placed at 1/3 span, and at the lower end of the lower cantilever, as shown in Figs. 10.4 and 10.5.

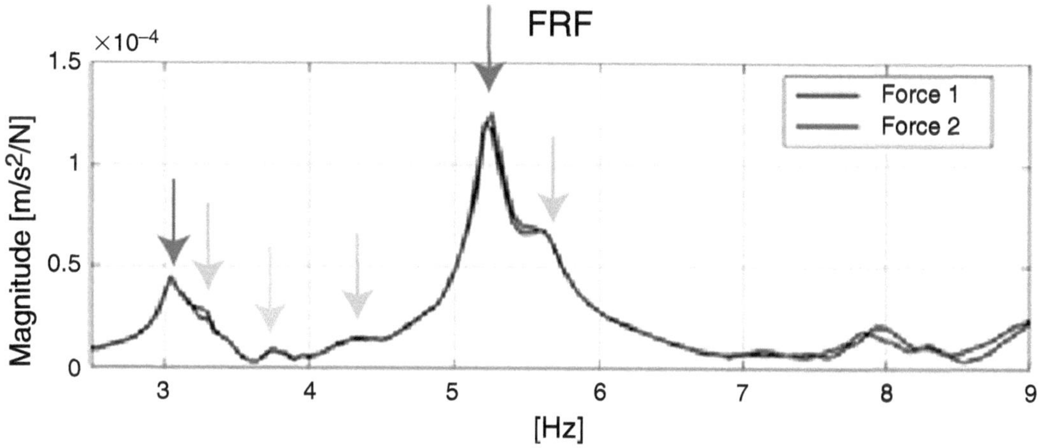


**Fig. 10.6** Vertical frequency response function of the stand of the third ring

**Table 10.1** Vibration modes

| N° mode | Frequency [Hz] | Damping [%] |
|---|---|---|
| 2 | 3.13 | 3.18 |
| (2b) | 3.29 | 2.31 |
| 3 | 3.73 | 1.25 |
| 4 | 4.36 | 2.48 |
| 5 | 5.21 | 1.94 |
| 6 | 5.58 | 1.78 |
| 7 | 7.13 | 1.73 |
| 8 | 7.82 | 2.18 |

**Table 10.2** Events considered

| ID case | Date | Event |
|---|---|---|
| 01 | 12th July 2012 | Take That's concert |
| 02 | 3rd June 2013 | Bruce Springsteen's concert |
| 03 | 16th June 2011 | Vasco Rossi's concert |
| 04 | 23rd February 2011 | Internazionale di Milano-Bayern Monaco Uefa football match |
| 05 | 19th June 2013 | Jovanotti's concert |

## 10.4 Data Analysis

This paper deals with the statistical characterization of the time histories of the stand between the fifth and sixth tower of the third ring. We studied the signal of the vertical accelerometer from 4th November 2010 to 6th June 2013. In Fig. 10.7, we show the time history of a single day to give a sample of the considered data, with a sampling frequency of 200 Hz, well above the needs for structural monitoring. Figure 10.8, gives the spectrum related to the same signal, in the frequency range from 0 to 9 Hz. The resolution of the spectrum is 0.01 Hz considered the threshold for meaningful changes in the natural frequencies to be monitored. The spectrum amplitude confirms that the vertical vibrations are mainly due to mode 2 and 5 already given in Table 10.1.

Over the considered period, 96 days were randomly selected in order to create a database of records representing the behaviour of the empty stadium, which is considered as the normal stand status. For each days, we considered 10 min and evaluate on these the AR parameters ($\Phi_i$). Thereafter, we statistically characterized the parameters as explain in the Sect. 10.4.2.

In the case of an existing civil structure, it is not possible to induce a relevant or even a controlled damage. So, it was looked for different operational situations to verify if it was possible to identify those as different behaviours of the stand. For this purpose, it has been decided to consider 4 concerts and 1 football match, those are deemed as extraordinary events. In the Table 10.2 are reported the events considered:

To be sure that the procedure does not create false positives, 3 days of empty stadium have been included in the validation data set, at the same time excluding them from the training database, to verify the robustness of the state evaluation of the stand. Table 10.3 shows the three select days.

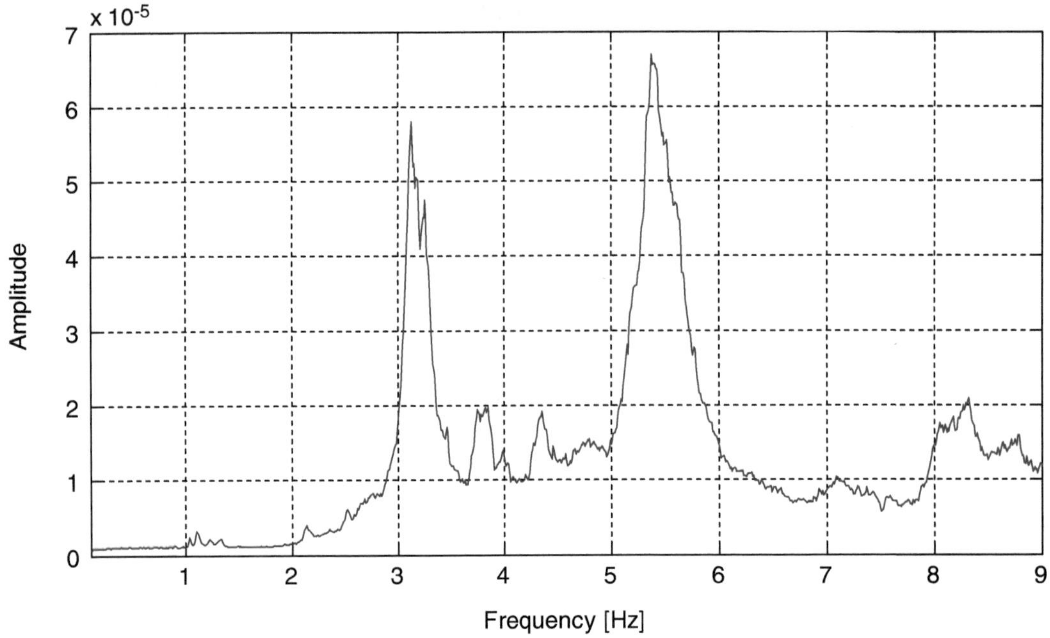


Fig. 10.7 Time history of the 1st January 2011

Fig. 10.8 Spectrum of the 1st January 2011

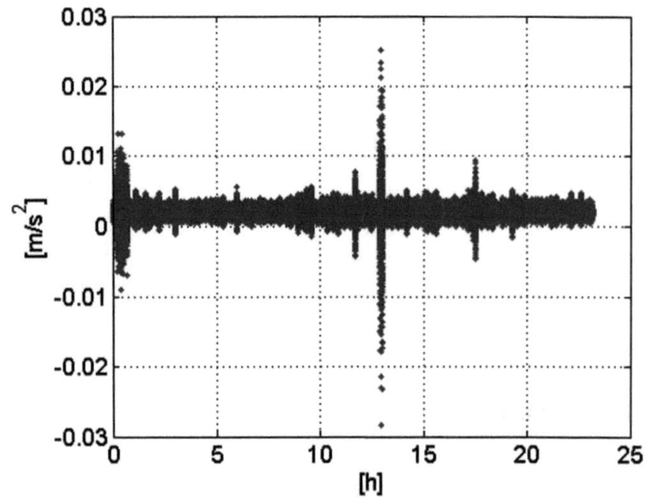


Table 10.3 Others days of the empty stadium

| ID case | Date |
|---|---|
| A | 18th March 2011 |
| B | 2nd February 2011 |
| C | 20th December 2012 |

In the following section, the adopted basic statistics to identify the behaviour of the structure related to different environmental and operational conditions is going to be described. Then, an explanation of the autoregressive model is provided to the purpose of identifying damage, also explaining our assumptions about the number of parameters, the duration of the considered data, and on the use of the Mahalanobis distance to manage the selected features.

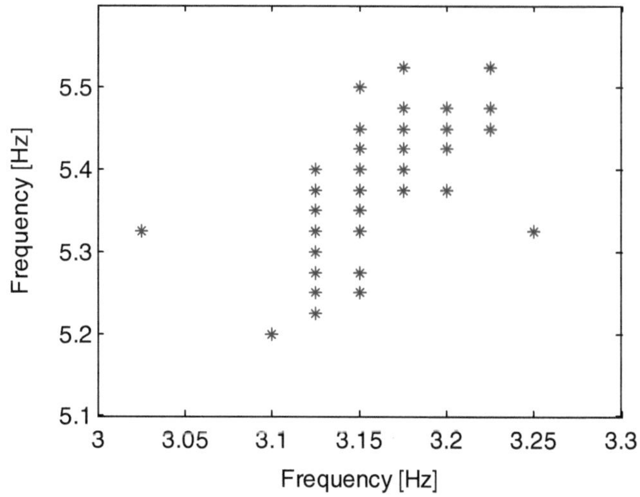


**Fig. 10.9** First versus second frequency at daytime

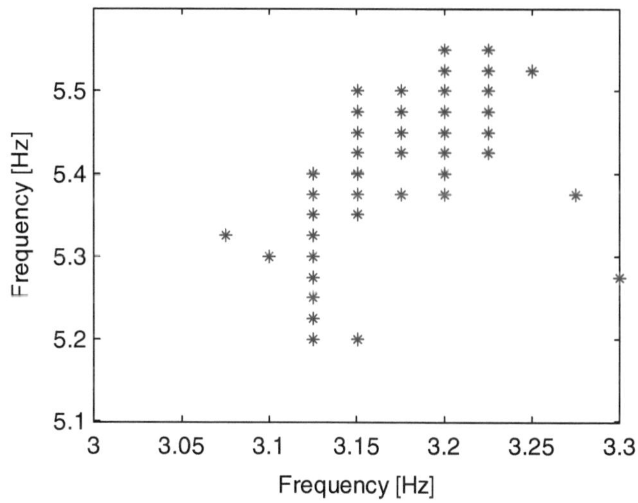


**Fig. 10.10** First versus second frequency at night-time

### *10.4.1 Basic Statistic*

At first, we divide the record between daytime and night-time to verify if it is possible to distinguish two different trends. We consider few hours for each day and night. The data considered in this section are form January 2011 to July 2011. During daytime the environmental conditions could be quite different from those recorded during night-time, for instance the maintenance staff might work on the stand with producing light forcing; conversely, during the night-time this is unlikely to happen. Moreover, close to the stadium there is the underground building site whose effect is more important during daytime. The same could be affirmed for road traffic. Finally, another possible contribution creating a gap between the two conditions (day and night), is temperature, not just having a seasonal trend but also a daily one.

First, we focus on the ratio between the two considered frequencies (2nd and 5th mode), in order to investigate if the different environmental and operational conditions can influence this relationship. In Figs. 10.9 and 10.10, although a not negligible dispersion is present, it is possible observe that, in the two considered situations, a similar trend is exhibited. Besides, it could be recognized a moderate linear trend, due to the linear behaviour of the structure under low load (although points are really concentrated).

If a damage occurs, the ratio between the first and second frequency changes, but an uncertainty band has to be fixed, so that if changes are within this band, no damage is supposed to be present. Unfortunately this band is expected to be rather wide.

Another aspect which has been considered is the relationship between temperature (provided by a weather station on the stadium roof) and the natural frequencies. For instance, Figs. 10.11 and 10.12 show the result for the frequency around 3 Hz (2nd mode).

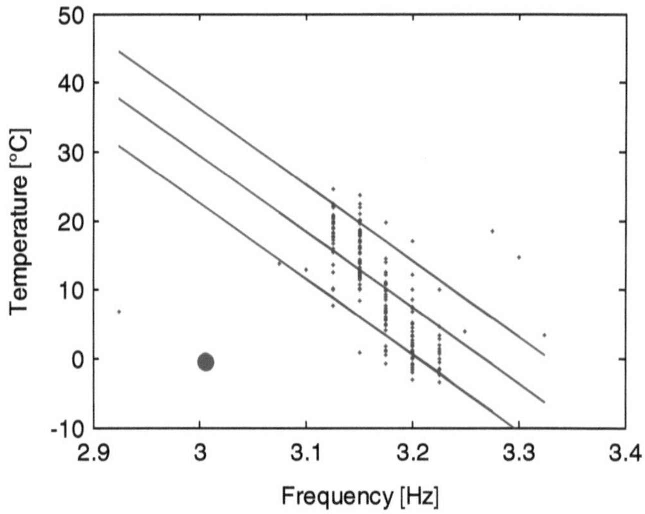


**Fig. 10.11** Temperature versus first frequency night-time

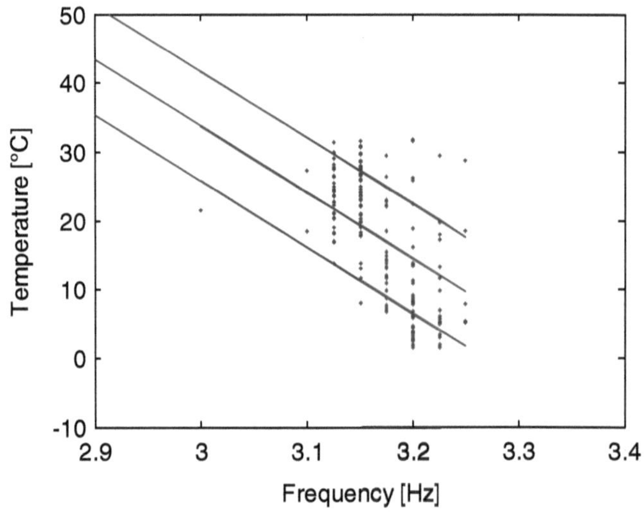


**Fig. 10.12** Temperature versus first frequency daytime

The behaviour during daytime is the same as that during night-time. In the first case, a wider spread is observed, because it has already been remarked that daytime can include different operating conditions: maintenance and traffic considerably change along the year. Conversely, during the night, the situation in and around the stadium is more stable and so the graph presents a lower data dispersion. In both the considered conditions, it is possible to define the uncertainty bands as those traced to define the seasonal thermal effects. Moreover, if we consider the data recorded during a football match (red point in Fig. 10.12) this is well outside the uncertainty bands. So it could be an evidence of the fact that during an event the dynamic behaviour changes.

### *10.4.2 Autoregressive Model*

The next step is to model the time histories by an appropriate autoregressive model, which means to make decisions about the records to be considered, their duration and model order to be adopted.

Primarily, a standardization has been applied to all signals, like Eq. (10.10) shows [6].

$$\widehat{x} = \frac{x - \mu_x}{\sigma_x} \tag{10.10}$$

**Table 10.4** Days considered with the corresponding orders predicted by the three selected methods

| Date | AIC | BIC | PACF |
|---|---|---|---|
| 11th January 2011 | 100 | 29 | 26 |
| 16th May 2011 | 100 | 46 | 36 |
| 2nd August 2011 | 100 | 33 | 36 |
| 30th September 2011 | 100 | 44 | 11 |
| Mean value | 100 | 38 | 27 |

where $\widehat{x}$ is the standardized signal, and $\mu_x$ and $\sigma_x$ are the mean and standard deviation of x, respectively. We use the parameters of the AR models ($\Phi_i$) as damage indicators. If damage occurs, the probability distribution function of the parameters of the autoregressive model changes.

The parameters of each time history have been estimated by means of a least square approach. The choice of the time range is based on simulations with different duration: 1, 5, 10, 30 and 60 min. After the model estimation, the root mean squares (RMS) of the residuals have been considered, and the 10 min approach appears as the minimum duration to obtain a good representation of the time histories. Then, to create the database, the empty stadium condition 8:00 am and 8:10 am has been considered, for the 96 days previously selected. The time band considered for the events (Table 10.2) has been between 9:30 pm to 9.40 pm (when the most important events take place). In the end, for the days used as a check for the procedure, both mornings and evenings have been considered.

The first step necessary to use the autoregressive models is the choice of the model order. In the next section, we report the result of the application of the three algorithms presented in Sect. 10.2.2.

#### 10.4.2.1 Model Order Estimation

The records used in this work are 96, so, the optimal model order estimation for each time history is time consuming, computationally speaking. In fact, the procedure consists of an iterative method with a fixed number of steps. Thereafter it could possible to identify the optimal order, in which one minimizing the function of the method adopted (Sect. 10.2.2). Therefore, we select four dates having a temperature range compatible with that of the considered events. The purpose is to ensure that the AR model is able to describe different environmental conditions. It has been decided to fix the maximum order to 100, which is a maximum that usually literature accepts as a limit the overfitting. In Table 10.4 the list of the selected days is shown, together with the corresponding predicted order by the three different methods, which are the most used in literature.

In literature, the Bayesian information criterion (BIC) is considered as the most stable method, so it has been decided to establish the model order as the mean value of the four values predicted by BIC for the four considered days. The mean value is 38, and then we decide to fix the model order at 40, as a factor of safety to taking into account the great variability of the time considered histories. As a next step the different parameters for all the time histories have been computed, considering an autoregressive model of order 40, AR(40). After this step it is possible to predict the new time histories by a proper autoregressive model.

The goodness of the data interpolation can be seen by considering the residuals between the original time history and the predicted one. Figure 10.13 shows the root mean square (RMS) of the residuals for all the 96 days considered.

The RMS value of the residuals is on the range between 0.1 and 0.7 (Fig. 10.13).

#### 10.4.2.2 Mahalanobis Distance on Parameters

Once the AR model has been estimated, the Mahalanobis distance has been adopted to distinguish between different conditions. To do that, it is necessary to characterize the normal behaviour of the stand using a database made of the 96 sets of AR parameters evaluated on the 96 records. Computing the mean vector and the covariance matrix of the 96 groups of parameters it is possible to define a threshold allowing one to discern if a time history, not considered in the database, is referred to the empty stadium or not. In Fig. 10.14 the result of the considered cases are shown, with the ID cases reported in the Tables 10.2 and 10.3.

The obtained threshold is 79.04. It is possible to see that the concerts and the football match are over the value of 79.04. Conversely, the other days of empty stadium are below this threshold. The evening of 20th December 2012 (C evening), appear as a fault, but it was discovered this was an outlier in the data catalogue of the stadium. In fact, not listed in the event agenda, a football match was played in that occasion, so the Mahalanobis index is correct, and it is comparable to the other

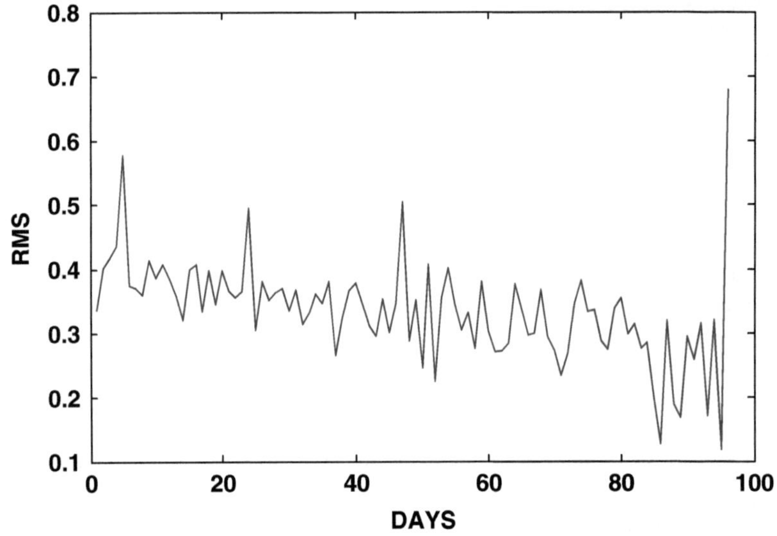


**Fig. 10.13** RMS values of the residuals between the original and predicted time histories

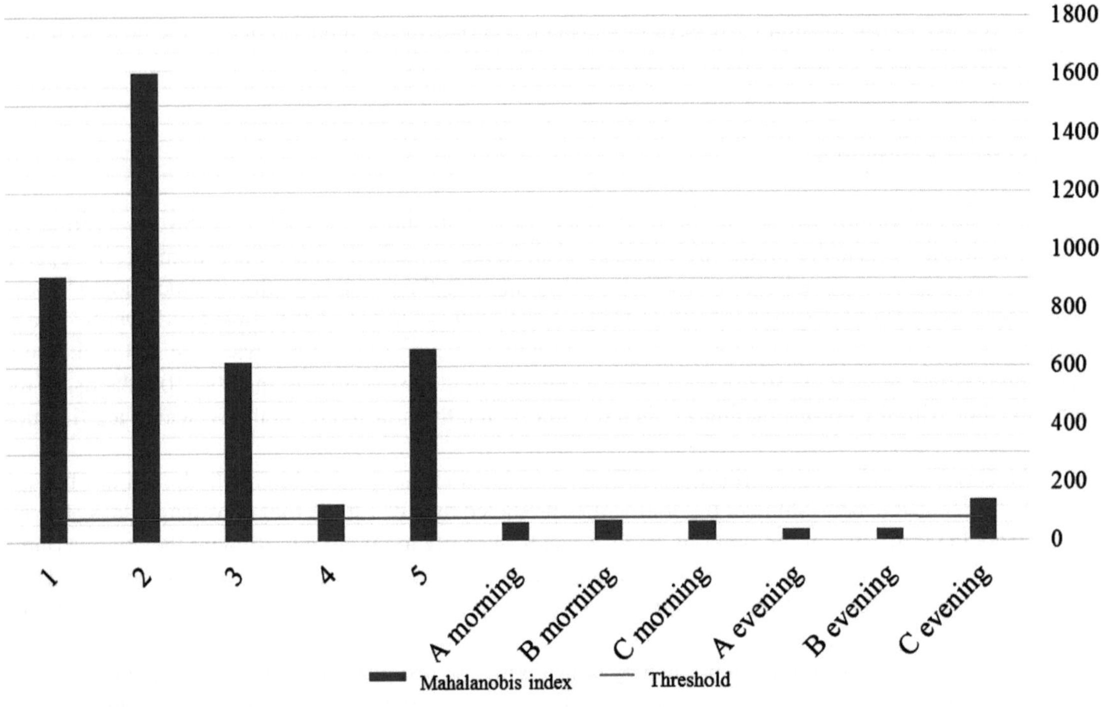


**Fig. 10.14** Mahalonobis distance on parameters

football matches. Taking into account also other events, not reported in this paper, it seems that for the considered stand, the Mahalanobis index can also distinguish a concert from a football match. Indeed it could be observed that for concerts the index reaches greater values than for the football matches.

## 10.5 Conclusions

This paper describes the application of some statistics tools for structural health monitoring. First, we proposed a basic analysis showing the dependence of the structural frequencies on temperature changes. As expected, the low vibration level makes the response of the structure in the linear field, that has a great dependence from temperature.

In the second part of the paper we applied an autoregressive model AR(40) to model the vibration signal acquired by an accelerometer fixed to a stand. This procedure allowed to properly describe the response of the structure to the ambient dynamic excitations. The residues demonstrated that the signal was correctly fitted when the structure is empty. It was also possible to identify some unusual working conditions of the stand, at least for the identified model, for instance during concerts and football matches, using the Mahalanobis distance applied to the coefficients of the AR model.

## References


1. Sohn H et al (2000) Damage diagnosis using statistical process control. Conference on recent advances in structural dynamics, Southampton, 22–25 July 2000
2. de Lautour OR et al (2009) Nearest neighbour and learning vector quantization classification for damage detection using time series analysis. Struct Control Health Monit, 17(6):614–631 October 2010
3. Box GEP et al (1994) Time series analysis: forecasting and control, 3rd edn. Prentice Hall, Englewood Cliffs
4. Wei WWS (1990) Time series analysis: univariate and multivariate methods. Addison-Wesley, Redwood City
5. Cremona E (2012) Analisi applicative di criteri per il Model Updating di strutture complesse il caso del modello di una tribuna rettilinea del terzo anello dello stadio Giuseppe Meazza. Master degree thesis, Politecnico di Milano
6. Sohn H et al (2005) Structural damage classification using extreme value statistics. J Dyn Syst Meas Control 127:125

# Chapter 11
# Output Only Functional Series Time Dependent AutoRegressive Moving Average (FS-TARMA) Modelling of Tool Acceleration Signals for Wear Estimation

**B.H. Aghdam, E. Cigeroglu, and M.H. Sadeghi**

**Abstract** In this paper, tool vibration signals obtained from a turning process are used for tool wear estimation purposes. During the cutting process, tool acceleration signals are recorded for different levels of wear. Due to non-stationarity of tool/holder system's response, Time dependent time series model of Functional Series Time dependent AutoRegressive Moving Average (FS-TARMA) type is used for modelling the signals and extraction of wear sensitive features that will be exploited in a wear estimation algorithm. Results of the analysis through FS-TARMA, reveals its higher accuracy with respect to stationary type models, since it captures time dependent properties as well, which can be used in an online tool wear estimation algorithm.

**Keywords** Tool wear • Turning • FS-TARMA • Time series

## 11.1 Introduction

On-line tool wear monitoring in machining processes has been focused by many researchers in the past two decades. Due to the fact that tool change policies, product quality, tooling costs, and productivity are all influenced by tool wear.

In addition these, tool wear is an important factor in assuring the quality of the machined products.

A cutting tool when worn, produces a poor surface quality. If the tool wear is not monitored continuously, the surface finish of the workpiece may be significantly degraded, which consequently can lead to loss of the workpiece and associated manufacturing time. Tool wear, if it is not detected, can also cause catastrophic failures and damage to the machine tool and workpiece which results in significant down time and loss in productivity.

Tool condition monitoring (TCM), detection of tool wear at the initial stages and estimating it during the growing level can be used as an aid to prevent unscheduled downtime and workpiece damage, especially in cutting process of high precision materials.

Different methods have been introduced and implemented in TCM including direct methods that measure tool wear directly and indirect methods that use secondary effects such as machined surface quality degradation to estimate the wear level [1].

Online tool wear estimation may be of advantage in different areas like tool change policy, adaptive control, economic optimization of machining operations and full automation of machining operations [2].

It is possible to use different types of signals for tool wear estimation. Cutting forces, acoustic emission (AE) and acceleration signals are the most commonly are used ones. Wear sensitive features can be extracted from time or frequency domain signals, each of which has limitations and advantages.

Pandit and Kashou [3] employed natural vibration mode power contribution for online tool wear estimation. Successful results have been reported by using AE signals [4, 5]. Liang and Dornfeld [6], estimated the AE signals and extracted wear sensitive features by an Auto Regressive (AR) time series model with time varying parameters.

B.H. Aghdam
Mechanical Engineering Department, K.N.T. University of Technology, Pardis St, Vanak Sq, Tehran, Iran

E. Cigeroglu (✉)
Mechanical Engineering Department, Middle East Technical University, Ankara, Turkey
e-mail: ender@metu.edu.tr

M.H. Sadeghi
Mechanical Engineering Department, Tabriz University, Tabriz, Iran

© The Society for Experimental Mechanics, Inc. 2015

C. Niezrecki (ed.), *Structural Health Monitoring and Damage Detection, Volume 7*, Conference Proceedings of the Society for Experimental Mechanics Series, DOI 10.1007/978-3-319-15230-1_11

The concept of energy in specific frequency bands has been used by many researchers. A specific method of this kind which considers the energy contribution of each single vibration mode has been applied by Yao and Fang to multi-dimensional force and acceleration signals [7].

Modal energy component in frequency domain was introduced as an effective wear sensitive feature by Roth and Pandit [8] for estimation of tool wear in milling.

Sick [9], performed an extensive review on online indirect tool wear estimation in turning with artificial neural network (ANN) methods and outlined possible directions for future studies. He concluded that certain features in time or frequency domain can be used for online tool wear estimation Features in time domain depend significantly on the values of cutting conditions; whereas, frequency domain features depend on fundamental resonant frequencies of the tool-workpiece system, chip lamination frequencies, chatter, etc.

In another review paper, Rehorn et al. [10] organized the results according to the type of the machining operation. They stated that time-frequency methods can be used to identify wear and breakage in machining operations but very few researches have been carried out on these items. Authors concluded that even though ANNs and similar methods can provide highly accurate results, simple force models can as well provide similar accuracies.

Another important aspect of tool wear estimation is the extraction of wear sensitive features from the recorded signals, which is independent from cutting conditions. Especially, Features in frequency domain have the property of being independent from cutting conditions. Some statistical features of tool bending and longitudinal vibration obtained by Singular Spectrum Analysis (SSA) in certain frequency bands has been employed for wear estimation [11, 12].

ANNs and its variants are among the more commonly used methods in tool wear estimation [13–15]. In many of the papers, different methods has been applied to train the networks to achieve higher performance in wear estimation [16–19]. Many researchers tried to exploit multiple sensors and sensor fusion techniques to enhance the accuracy and robustness of the wear estimation algorithm [20–25], however in some cases it did not guarantee the enhancement of the estimations. Simultaneous use of Back-propagation neural networks (BPNs) and adaptive neuro-fuzzy inference system (ANFIS) was investigated by Liu et al. [26] and it was shown that the error is lower when ANFIS method is used.

In a more recent review paper [27], based on a comprehensive investigation of research works it has been concluded that the use of frequency or time-frequency features is more advantageous.

According to the review papers [9, 27] it can be said that very often researchers employed ANNs and its variants for modelling tool wear and features relation for TCM researches and have proposed different procedures for neural network training. These methods have important disadvantages such as sensitivity to cutting conditions and the need for too many training samples.

Time invariant (TI) systems under stationary input produce stationary vibration signals with TI statistical characteristics, while time varying (TV) systems have non-stationary responses with time varying statistical characteristics [28]. However, TI structures which are subjected to non-stationary input can also produce non-stationary response, for instance bridges and tool-holder set of a lathe can be considered as TI structures that have non-stationary response due to ambient excitations and cutting forces, respectively.

ANN's that are capable of modelling non-stationary signals has been extensively used for estimation of tool acceleration signals. However, ANN's do not have a parametric structure and system dynamic parameters such as damping ratios and natural frequencies cannot be extracted using them.

In certain dynamic systems, the input to the system cannot be measured or it is unobservable. In such cases, the identification methods that only rely on system output are attractive. These methods that are known as output-only identification problems in the focus of this study and they are used for modelling the recorded tool vibration. In estimation of tool/holder system non-stationary response measured by an accelerometer, output-only FS-TARMA method can be used. Although it is a very effective method, it has never been used in identification of tool-holder dynamics and tool wear estimation.

For the modelling and analysis of non-stationary signals, two types of methods can be used, non-parametric methods and parametric methods.

Parametric methods are mainly consisting of parameterized versions of Time-Dependent AutoRegressive Moving Average (TARMA). The difference between these versions and ARMA counterparts is that in the former models, the parameters are time-dependent. Parametric methods have some advantages over the non-parametric counterparts such as representation parsimony, improved accuracy, improved frequency resolution, improved tracking of time-dependent dynamics, and flexibility in analysis [28].

Parametric methods can be classified as unstructured and structured, stochastic and deterministic parameter evolution model considering the "structure" that is imposed upon time evolution of the model parameters [28].

Deterministic parameter evolution methods impose deterministic "structure" upon the evolution of TV parameters. Functional Series TAR and TARMA (FS-TAR and FS-TARMA) types are of the main version of these methods and employ deterministic basis functions belonging to specific functional subspaces for estimation of model parameters [28]. Since TV parameters often change in a non-random way, deterministic methods are suitable for capturing deterministic parameter evolution.

Based on our knowledge upon the survey of the related literature published since now, it can be said that there is no research work done by non-stationary output-only FS-TARMA method in the field of tool wear estimation. Also, most of the studies have focused on mathematical development of signal processing/modelling techniques and a method that considers the physics of tool/holder assembly during tool wear still needs to be developed.

The aim of this paper is to develop a tool wear estimation algorithm in turning process that employs FS-TARMA method for identification of tool/holder system dynamics based on the vibration output. To this aim, wear sensitive features extracted using FS-TARMA models of signals will be related to major flank wear values. By the use of the foresaid method, higher accuracy is obtained in comparison to stationary signal modelling methods.

## 11.2 Functional Series TARMA Models of Signals and Their Estimation

Estimation of non-stationary tool vibration signals based on Functional Series TARMA models is considered in this section.

### *11.2.1 FS-TARMA Models*

An FS-TARMA $(n_a, n_c)_{[p_a, p_c, p_s]}$ model, with $n_a, n_c$ indicating AR and MA orders respectively, $p_a, p_c$ their corresponding functional basis dimensionalities, and $p_s$ dimensionality of the respective innovations variance, can be expressed in the following form [28]:

$$x[t] + \sum_{i=1}^{n_a} a_i[t].x[t-i] = w[t] + \sum_{i=1}^{n_c} c_i[t].w[t-i], \quad \mathrm{w[t]} \sim \mathrm{NID}\left(0, \sigma_w^2[t]\right), \quad t = 1, \ldots, N, \tag{11.1}$$

in which $x[t]$ designates the estimated non-stationary signal, $w[t]$ an innovations sequence with zero mean and time-dependent variance $\sigma_w^2[t]$. Supposing that the system input is unobservable, it can be considered as to be a zero mean innovations (uncorrelated) sequence $w[t]$ $(f[t] \equiv w[t])$ with time-dependent variance. $\sigma_w^2[t]$. $a_i[t]$ and $c_i[t]$ represent AR and MA time-dependent parameters respectively, that as well as the innovations variance $\sigma_w^2[t]$ can be estimated by the related functional subspaces with respective bases:

$$\begin{aligned}
\mathcal{F}_{AR} &\triangleq \left\{G_{b_a(1)}[t], G_{b_a(2)}[t], \ldots, G_{b_a(p_a)}[t]\right\}, \\
\mathcal{F}_{MA} &\triangleq \left\{G_{b_c(1)}[t], G_{b_c(2)}[t], \ldots, G_{b_c(p_c)}[t]\right\}, \\
\mathcal{F}_{\sigma_w^2} &\triangleq \left\{G_{b_s(1)}[t], G_{b_s(2)}[t], \ldots, G_{b_s(p_s)}[t]\right\}.
\end{aligned} \tag{11.2}$$

In the above expressions "$\mathcal{F}$" designates functional subspace of the indicated quantity and $\{G_j[t] : j = 0, 1, \ldots\}$ a set of orthogonal basis functions which are selected from a suitable family (such as Chebyshev, Legendre, polynomial, trigonometric, etc. functions). The indices $b_a(i)$ $(i = 1, \ldots, p_a)$, $b_c(i)$ $(i = 1, \ldots, p_c)$ and $b_s(i)$ $(i = 1, \ldots, p_s)$ designate the specific basis functions of a particular family that are included in each subspace.

For an FS-TARMA model, the time-dependent AR and MA parameters and innovations variance, can be expressed based on basis functions as follows [28]:

$$a_i[t] \triangleq \sum_{j=1}^{p_a} a_{i,j}.G_{b_a(j)}[t], \quad c_i[t] \triangleq \sum_{j=1}^{p_a} c_{i,j}.G_{b_c(j)}[t], \quad \sigma_w^2[t] \triangleq \sum_{j=1}^{p_a} s_j.G_{b_s(j)}[t], \tag{11.3}$$

where $a_{i,j}$, $c_{i,j}$ and $s_j$ represent the AR, MA and innovations variance coefficients of projection, respectively. By the definitions of Eq. (11.3) the model parameters consist of projection coefficients $a_{i,j}$, $c_{i,j}$ and $s_j$, while a specific model structure, like $\mathcal{M}$ is defined by model orders $n_a, n_c$ and the functional subspaces $\mathcal{F}_{AR}, \mathcal{F}_{MA}, \mathcal{F}_{\sigma_w^2}$:

$$\mathcal{M} \triangleq \left\{n_a, n_c, \mathcal{F}_{AR}, \mathcal{F}_{MA}, \mathcal{F}_{\sigma_w^2}\right\}. \tag{11.4}$$

The complete identification problem can be expressed as follows [28]: "Given N observations of the vibration response, say $x^N = \{x[1] \ldots x[N]\}$ and the FS-TARMA model set:

$$\mathbb{M} \triangleq \left\{ M(\theta) : x[t] + \sum_{i=1}^{n_a} a_i[t,\theta] . x[t-i] = e[t,\theta] + \sum_{i=1}^{n_c} c_i[t,\theta] . e[t-i] ;\ \sigma_w^2 = \mathrm{E}\left\{e^2[t,\theta]\right\},\ t=1,\ldots,N,\ \theta \in \Re^{\dim(\theta)} \right\}, \tag{11.5}$$

select an element of $\mathbb{M}$ that best fits the observations".

In this expression, $\dim(\theta)$ represents the dimension of the parameter vector $\theta$ which consists of the projection coefficients $a_{i,j}$, $c_{i,j}$ and $s_j$, while $e[t, \theta]$ designates the model's one-step-ahead prediction error (residual) sequence, that coincides with the model's innovations sequence $w[t]$.

The complete identification problem is divided into two subproblems: (a) the parameter estimation subproblem (for a given structure $\mathscr{M}$) and (b) the model structure selection subproblem.

### 11.2.2 FS-TARMA Parameter Estimation

In this subsection the projection coefficient vector $\theta$ estimation is considered. The vector $\theta$ which consists of AR, MA and innovations variance projection coefficients, is estimated for a given structure by employing the non-stationary vibration samples [28].

$$\begin{aligned}
\theta &\triangleq \left[\vartheta^T \middle| s^T\right]_{(n_a \cdot p_a + n_c \cdot p_c + p_s)\times 1},\ \vartheta \triangleq \left[a^T \middle| c^T\right]_{(n_a \cdot p_a + n_c \cdot p_c)\times 1},\ s \triangleq \left[s_1, \ldots, s_{p_s}\right]_{p_s \times 1}, \\
a &\triangleq \left[a_{1,1}, \ldots, a_{1,p_a} \middle| \cdots \middle| a_{n_a,1}, \ldots, a_{n_a,p_a}\right]^T_{(n_a \cdot p_a)\times 1}, \\
c &\triangleq \left[c_{1,1}, \ldots, c_{1,p_c} \middle| \cdots \middle| c_{n_c,1}, \ldots, c_{n_c,p_c}\right]^T_{(n_c \cdot p_c)\times 1},
\end{aligned} \tag{11.6}$$

The AR/MA projection coefficient vector $\vartheta$ is estimated based on minimization of the Ordinary Least Squares (OLS) criterion which consists of the sum of squares of the model's one step ahead prediction errors (Residual Sum of Squares, RSS) [28]:

$$\widehat{\vartheta} = \arg\min_{\vartheta} \sum_{t=1}^{N} e^2[t, \vartheta], \tag{11.7}$$

where arg min designates "argument minimizing" and $e[t, \vartheta]$ is the one-step-ahead prediction error of the model.

The estimation of $\vartheta$ based on OLS results in a non-linear optimization problem. Therefore, iterative optimization methods must be used to estimate the $\vartheta$. These methods can be based on linear multi stage procedures which tackle the difficulties intrinsic to non-linear optimization by a sequence of linear optimization techniques. The Two Stage Least Squares (2SLS) method [29] is of linear multi stage methods that is based on an infinite order FS-TAR representation of the original FS-TARMA model.

### 11.2.3 Distance Between the FS-TARMA Models

After estimation of an FS-TARMA model for each cutting condition and different wear values, a metric should be introduced to compare the models and obtain a measure for prediction of tool wear. For this comparison, different metrics can be used. Here, a metric introduced by Martin [30] which is originally developed for ARMA models, is employed.

Is has been shown that the distance between two stable ARMA models $M$, $M'$ with AR orders $p$, $p'$ and poles $(\alpha_j)$, $(\alpha'_j)$ can be expressed in the form [30]:

$$d\left(M, M^{'}\right) = \left[ \ln \frac{\prod_{i=1}^{p}\prod_{j=1}^{p^{'}}\left(1-\alpha_i\alpha_j^{'*}\right)\prod_{i=1}^{p^{'}}\prod_{j=1}^{p}\left(1-\alpha_i^{'}\alpha_j^{*}\right)}{\prod_{i=1}^{p}\prod_{j=1}^{p}\left(1-\alpha_i\alpha_j^{*}\right)\prod_{i=1}^{p^{'}}\prod_{j=1}^{p^{'}}\left(1-\alpha_i^{'}\alpha_j^{'*}\right)} \right] \tag{11.8}$$

this distance can also be used for classification of signals and fault diagnosis purposes.

In this study, signals of the same system are obtained under several different working conditions with varying wear (fault) levels and modelled using FS-TARMA models. Then the distance between each of these models and the baseline model with no wear is evaluated. An important advantage of this metric is that it is expressed in terms of eigenvalues of the system and can be directly related to the natural frequencies of the tool/holder set. Since the original metric is developed for stationary ARMA models, here, FS-TARMA model parameters at a specific time are used considering the stationarity conditions.

## 11.3 Major Flank Wear Estimation

In tool wear experiments, 3-D tool acceleration signals are recorded. In the following subsections the procedure of wear estimation with the use of acceleration signals is presented.

### *11.3.1 Description of Experiments*

During a tuning process, forces exerted on different faces of the tool, cause it to vibrate in all three directions. Before starting any measurements on a working machine, first, the approximate natural frequencies of the tool/holder assembly were obtained using an impact hammer (type 8202) and a 3-D accelerometer (Deltatron code No. 4504). At each direction, hammer tests were repeated several times to get the best results in terms of good coherence.

Experimental test setup is shown in Fig. 11.1, where a 3-D accelerometer is installed on the tool holder. In order to acquire signals that are highly affected by tool wear, the accelerometer needs to be as close as possible to the tool yet far enough to prevent damage. Five typical cutting conditions were selected for tool wear experiments, as shown in Table 11.1. Different directions on the tool/holder system is shown in the Fig. 11.2.

**Fig. 11.1** Experimental test setup [31]

**Table 11.1** Machining conditions used in tool wear experiments

| | |
|---|---|
| Machine tool | Universal Lathe *TN*50 |
| Tool insert type | TNMG (SANDVIK) |
| Work material | AISI 1045 |
| Workpiece size | Length = 600 *mm*, Diameter = 50 *mm* |
| Cutting conditions | 1. $n = 500\ rpm, f = 0.2\ mm/rev,\ d = 0.5\ mm$ |
| | 2. $n = 710\ rpm, f = 0.1\ mm/rev,\ d = 0.5\ mm$ |
| | 3. $n = 1{,}000\ rpm, f = 0.2\ mm/rev,\ d = 0.5\ mm$ |
| | 4. $n = 1{,}000\ rpm, f = 0.2\ mm/rev,\ d = 0.1\ mm$ |
| | 5. $n = 1{,}000\ rpm, f = 0.1\ mm/rev,\ d = 0.1\ mm$ |
| Cutting fluid | No |

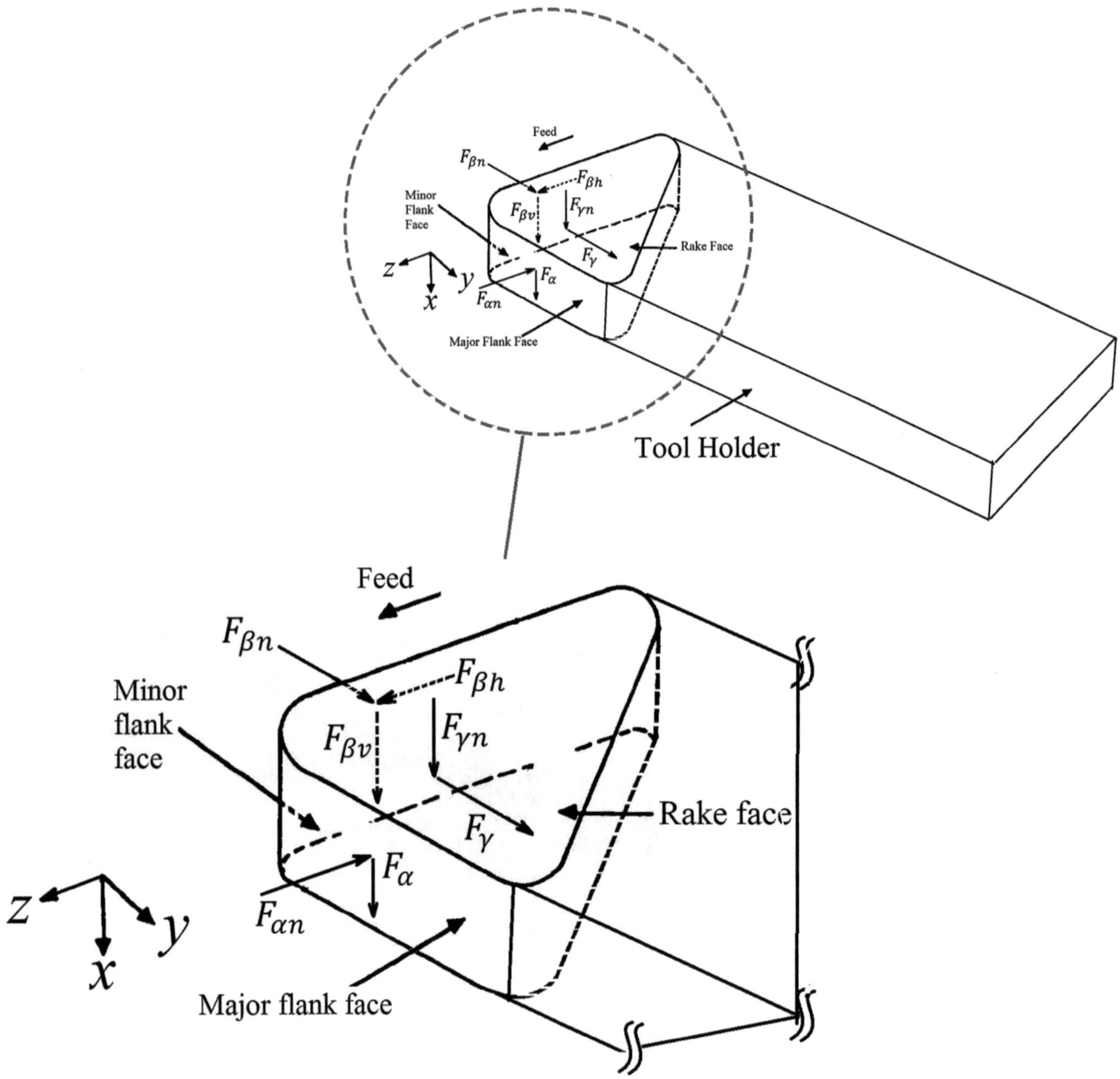


**Fig. 11.2** Tool/holder assembly and cutting axes [31]

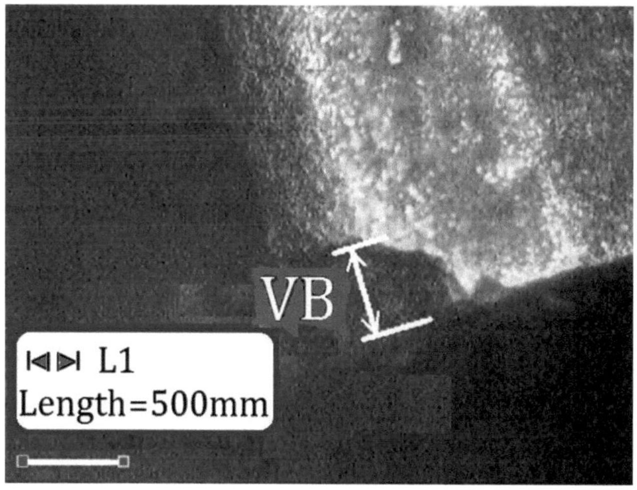


**Fig. 11.3** Major flank wear observed under optical microscope [31]

### 11.3.2 *Tool Wear Measurement*

An optical microscope with camera attachment, a stereo microscope and a digital camera equipped with a precise lens that provides the ability of high zoom rates, are used jointly to measure major flank wear *VB* of the tools. In the Fig. 11.3, a typical tool wear photo taken by optical microscope indicating major flank wear has been presented.

### 11.3.3 *FS-TARMA Estimation*

In this section, estimation of FS-TARMA models for acceleration signals is considered. The basis functions are chosen from Walsh functions. Walsh functions have been shown to be effective in capturing the instantaneous frequencies in burst type signals modeled by AutoRegressive time series models [32]. The parameter estimation is done based on 2SLS estimation method and then the nonlinear refinement. The orders used for the model are as follows:

$n_i = 48,\ p_i = 50,\ n_a = n_c = 30,\ p_a = p_c = 48$. The achieved non-stationary response modelling accuracy is very good in comparison with ARMA counterpart with a 6% Residual Sum of Squares normalized by the Series Sum of Squares (RSS/SSS).

A segment of a sample of estimated signals is presented in Fig. 11.4. The residual series and the variance for the estimated signal is shown in Figs. 11.5 and 11.6 respectively.

### 11.3.4 *Analysis Based on the Martin's Distance*

In a previous paper by Aghdam et al. [31], with the use of ARMA models and dispersion ratio of specific vibration modes [33], it has been shown that the phenomenon happening in a major flank wear process is the change of natural vibration mode of tool/holder assembly from the second bending mode about $z$ axis to first bending mode about $x$ axis (shown in Fig. 11.7).

Also in [32], the distances calculated based on tool acceleration, with ARMA models, in the main cutting direction $x$ and feed direction $z$, were plotted versus major flank wear. It was observed that the fitted curves possess a local minimum in the range (250–400) $\mu m$ and was revealed that this point is the same point at which the change in the vibration mode occurs. Here, the results obtained based on FS-TARMA models are shown in the Figs. 11.8 and 11.10. since different cutting conditions provide the same results, only the results for fifth cutting conditions (according to Table 11.1) is presented.

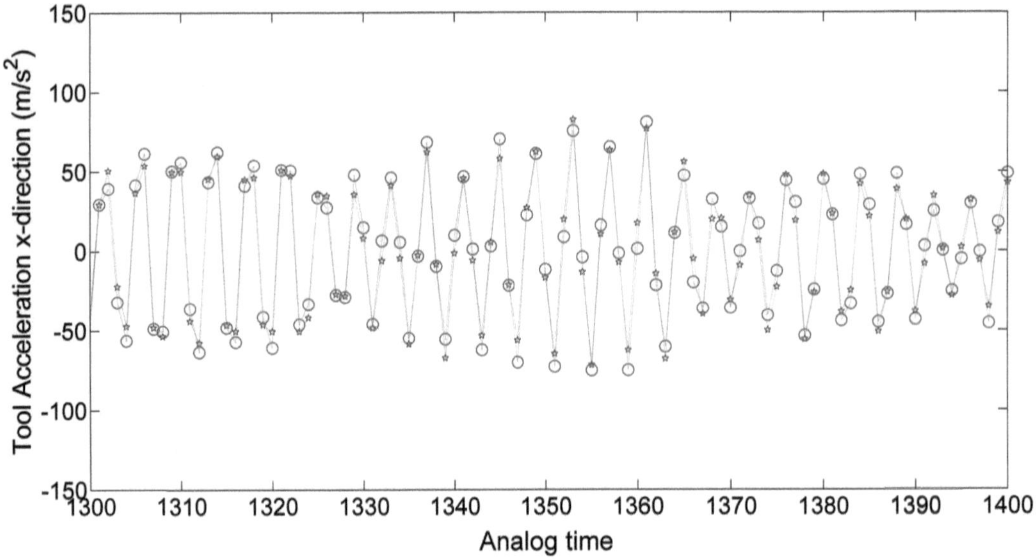


**Fig. 11.4** A segment of non-stationary tool vibration signal in x direction based on FS-TARMA predictions

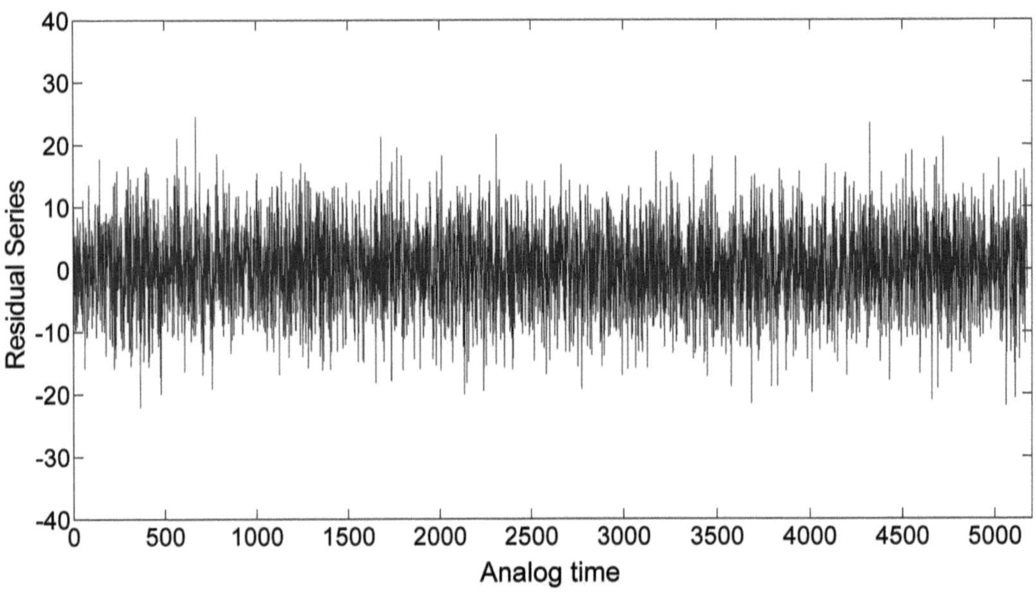


**Fig. 11.5** FS-TARMA residual sequence for tool vibration signal in x direction

As it can be seen from Figs. 11.8 and 11.10, for the distances obtained using FS-TARMA models, there is a local maximum for the distance in $x$ direction, while in $z$ direction there exist a local minimum point in (300–400) $\mu m$ of wear. This point can be associated with the point at which the tool mode of vibration changes and the tool should be replaced. However, the results of the analysis based on ARMA model of signals and the same metric [30], provide another measure for the identification of this point. As shown in the Figs. 11.9 and 11.11, (300–400) $\mu m$ there is a local minimum point. Comparing the results from the two methods, i.e. FS-TARMA and ARMA, reveals that FS-TARMA model gives smoother curves for the distance between the models of signals in tool wear progress.

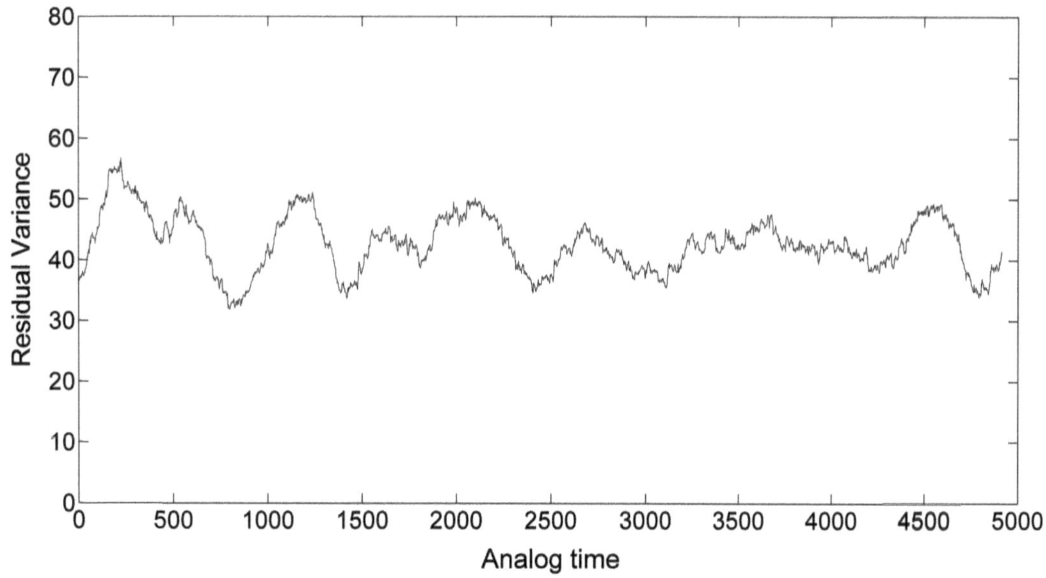


**Fig. 11.6** Estimated variance of residual

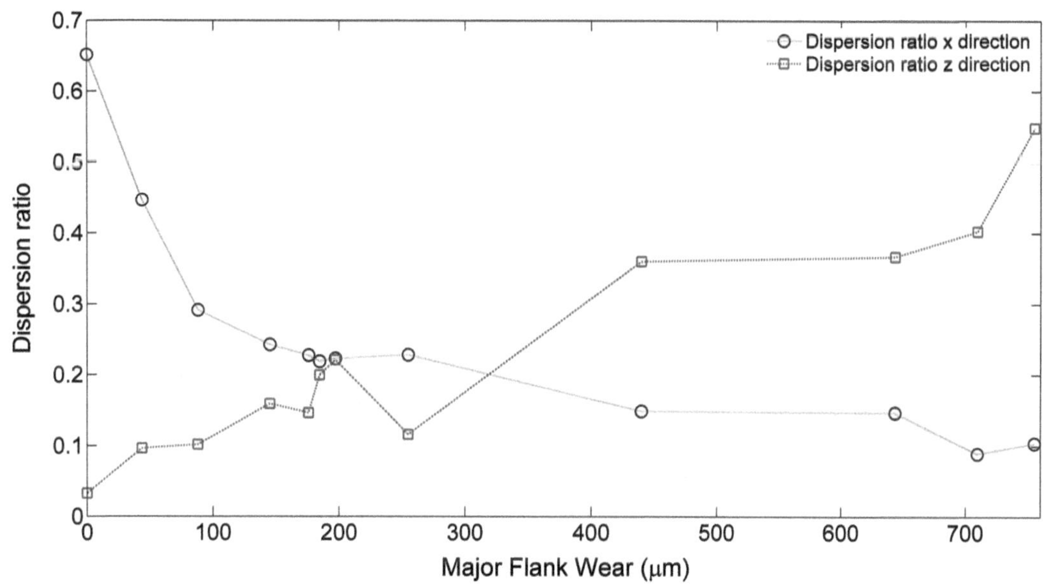


**Fig. 11.7** Dispersion pattern in feed direction $D_z$, in (1.7–3.0) *kHz* (*squares*) and main cutting direction $D_x$, in (8.0–9.6) *kHz* (*circles*) [31]

Since the Martin's metric, calculates the distance between the models based on system eigenvalues (which can be related to natural frequencies), it can be said that the change in distance is due to appearance of new poles (frequencies) in the model. According to Aghdam et al. [31], the tool vibrates mainly in second bending mode about $z$ axis when it is sharp and in the first bending mode about $x$ axis when is worn. Vibrating in a mode, can also excite the other frequencies lying in vicinity of the associated natural frequency, which in case of Martin's metric, it implies appearance of other poles near to the poles associated with the corresponding natural frequencies. When the tool is sharp, the distance curve in $x$ direction possesses a minimum, and when the tool reaches to the severe condition of wear, in (300–400) $\mu m$, the distance approaches a maximum. This means appearance of other frequencies which belong to a frequency band in vicinity of bending natural frequency about $x$ axis. In case of vibration in $z$ direction, the maximum point is in (300–400) $\mu m$ of wear, since in this condition the tool vibrates in another frequency band associated with first bending mode about $x$ axis.

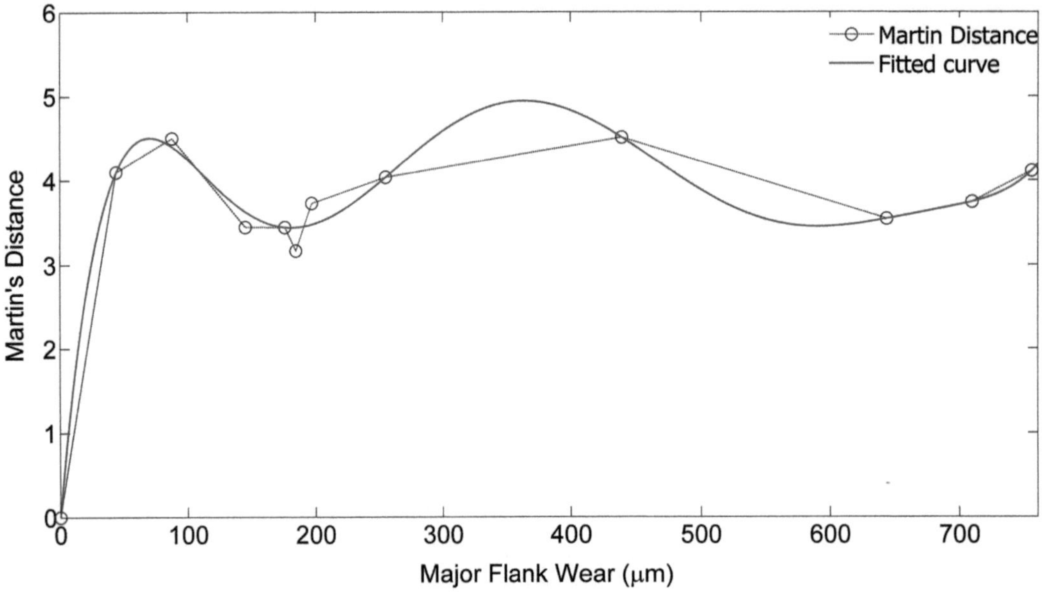


**Fig. 11.8** Martin distance of worn tools from sharp one in $x$ direction by FS-TARMA, fifth experiment

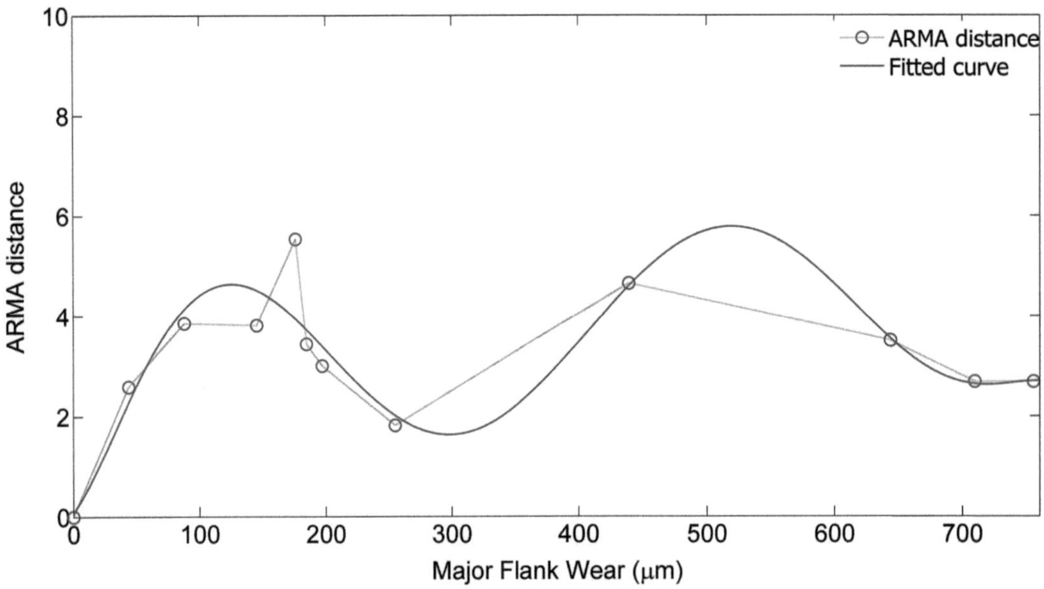


**Fig. 11.9** Martin distance of worn tools from sharp one in $x$ direction by ARMA, fifth experiment [31]

## 11.4 Conclusion

In this paper, recorded vibration signals of the tool in a turning process are modelled employing FS-TARMA models, through which the tool/holder system vibrational properties are obtained. Since FS-TARMA model is capable of modelling non-stationary signals, it provides higher accuracy in capturing the non-stationary signal behaviour. For comparison of FS-TARMA models, a metric which was applied in a previous work, is employed. The results of the metric shows a smoother behaviour in comparison with the results obtained via conventional ARMA models. Although the same metric is applied for both of the models, the distances obtained based on the models have different behaviour, while each of them give a criterion for the identification of the point where the tool should be replaced.

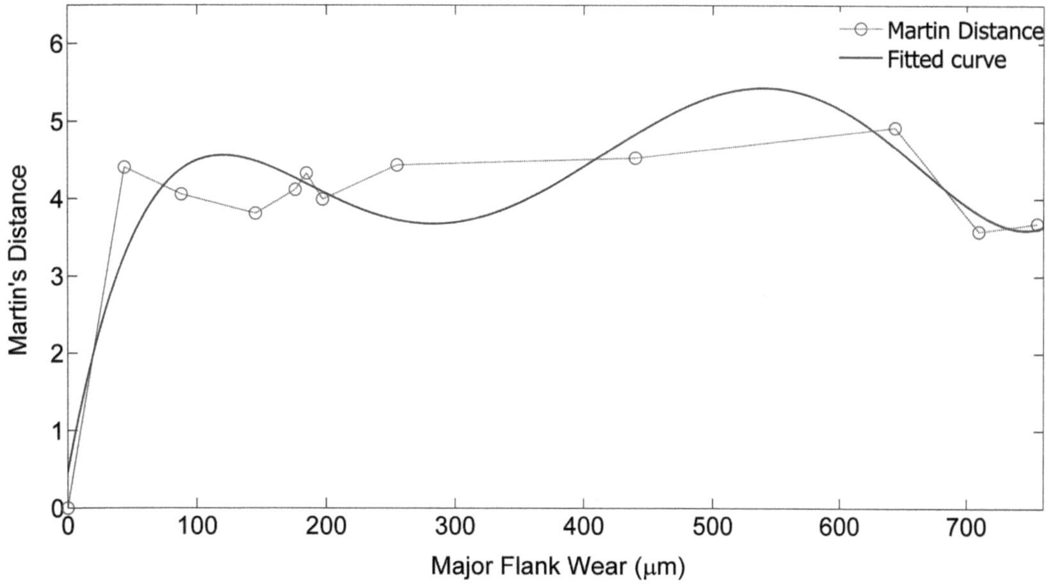


**Fig. 11.10** Martin distance of worn tools from sharp one in $z$ direction, FS-TARMA, fifth experiment

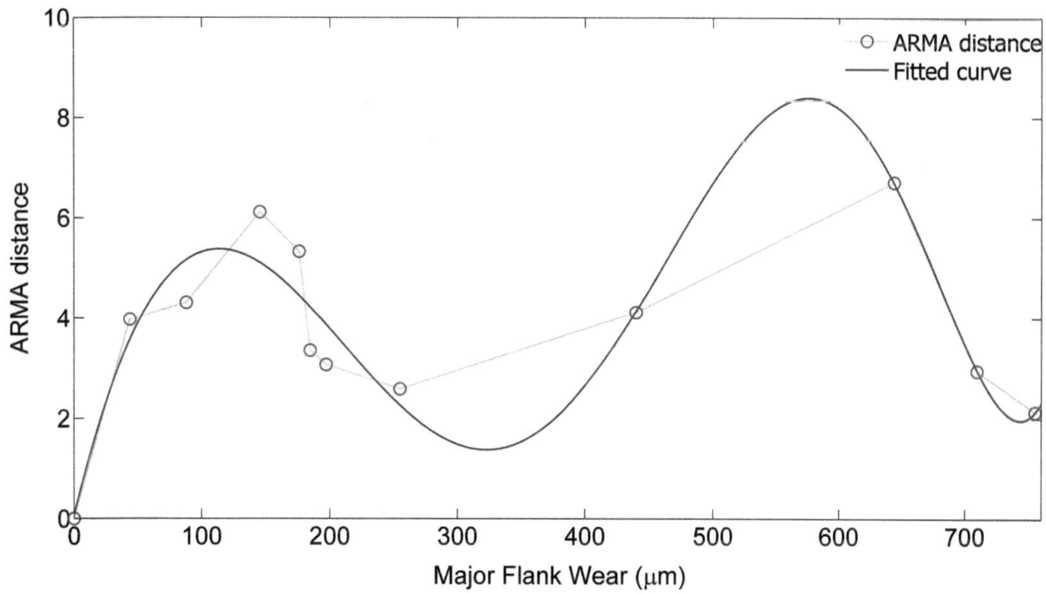


**Fig. 11.11** Martin distance of worn tools from sharp one in $z$ direction, ARMA, fifth experiment [31]

## References

1. Martin KF (1994) A review by discussion of condition monitoring and fault diagnosis in machine tools. Int J Mach Tools Manuf 34(4):527–551
2. Ghasempoor A, Moore TN, Jeswiet J (1998) On-line wear estimation using neural networks. Proc Inst Mech Eng B 212:105–112
3. Pandit SM, Kashou S (1982) A data dependent systems strategy of on-line tool wear sensing. J Eng Ind 104:217–223
4. Jemielniak K, Kossakowska J, Urbanski T (2011) Application of wavelet transform of acoustic emission and cutting force signals for tool condition monitoring in rough turning of Inconel 625. Proc Inst Mech Eng Part B J Eng Manuf 225:123–129
5. Li X (2002) A brief review: acoustic emission method for tool wear monitoring during turning. Int J Mach Tools Manuf 42:157–165
6. Liang SY, Dornfeld DA (1989) Tool wear detection using time series analysis of acoustic emission. J Eng Ind 111:199–205
7. Yao Y, Fang XD (1992) Modelling of multivariate time series for tool wear estimation in finish-turning. Int J Mach Tools Manuf 32(4):495–508
8. Roth JT, Pandit SM (1999) Monitoring end-mill wear and predicting tool failure using accelerometers. J Manuf Sci Eng 121:559–567
9. Sick B (2002) Online and indirect tool wear monitoring in turning with artificial neural networks: a review of more than a decade of research. Mech Syst Signal Process 16(4):487–546

10. Rehorn AG, Jiang J, Orban PE (2005) State-of-the-art methods and results in tool condition monitoring: a review. Int J Adv Manuf Technol 26:693–710
11. Alonso FJ, Salgado DR (2008) Analysis of the structure of vibration signals for tool wear detection. Mech Syst Signal Process 22:735–748
12. Kilundu B, Dehombreux P, Chiementin X (2011) Tool wear monitoring by machine learning techniques and singular spectrum analysis. Mech Syst Signal Process 25:400–415
13. Wang X, Wang W, Huang Y et al (2008) Design of neural network-based estimator for tool wear modelling in hard turning. J Intell Manuf 19:383–396
14. Aliustaoglu C, Ertunc H, Ocak H (2009) Tool wear condition monitoring using a sensor fusion model based on fuzzy inference system. Mech Syst Signal Process 23:539–546
15. Purushothaman S (2010) Tool wear monitoring using artificial neural network based on extended Kalman filter weight updation with transformed input patterns. J Intell Manuf 21:717–730
16. Sharma VS, Sharma SK, Sharma AK (2008) Cutting tool wear estimation for turning. J Intell Manuf 19:99–108
17. Srikant RR, Krishna PV, Rao ND (2011) Online tool wear prediction in wet machining using modified back propagation neural network. Proc Inst Mech Eng Part B J Eng Manuf 225:1009–1018
18. Brezak D, Majetic D et al (2012) Tool wear estimation using an analytic fuzzy classifier and support vector machine. J Intell Manuf 23:797–809
19. Penedo F, Haber RE, Gajate A, Del Toro RM (2012) Hybrid incremental modeling based on least squares and Fuzzy-NN for monitoring tool wear in turning processes. IEEE Trans Ind Inf 8(4):811–818
20. Leem CS, Dornfeld DA, Dreyfus SE (1995) A customized neural network for sensor fusion in on-line monitoring of cutting tool wear. J Eng Ind 117:152–159
21. Chen SL, Chang TH (2001) Using a data fusion neural network in the tool wear monitoring of a computer numerical control turning machine. Proc Inst Mech Eng 215:1241–1255
22. Segreto T, Simeone A, Teti R (2012) Sensor fusion for tool state classification in Nickel superalloy high performance cutting. Proc CIRP 1:593–598
23. Segreto T, Simeone A, Teti R (2013) Multiple sensor monitoring in nickel alloy turning for tool wear assessment via sensor fusion. Proc CIRP 12:85–90
24. Paul PS, Varadarajan AS (2012) A multi-sensor fusion model based on artificial neural network to predict tool wear during hard turning. Proc Inst Mech Eng Part B J Eng Manuf 226(5):853–860
25. Jemielniak K, Urbanski T et al (2012) Tool condition monitoring based on numerous signal features. Int J Adv Manuf Technol 59:73–81
26. Liu TI, Song SD et al (2013) Online monitoring and measurements of tool wear for precision turning of stainless steel parts. Int J Adv Manuf Technol 65:1397–1407
27. Siddhpura A, Paurobally R (2013) A review of flank wear prediction methods for tool condition monitoring in a turning process. Int J Adv Manuf Technol 65:371–393
28. Poulimenos AG, Fassois SD (2009) Output-only stochastic identification of a time-varying structure via functional series TARMA models. Mechl Syst Signal Process 23:1180–1204
29. Grenier Y (1983) Time-dependent ARMA modeling of nonstationary signals. IEEE Trans Acoust Speech Signal Process 31:899–911
30. Martin RJ (2000) A metric for ARMA processes. IEEE Trans Signal Process 48(4):1164–1170
31. Aghdam BH, Vahdati M, Sadeghi MH (2014) Vibration-based estimation of tool major flank wear in a turning process using ARMA models. Int J Adv Manuf Technol. doi:10.1007/s00170-014-6296-3
32. Reddy GRS, Rao R (2014) Performance analysis of basis functions in TVAR model. Int J Signal Process Image Process Pattern Recog 7(3):317–338
33. Pandit SM, Wu SM (1983) Time series and system analysis with applications. Wiley, USA

# Chapter 12
# A High-Speed Dual-Stage Ultrasonic Guided Wave System for Localization and Characterization of Defects

**Adam Gannon, Elizabeth Wheeler, Kyle Brown, Eric Flynn, and Will Warren**

**Abstract** This paper presents a joint approach to the detection, localization and characterization of structural defects. Our technique, termed Dual-Stage Structural Health Monitoring (DSSHM), combines an online, embedded ultrasonic structural health monitoring system with acoustic wavenumber spectroscopy data gathered from a scanning laser Doppler vibrometer. The combined system provides non-disruptive monitoring and high measurement fidelity necessary to drive critical decisions. Stage 1 is performed in situ (while the structure is in operation): an array of embedded ultrasonic transducers is used to excite and measure ultrasonic wave pulses in order to detect and locate irregularities within the test structure. In Stage 2, a subset of the same transducer array produces a steady-state ultrasonic response in the immediate vicinity of each structural irregularity. A laser Doppler vibrometer is then introduced to measure the full field response around that particular location in order to verify the presence of damage and provide high fidelity localization and characterization of the affected area. This document explores both the application of ultrasonic guided wave physics and the integration of sensing hardware and signal processing algorithms in implementing DSSHM.

**Keywords** Structural health monitoring • Non-destructive testing • Ultrasonic guided waves • Laser Doppler vibrometer • Embedded transducers • Hybrid monitoring

## 12.1 Introduction

The need for damage detection in aging structures is of major concern in the fields of aerospace, mechanical, and civil engineering [1]. There are three questions of primary importance regarding damage analysis for any engineering structure: First, has any damage occurred? Second, where has the damage occurred? Third, how severe is the damage? In more concise terms, relevant information regarding a structure's condition comes down to detection, localization, and characterization of damage.

The world of structural integrity analysis is currently somewhat divided between two fields, known respectively as Structural Health Monitoring (SHM) and Non-Destructive Testing (NDT). Both techniques are concerned with the three aforementioned fundamental questions regarding structural integrity. At present, Structural Health Monitoring concentrates most heavily on damage detection. The general idea is to instrument a structure with online embedded sensors that can monitor a structure's condition in situ. One example is a sparse array of ultrasonic transducers that can both actuate the structure and sense the vibrational response of the original pulse as it scatters and reverberates from the various structural features [2]. A change in the way that the ultrasonic waves propagate through the structure indicates that damage has likely occurred somewhere in the structure. The primary advantage of Structural Health Monitoring is that the system being evaluated does not need to be brought offline in order for an analysis to be performed. The main disadvantage of SHM lies in its inability to provide high fidelity localization and characterization of damage because of a fundamental lack of sensor density. As sensors are added to the structure the disadvantage of their additional cost and weight outweighs their potential benefit.

A. Gannon
Department of Electrical Engineering, State University of New York at Buffalo, Buffalo, NY 14260, USA

E. Wheeler
Department of Civil Engineering, Clemson University, Clemson, SC 29634, USA

K. Brown
Department of Mechanical Engineering, Brigham Young University, Provo, UT 84602, USA

E. Flynn (✉) • W. Warren
Engineering Institute, Los Alamos National Laboratory, Los Alamos, NM 87545, USA
e-mail: eflynn@lanl.gov

© The Society for Experimental Mechanics, Inc. 2015

C. Niezrecki (ed.), *Structural Health Monitoring and Damage Detection, Volume 7*, Conference Proceedings of the Society for Experimental Mechanics Series, DOI 10.1007/978-3-319-15230-1_12

In contrast, Non-Destructive Testing provides highly accurate characterization of damage, thanks to a high measurement density over the location being tested. However, thoroughly inspecting a structure using NDT techniques requires the system to remain offline during inspection. The high-fidelity damage characterization provided by NDT comes with the drawback of extended amounts of time and effort needed for inspection.

At present, Non-Destructive Testing is used predominantly in industry because its data density provides the information necessary to drive critical decisions regarding a structure's condition. However, current research aims to improve the reliability and accuracy of Structural Health Monitoring in order to utilize SHM as the sole means of evaluating structures in wide-scale commercial applications. Essentially, there is a need for a system than can combine the strengths of SHM and NDT. Recent efforts have been made in [3] to find a solution to this challenge using a combination of embedded transducers and scanning systems. This paper follows a similar approach but uses a different set of technologies that we believe are especially conducive to use together as a dual-stage system.

First a typical ultrasonic guide wave measurement is performed for baseline acquisition. Next, the likely area of defect is determined using Rayleigh Maximum Likelihood Estimation. Once the suspected area of defect is found, local transducers in the vicinity of the defect produce a steady-state ultrasonic response which is measured by a laser Doppler vibrometer. Finally, local wavenumber estimates are extracted from the measurements in order to localize and characterize the damage. Our approach, known as Dual-Stage Structural Health Monitoring (DSSHM), combines the constant monitoring capability of embedded piezoelectric sensors and the high resolution of a laser Doppler vibrometer. We present herein the application of ultrasonic guided wave physics and the associated integration of sensing hardware and signal processing algorithms to capitalize on the respective strengths of both SHM and NDT. The resulting scanning system is both faster and less invasive than other existing technologies while providing an accurate characterization of structural integrity.

## 12.2 Background

Guided waves are useful in structural health monitoring because of the way in which they interact with material defects and the distance they travel in a structure. Our DSSHM research concentrates on plate-like structures (i.e., the sheet metal skin of an airplane or the composite skin of a wind turbine blade), for which guided wave analysis is particularly useful. Stress waves that propagate through plate-like structures are classified as lamb waves, which are named for the famous mathematical physicist Horace Lamb. The original text of Lamb's paper can be found in [4].

Lamb wave modes are classified as either symmetric or anti-symmetric, depending on whether their displacement is either symmetric or anti-symmetric about the centerline of the plate [2]. At low frequencies the particle wave motion of the zero-order symmetric wave mode ($S_0$) is primarily parallel to the wave propagation direction, while the zero-order anti-symmetric or flexural mode ($A_0$), exhibits particle motion primarily perpendicular to the direction of wave propagation (Fig. 12.1).

The two distinct modes of propagation are significant in the application of structural health monitoring because they each interact differently with different kinds of structural defects. For example, the $S_0$ mode passes through surface or in-plane defects (like corrosion or delamination) with little scattering. The $A_0$ mode, on the other hand, generally experiences significant reflection/scattering upon reaching such a defect. The $S_0$ wave mode tends to travel through the structure approximately twice as fast as the $A_0$ mode of the same frequency [5]. Our scanning laser Doppler vibrometer is also primarily sensitive to the A0 mode. Because this paper only considers damage caused by corrosion, and our damage localization techniques are based on wave scattering, the wave mode of principal interest in this study is the $A_0$ mode.

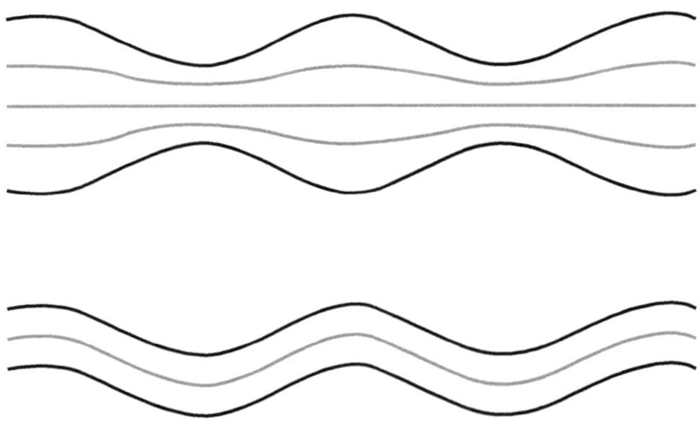

**Fig. 12.1** The symmetric wave mode, $S_0$ (*top*) and the antisymmetric mode, $A_0$ (*bottom*)

## 12.3 Experimental Procedures

### *12.3.1 Equipment and Preparation Considerations*

The DSSHM system includes a National Instruments Data Acquisition system for data capture and signal generation. The NI system includes an $8 \times 16$ matrix switch to make connections between any pair of transducers. Other hardware includes a set of ultrasonic transducers and a Scanning Laser Doppler Vibrometer (LDV). All of the control and data processing are performed using Matlab.

In order to successfully perform a DSSHM test, several preparatory steps must be taken. A test structure is first identified and instrumented with the piezoelectric transducers, each of which will be used both as an actuator and a sensor at different times in the testing process.

#### 12.3.1.1 Transducer Attachment

The manner in which the transducers are attached to the test surface can cause significant variation in the shape and mode of the waveforms that are produced and sensed. For example, when a transducer is attached with accelerometer wax, sensor actuation produces far more pronounced $A_0$ waves than $S_0$. This is because the transducer is well coupled only in the direction normal to the plane. The lack of shear coupling prevents the transducer from inducing longitudinal (in-plane) strain in the test medium. Thus, the $S_0$ wave mode (where particle motion occurs in the same direction as wave propagation) is not nearly as pronounced as the anti-symmetric mode (where particle movement is perpendicular to the plane of the surface). On the other hand, attaching transducers with superglue or an epoxy imparts far better shear coupling to the transducer-surface interface, meaning that both the $S_0$ and $A_0$ wave modes are generated by actuation.

#### 12.3.1.2 Frequency Selection

Regardless of what method is used to attach the transducers, it is important to pick a test frequency that minimizes (to the extent possible) the amplitude of the $S_0$ mode produced. As mentioned before, the $S_0$ mode travels faster than the $A_0$ mode. This property can be useful in seeking out an ideal test frequency for a structure of given material properties. Pulses of varying frequency are generated by an actuator and sensed by a nearby transducer, and the relative amplitudes of the first arrival ($S_0$) and second arrival ($A_0$) waves can be visually inspected to see which frequency minimizes (to the extent possible) the $S_0$ mode and maximizes the $A_0$ mode. After taking a measurement of the material's thickness the corresponding lamb wave velocity can be estimated.

### *12.3.2 Baseline Acquisition*

DSSHM takes a supervised learning approach to damage detection and localization. Following the detection, in order to locate damage in a structure, the DSSHM system compares data corresponding to the damaged state against data corresponding to the undamaged state. It is in comparing the pre- and post-damage data that the system is able to extract the information necessary to determine the approximate location of the defect. Thus, the stage-one localization and characterization can only be performed after appropriate "baseline" time history data has been obtained. The baseline time history corresponding to the undamaged state of a structure is obtained in the following manner:

Assume that a structure to be tested is instrumented with $P$ piezoelectric transducers. One by one, the transducers take turns acting as the actuator node in a pitch-catch setup. The actuator emits a narrowband high-frequency Gaussian-envelope pulse. Every other transducer acts as a sensor while the pulse propagates radially through the structure and reverberates from structural boundaries and features. Thus, we can say that there are M unique actuator-sensor pairs, where $M = P \cdot (P - 1)$. The data acquisition system records a time-history of the voltage across the sensor transducer for each of the M actuator-sensor pairs. Each time-history consists of $N$ samples. Several example baseline time-histories are shown in Fig. 12.2. The pulse which appears on the left end of each time-history plot is the electromagnetic interference from the actuation signal.

Recall that the baseline waveforms reflect the condition of the test structure in its undamaged state. That said, it is important to recognize that the acoustic properties of a structure tend to vary significantly according to environmental

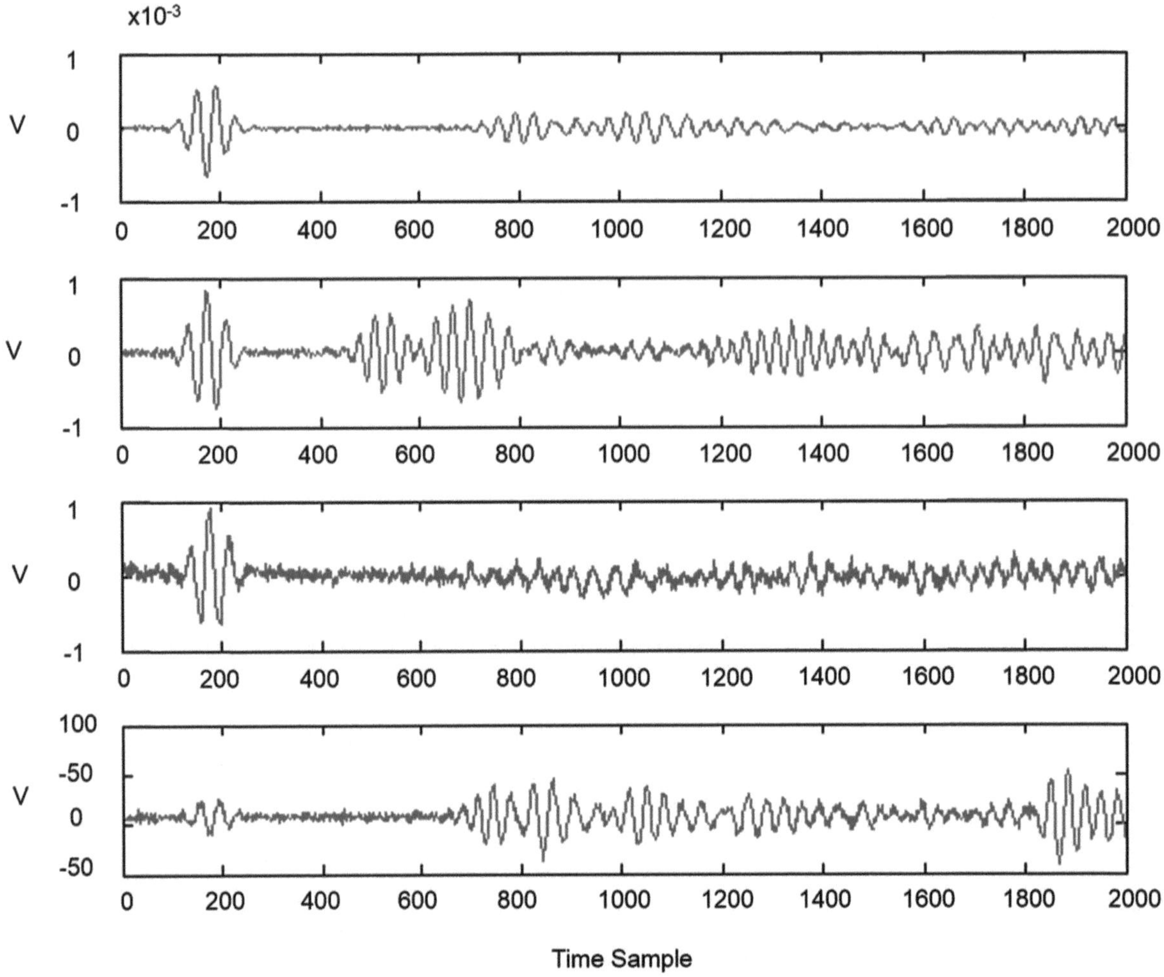


**Fig. 12.2** Four example time histories. The x-axis represents time samples, captured at rate of 1 MS/s. The y=axis represents the instantaneous voltage across the sensor transducer

factors [6]. For example, waves will travel slower through a plate at 18 °C than at 20 °C or 22 °C. Thus, it is necessary to establish a database of baselines that characterize the undamaged state of the structure across a full range of environmental conditions in which the structure may operate. In our study, the only environmental factor considered to have a significant effect on the test structure's properties is temperature. Accordingly, we populate our baseline database by continuously performing baseline tests throughout a 24 h period (or longer) to obtain baseline time histories that cover the entire temperature range generally experienced by the test structure.

### *12.3.3 Sensor Auto-Localization*

Although not directly related to the DSSHM process, sensor auto-localization describes a tool developed to increase the efficiency of our experimental procedures. The ability of the DSSHM system to localize damage is predicated on prior knowledge of the location of each sensor on the structure. The simple (albeit tedious) way of obtaining this information is to define a coordinate system and measure (by hand) the location of each sensor in the array relative to that coordinate system. This process is not particularly inconvenient, especially considering that plausible future commercial applications of DSSHM would incorporate a permanently embedded array of transducers. In such a case, determining the relative placement of transducers would be a one-time activity. However, in the context of our experimentation, or cases where some transducers are used temporarily, attaching and reattaching sensor arrays to various structures means a multitude of hand-measurements. Thus, we decided it would be worthwhile to write code that utilizes the baseline readings for a given sensor array to automatically determine the location of each sensor in the array based on time-of-flight information that can be extracted from each sensor-pair waveform. Using three or four reference nodes at known positions, the computer can compare the time-of-flight information from each sensor to each other sensor in order to determine the maximum likelihood location of each other transducer in the array.

### 12.3.4 *Stage 1 – Detection and Rough Localization Using Rayleigh Maximum Likelihood Estimation*

Once all of the aforementioned tasks have been performed, DSSHM can be implemented. The following material assumes that damage has only occurred at a single location in the inspection area since the baseline readings were recorded.

#### 12.3.4.1 "Ping" Mode

In stage one, the initial procedure is identical to the baseline acquisition phase. Each transducer will take a turn to act as actuator, delivering a narrowband high-frequency Gaussian pulse to the structure. Every other transducer acts as a sensor to record the structural response at that location. The result is another set of $M$ time-histories of the voltage produced at the sensor for each actuator-sensor pair. The test waveforms produced during what we refer to as the "Ping" mode will appear similar to the baseline waveforms. The differences between corresponding baseline and test waveforms are due to the scattering of waves from the damage location. Once the test waveforms have been produced, the next step is to extract the variations caused by the damage. This is accomplished by baseline subtraction.

#### 12.3.4.2 Optimal Baseline Subtraction and Damage Detection

In this step, the baseline waveform for each sensor pair is subtracted from the corresponding test waveform to reveal the changes caused by scattering from the damage location. The idea is that the baseline time-history and the test time-history for any given actuator-sensor pair will be identical until the arrival of the first scattered wave from the damage location. The time sample corresponding to the arrival of this first reflection will be the first non-noise feature of the difference between baseline and test data (Fig. 12.3).

Transducers are shown in blue and the defect is shown in dark gray In order for this subtraction to be effective, however, the correct baseline must be selected. Recall that each baseline in the database corresponds to a different structure temperature. A quick way of choosing which baseline to use for the baseline subtraction would be to simply select the baseline taken at the temperature closest to the temperature at the time of testing. However, it is possible (indeed quite likely) that the temperature throughout the structure is not uniform. Thus, the baseline with the corresponding temperature may not really be the best one to use in the subtraction process.

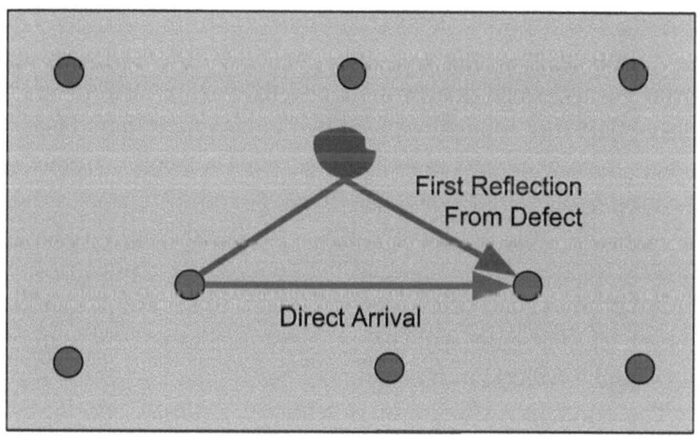


**Fig. 12.3** Representation of wave propagation through plate with damage present

A better metric to compare baselines is mean squared error. We follow a method baseline selection using mean squared error introduced in [7]. First the test data, $y[n]$, is normalized to produce $y'[n]$

$$y'[n] = \frac{y[n]}{\sqrt{\sum_{n=0}^{N-1} y^2[n]}} \tag{12.1}$$

and the baselines $x_l[n]$ for baselines $l = 1,2,\ldots,L$

$$x'_l[n] = A \cdot x_l[n] \tag{12.2}$$

where

$$A = \frac{\sum_{n=0}^{N-1} x_l[n]\, y'[n]}{\sum_{n=0}^{N-1} x_l^2[n]}. \tag{12.3}$$

The optimal baseline is that which minimizes the mean squared error between the test data and that baseline

$$l_{opt} = \underset{l}{\operatorname{argmin}} \sum_{n=0}^{N-1} \left(y'[n] - x'_l[n]\right)^2. \tag{12.4}$$

Once the optimal baseline set has been selected, we must determine whether or not damage has actually occurred. If the average magnitude of the time history difference exceeds a detection threshold, the system deduces that damage has indeed occurred. This detection threshold is determined in the following manner: each baseline set in the database is compared to each other baseline set. The threshold is equal to the average magnitude of the differences between each pair of baseline time-histories, plus two standard deviations of those differences.

Figure 12.4 shows an example of a test waveform (top), the corresponding baseline waveform (middle), and the resultant waveform from the baseline subtraction (bottom). Note that the resultant waveform is plotted on a smaller y-scale to more clearly exhibit its shape.

#### 12.3.4.3 Rayleigh Maximum Likelihood Estimate

As mentioned in Sect. 12.3.4.2, the time-sample at which the difference time-history first becomes anything more than noise should represent the time of flight corresponding to the shortest path from actuator to damage to sensor. This makes sense intuitively, because the sensor shouldn't record anything different until the wave has had time to scatter from the damage location.

The task is now to determine which time sample corresponds to the first scattered arrival. This will be accomplished using a Rayleigh Maximum Likelihood Estimate filter, which assigns to each time sample in the feature vector a relative likelihood that the first scattered arrival occurs at that time-sample. Our first step is to obtain an analytical signal from the difference waveform by applying a complex matched filter with the actuation waveform. In the figure below, the red waveform is the difference between the test data and the baseline data for a given actuator-sensor pair, while the blue curve is the envelope of that data difference produced by the matched filter. The envelope for each actuator-sensor pair becomes the feature vector that will be analyzed in order to provide a good estimate of the time-of-flight for the first scattered arrival at the sensor. Assuming one single damage location, the difference time-history for each actuator-sensor pair can be divided into two parts: (i) the time before any waves scattered from the damage reach the sensor, and (ii) the arrival at the sensor of the first directly scattered wave and all the subsequent echoes of the scattering. For the waveform below, the dividing point can be identified by visual inspection to be near time-sample 2,500. This location coincides with the peak of the green curve at the bottom of the figure. This green curve is obtained by applying a Rayleigh Maximum Likelihood Estimate (RMLE) filter to the signal envelope in blue (Fig. 12.5).

The Rayleigh Maximum Likelihood Estimate works as follows: for each time-sample $\eta_m$ on the feature vector, two Rayleigh parameters ($\sigma_1$ and $\sigma_2$) are estimated based on the distribution of signal magnitudes $\upsilon_m$ according to

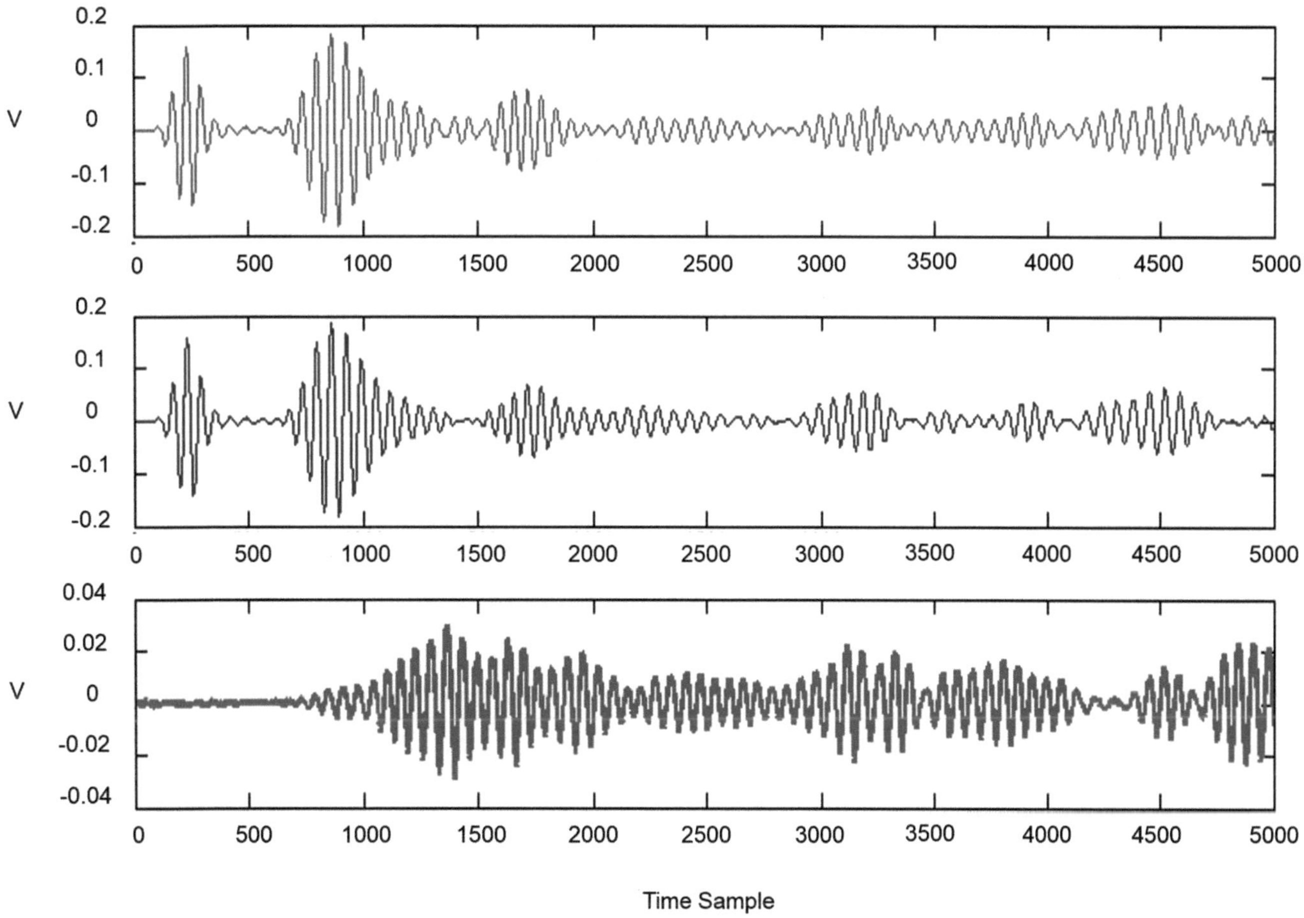


**Fig. 12.4** A test waveform (*top*), baseline waveform (*middle*), result of baseline subtraction (*bottom*)

$$\sigma_1(\eta_m) = \sqrt{\frac{1}{2(\eta_m)}\sum_{n=1}^{\eta_m} v_m^2[n]} \tag{12.5}$$

$$\sigma_2(\eta_m) = \sqrt{\frac{1}{2(N-\eta_m)}\sum_{n=\eta_m+1}^{N} v_m^2[n]} \tag{12.6}$$

Thus, $\sigma_1$ describes the distribution of signal magnitudes to the left of sample $\eta_m$ while $\sigma_2$ describes the distribution to the right of sample $\eta_m$. The likelihood used to compare the two Rayleigh distribution parameters reduces to form

$$w_m[\eta] = -\eta \log\left\{\sigma_1^2(\eta)\right\} - (N-\eta)\log\left\{\sigma_2^2(\eta)\right\} \tag{12.7}$$

This likelihood function is used to assign a relative to each time-sample in the analytical signal that the first scattered arrival occurred at that time signal. The result is the green curve shown above. It is clear from this curve that the likelihood function is maximized at the dividing point mentioned above, which separates the pre-first arrival data from the post-first arrival data.

The RMLE filter results for each actuator-sensor pair can now be combined to give a maximum likelihood estimate for the damage location on the structure. The test area is divided up into pixels, each of which will be assigned a test statistic value corresponding to the likelihood that the damage occurred at that particular pixel. For each pixel $x$ on the test structure, there exists a time-sample $\eta(x)$ for a given actuator-sensor pair that corresponds to the time-of-flight from the actuator to $x$ to the sensor. The total likelihood that damage has occurred at $x$ is the sum of the likelihoods calculated for each actuator-sensor pair by Eq. 12.8.

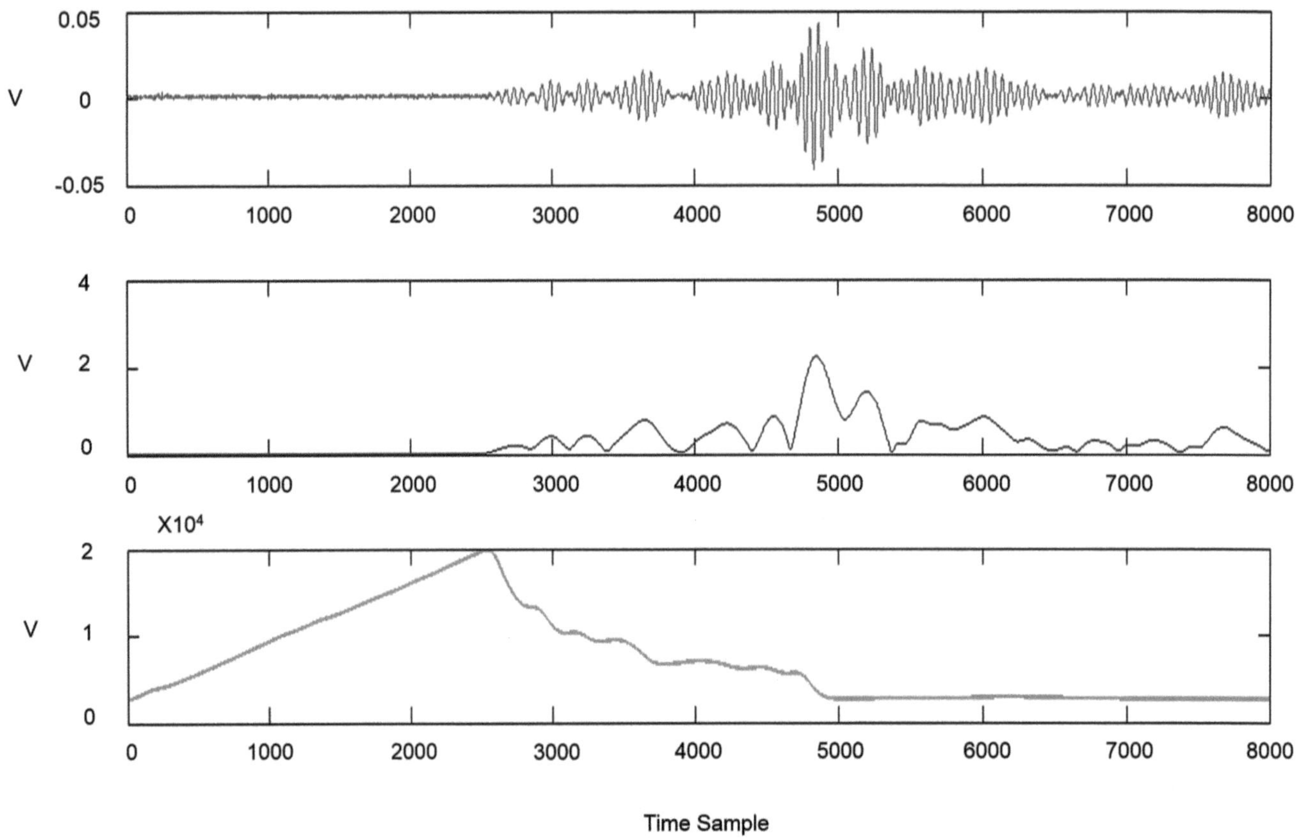


**Fig. 12.5** Baseline subtracted waveform (*top*), matched filter result (*middle*), RMLE result (*bottom*)

$$I(x) = \sum_{m} w_m \left[\eta_m(x)\right] \tag{12.8}$$

The result of applying Eq. 12.8 to every pixel on the test structure is a heat map that gives an estimate for the most likely location of damage. The rate at which the likelihood values decrease in moving away from the hottest location gives an indication of the confidence associated with the estimate of damage location (Fig. 12.6).

For a more comprehensive overview of RMLE, see Flynn et al. in [8].

### *12.3.5 Stage 2 – High Fidelity Characterization Using Acoustic Wavenumber Spectroscopy*

While the rough localization in Stage 1 provides an approximate damage location, it would be ideal to produce an image of the defect to provide the operator with information about its size, shape, and depth to help drive decisions about replacement and mitigating actions.

We use Acoustic Wavenumber Spectroscopy to produce such images from scans of the structure taken using a scanning Laser Doppler Vibrometer. The embedded transducers in the immediate vicinity of the damage are used to excite a steady state response in the structure. We take advantage of the distributed array of transducers and energize only those closest to the damage. After processing, we arrive at an estimate of wavenumber, the spatial equivalent of frequency, which corresponds to the thickness of the vibrating structure at each pixel scanned.

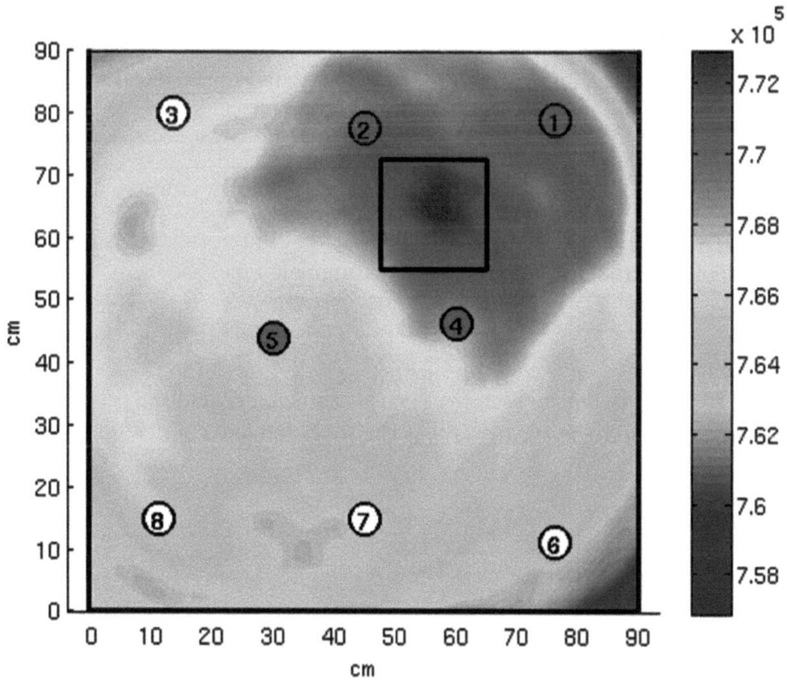


**Fig. 12.6** Example of generated heat map

#### 12.3.5.1 Scan Area Selection

The localization results obtained in Stage 1 are used to guide the laser to an area to scan. Reducing the size of the scan area will reduce the amount of time needed to image the defect and provide a higher resolution scan. We would like to choose the smallest scan area possible while still being reasonably confident that our scan area includes the damage.

We examine the range of local maximum likelihood estimates produced in Stage 1, and define the scan area to be a rectangle that includes all pixels whose likelihood estimates exceed a threshold. After some experimentation our threshold was chosen as 90 %. In future work, a more rigorous statistical analysis could be used to determine scan area.

#### 12.3.5.2 Active Transducer Selection

In order to produce the steady-state excitation necessary for a high-fidelity laser scan, we take advantage of our distributed array of transducers by selecting the transducers nearest to the estimated damage location.

The distributed nature of our transducer array allows us to route energy close to the scan area. Experimental testing has shown that the best scan images are produced when the area of suspected damage is surrounded with a transducer on each side. Intuitively this makes sense because we are able to route energy around structural features that may damp out waves coming from one direction. Thus, we look for a subset of transducers that are near the damage location and that effectively surround it.

Our set of active transducers will include four transducers. It is possible that one or more of them will be within the scan region, while others may be outside the scan region. We begin by selecting all transducers contained within the scan bounds. These transducers belong to set $\tau_I$. For each sensor inside the scan bounds we calculate the closest side to the transducer by finding the side that minimizes the distance to the midpoint of the side. There are then S remaining sides where S can be any integer in the range [0,4].

The set $\tau_A$ contains all the transducers outside the scan area. The set of transducers outside the scan area to energize, $\tau_O$, are chosen as those that minimize the sum of the distances to the remaining sides

$$\tau_O = \underset{i \in \tau_A}{\operatorname{argmin}} \left( \sum_{j=1}^{S} d_{ij} \right) \tag{12.9}$$

Where $d_{ij}$ is the Euclidean distance between sensor $i$ and the midpoint of remaining side $j$. The union of sets $\tau_I$ and $\tau_O$ give the indices of the four transducers to energize to provide the excitation. All selected transducers are connected in parallel. An 80 V peak-to-peak sine wave at 40 kHz is sent to the selected transducers to set up a steady state response in the structure.

#### 12.3.5.3 Laser Scan

The LDV measures the response of the structure to the steady state excitation using the light from a laser reflected off the surface of the structure. The velocity measurement takes the form

$$v\left[x,y,t\right] = a\left[x,y\right]\cos\left(2\pi ft\right) + b\left[x,y\right]\sin\left(2\pi ft\right) + n\left[x,y,t\right] \tag{12.10}$$

where $n[x,y,t]$ is the measurement noise and $a[x,y]$ and $b[x,y]$ denote the quadrature components of the steady state response measurement, respectively. The time-domain Fourier transform of the measurement is computed at each spatial point according to

$$c\left[x,y\right] = \frac{1}{T}\sum\nolimits_{i=0}^{T} v\left[x,y,t\right]\exp\left(-j2\pi ft\right) \tag{12.11}$$

Processing on the complex measurement matrix $c[x,y]$ is performed according to the process described in [9]. We start by performing a two-dimensional Fourier transform on the complex measurement matrix in order to obtain a wavenumber-domain representation of the data. We then determine the radial wavenumber $k_R = \sqrt{k_X + k_Y}$ of the peak in the wavenumber intensity diagram which will almost always correspond to the A0 mode out-of-plane sensitivity of the LDV. We pass the measurement matrix through a high pass filter with a cutoff equal to half this wavenumber. Next, the filtered measurement matrix $c\left[x,y,t\right]$ is passed through a bank of narrowband wavenumber filters

$$Z\left[k_X k_Y, k_C\right] = \tilde{C}\left[k_X, k_Y\right] W\left[k_X, k_Y, k_C\right] \tag{12.12}$$

$$W\left[k_X, k_Y, k_C\right] = \exp\left(-\left(\sqrt{k_X^2 + k_Y^2} - k_C\right)^2 / 0.72 B_K^2\right) \tag{12.13}$$

After transforming back to the spatial domain, we compute the envelope of each filter bank result. Finally, the wavenumber estimate at each coordinate is the wavenumber that maximizes the envelope at that point or

$$\widehat{k}\left[x,y\right] = \underset{k_C}{\operatorname{argmax}} \left| z\left[x,y,k_C\right] \right|_E \tag{12.14}$$

An image is generated from the local wavenumber estimates. Locations where corrosion damage has occurred will generally have a higher wavenumber. The shape and depth of the damage can be inferred from these images (Fig. 12.7).

## 12.4 Experiments

### *12.4.1 Laboratory Experiments*

Initial experiments were performed in a laboratory setting at the Los Alamos Engineering Institute. The test subject was a thin aluminum plate with dimensions 90 × 90 cm and thickness 1.5 mm.

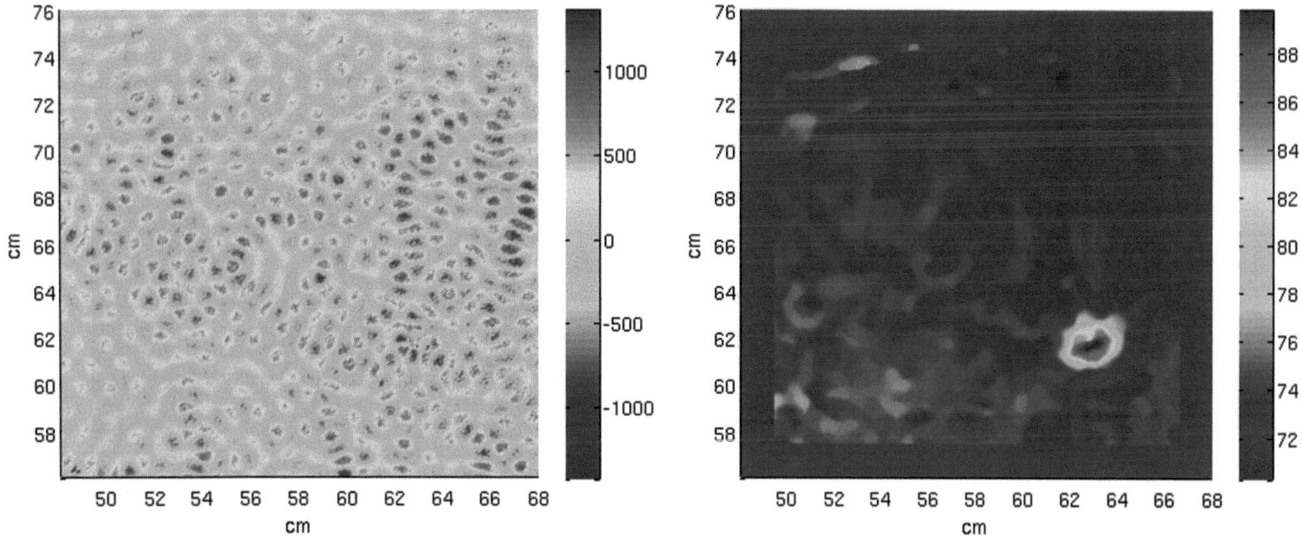


**Fig. 12.7** Raw laser measurement (*left*) and wavenumber estimate (*right*)

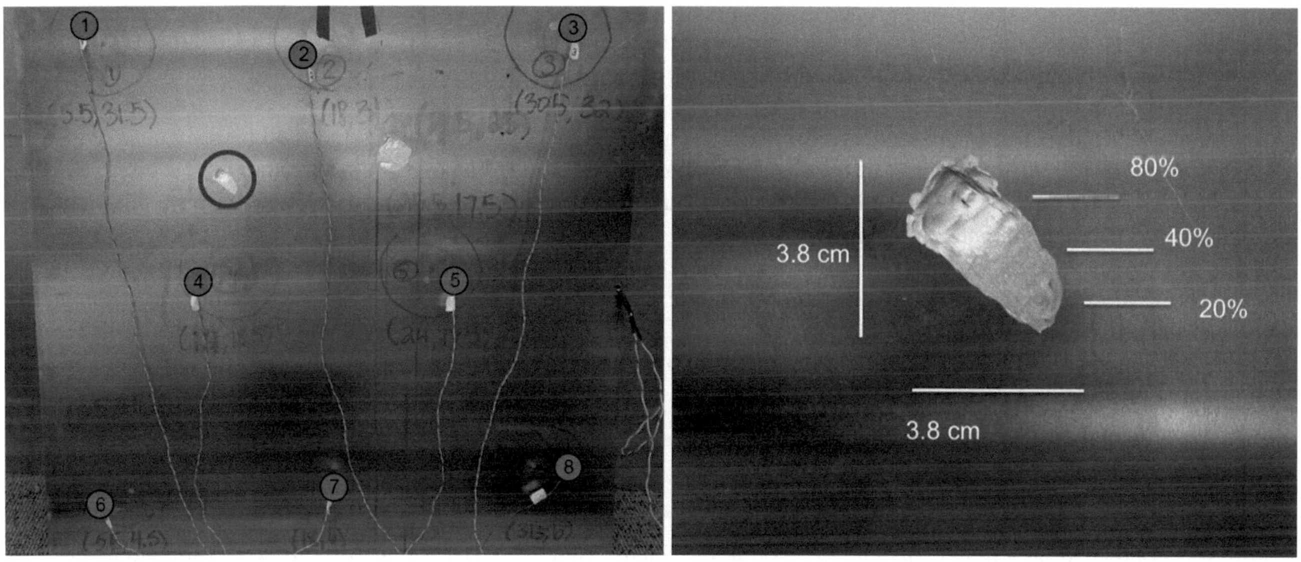


**Fig. 12.8** Rear view of test plate with transducers numbered 1–8 and damage indicated with circle (*left*) and detailed view of simulated corrosion (*right*)

#### 12.4.1.1 Test Setup

Eight piezoelectric transducers were attached to the aluminum plate using commercial superglue (Fig. 12.8, left). Exfoliation corrosion was simulated over an area of approximately 3.8 × 3.8 cm by removing 20–80 % of the original plate thickness using a Dremel rotary tool, Fig. 12.8 (right).

#### 12.4.1.2 Results and Discussion

Before creating the damage, a total of three baselines of the undamaged plate were recorded. After this database was obtained and the simulated corrosion was produced, the "ping" mode for Stage 1 began. The test waveforms produced during this stage were compared to the recorded baselines, baseline subtraction was performed, and the results were analyzed. A map showing

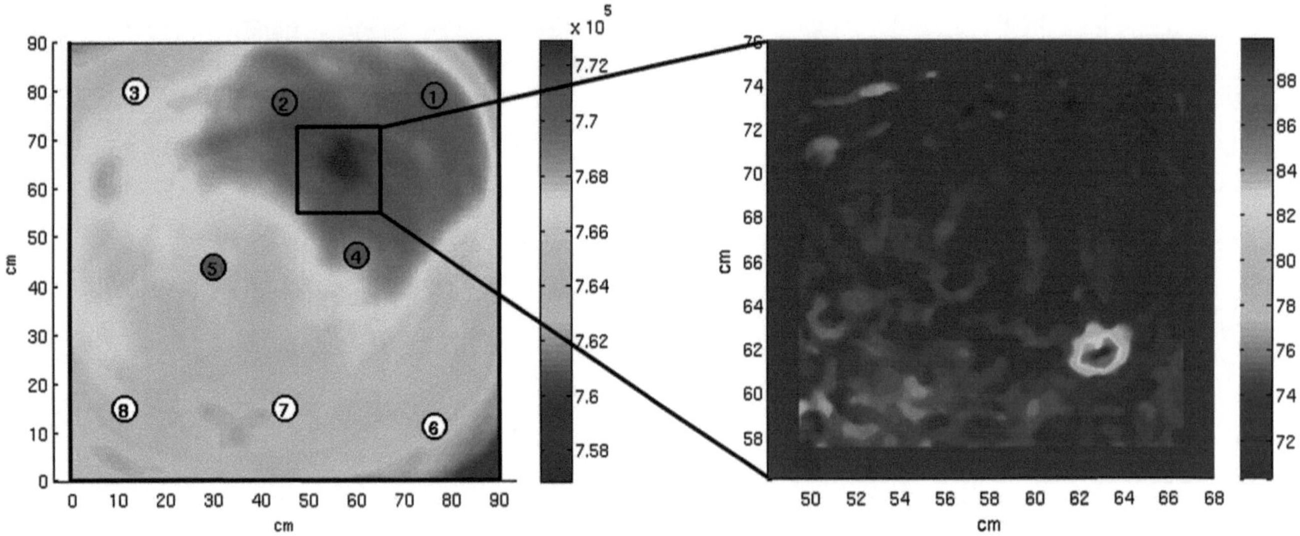


**Fig. 12.9** Generated map of suspected damage location using RMLE filter (*left*) and generated image of damage from wavenumber data (*right*)

the likelihood of damage at each location on the plate was generated from the results (Fig. 12.9, left). Note that this map was generated from the front view of the plate and the sensor layout and damage location are a mirrored from their locations in Fig. 12.8 (left).

The smaller black box shows the scan area for Stage 2, chosen to be a rectangle that includes all pixels whose likelihood estimates are in the top 90 %. A subset of transducers that are closest to the damage while surrounding it were selected to provide the excitation for the LDV according to the procedure in Sect. 12.3.5.2. Sensors 1, 2, 4, and 5 were automatically selected and are shown highlighted in red in Fig. 12.9 (left).

Stage 2 of the DSSHM methodology then started. The scanning laser Doppler vibrometer scanned the area where Stage 1 determined the most likely location of damage to be. Figure 12.9 (right) shows the image generated from the wavenumber data. The image provides a better representation of the actual size and shape of the damage than Fig. 12.9 (left) from Stage 1. From the wavenumber estimate we can also infer the depth of the damage.

### *12.4.2 Field Experiments*

Testing was conducted on a retired Boeing 737–200 at the Sandia National Labs Aging Aircraft NDI Validation Center in Albuquerque, New Mexico.

#### 12.4.2.1 Test Setup

Thirteen piezoelectric transducers were attached to the Boeing 737, once again using superglue (Fig. 12.10, left). To simulate exfoliation corrosion, a 6.35 × 4.45 cm area of the aircraft fuselage had 20 % of its skin depth removed, Fig. 12.10 (right). The damage was created with a Dremel rotary tool and the depth verified using Eddy Current.

#### 12.4.2.2 Results and Discussion

Similar to the procedure described in the laboratory testing (Sect. 12.4.1), baselines were recorded, damage was introduced, and the DSSHM methodology was performed. Due to time restraints, a total of three baselines were recorded over a time period with an average temperature of 28 °C. Figure 12.11 shows the generated map of suspected damage along with the determined scan area of the laser Doppler vibrometer.

**Fig. 12.10** Sensor layout with simulated damage identified (*left*) and simulated corrosion (*right*)

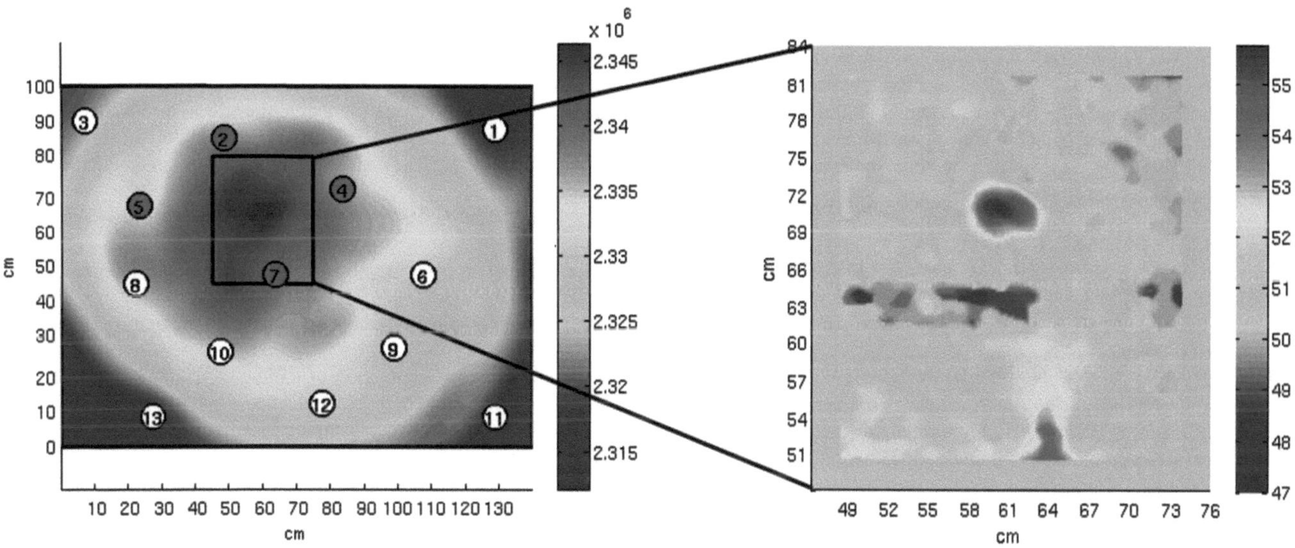


**Fig. 12.11** Boeing 737 damage localization results (*left*) and scan results (*right*)

Transducers 2, 4, 5, and 7 were selected to provide the needed excitation for the laser Doppler vibrometer. The generated image based on the scan area determined in Stage 1, is shown in Fig. 12.11 (left). Once again, Stage 2 (result shown in Fig. 12.11, right) was able to produce an accurate characterization of the damage, which is unable to be achieved with just Stage 1 alone.

## 12.5 Conclusion

In this work we have described a two-stage process for the detection, localization, and characterization of structural defects. An array of embedded transducers is used to measure the response of the structure to ultrasonic guided waves. In Stage 1, a test measurement is compared to a set of baseline measurements of the undamaged structure taken over a range of environmental conditions. Using the results from Stage 1, we then isolate an area of the structure where the damage is most likely to be located. This area is the target for the laser scan of Stage 2 where the same nearby embedded transducers to set up a steady state vibration. A scanning laser Doppler vibrometer measures this response and the data is processed to yield an estimate of wavenumber. The high-resolution image created from this wavenumber data gives information on the shape and depth of the damage and can be used to make informed decisions on the structure's repair and maintenance. In this way Stage 1 uses the information it has gathered and its existing hardware to help Stage 2 produce more detailed information.

Our system was validated with tests in the laboratory and in the field. Due to time constraints, only a few field tests could be performed. It would be valuable for future work to include tests on larger areas of the aircraft and run for longer durations. As mentioned above, a more rigorous statistical analysis could be developed for selecting the size of the scan area. We would like to expand future testing to include identification of different types of damage in addition to simulated corrosion. The same system could also be tested for use with other structures such as composite-based aircraft and wind turbines. Our experience with field testing highlights the importance of measuring baselines over a wide arrange of temperatures and environmental conditions to allow for an accurate comparison with test data. A temperature variation of only a few degrees between baseline and test measurements can produce inaccurate results.

Through this research our system, DSSHM, demonstrates the advantage of a two-stage approach for the detection and localization of structural defects. The two stages working together combine the advantages of both SHM and NDT into one system with the benefits of non-disruptive monitoring and high measurement fidelity. In this paper we have described the design of such a system and demonstrated its ability to identify structural defects.

**Acknowledgments** The authors would like to thank Mike Bode and the employees at the Aging Aircraft Facility at Sandia for allowing us to perform our testing on the Boeing 737. The hospitality and information you provided for us are greatly appreciated.

## References

1. Farrar CR, Worden K (2007) An introduction to structural health monitoring. Philos Trans Roy Soc A Math Phys Eng Sci 365(1851):303–315
2. Raghavan A, Cesnik CE (2007) Review of guided-wave structural health monitoring. Shock Vib Dig 39(2):91–116
3. Michaels TE, Michaels JE, Mi B, Ruzzene M (2005) Damage detection in plate structures using sparse ultrasonic transducer arrays and acoustic wavefield imaging. In: AIP conference proceedings. IOP Institute of Physics Publishing Ltd., Golden, Colorado, USA, vol 760, no 1, p 938
4. Lamb H (1917) On waves in an elastic plate. In: Proceedings of the royal society of London. Series A, Containing papers of a mathematical and physical character, pp 114–128
5. Staszewski WJ, Lee BC, Mallet L, Scarpa F (2004) Structural health monitoring using scanning laser vibrometry: I. Lamb wave sensing. Smart Mater Struct 13(2):251–260
6. Croxford AJ, Moll J, Wilcox PD, Michaels JE (2010) Efficient temperature compensation strategies for guided wave structural health monitoring. Ultrasonics 50(4):517–528
7. Lu Y, Michaels JE (2005) A methodology for structural health monitoring with diffuse ultrasonic waves in the presence of temperature variations. Ultrasonics 43(9):717–731
8. Flynn EB, Todd MD, Wilcox PD, Drinkwater BW, Croxford AJ (2011) Maximum-likelihood estimation of damage location in guided-wave structural health monitoring. In: Proceedings of the royal society A: mathematical, physical and engineering science, rspa20110095
9. Flynn EB, Jarmer GJ (2013) High-speed, non-contact, baseline-free imaging of hiddden defects using scanning laser measurements of steady-state ultrasonic vibration. In: 9th international workshop on structural health monitoring, Stanford, California, USA, vol 1

# Chapter 13
# Vibration-Based Scour Monitoring: Prototype Design, Laboratory Experiments and Field Deployment

**Sez Atamturktur and Abdul Khan**

**Abstract** Scouring is caused by the removal of bed material surrounding the piers and abutments and can rapidly compromise the integrity of a bridge structure and cause catastrophic failure. Such failures, which occur most frequently during peak flow periods, such as flooding, are a hindrance to emergency personnel trying to enter affected areas and to individuals trying to evacuate. As a countermeasure to mitigate the effects of the bed degradation, this article presents a vibration-based scour monitoring technique. This novel approach exploits the differences between the measured low-frequency ambient excitations of a thin, flexible plate located in the flow and the same plate located in the sediment. The underlying principle is that a flexible plate excited by the turbulent flow vibrates at significantly higher amplitude compared to an identical plate placed within sediment. Laboratory and field results obtained at various flow conditions indicate that the vibration-based scour monitoring concept is able to supply reliable information regarding both scour and refill processes. This article details the underlying monitoring concept, the design and optimization of sensors, the evaluation of sensitivity of the developed sensors to environmental conditions, and their long-term field deployment.

**Keywords** Scour monitoring • Riverbed detection • Pier protection from scour • Structural health monitoring • VTP sensors

## 13.1 Introduction

Riverbed scouring entails the abrasion of the material surrounding the bridge piers and abutments by high velocity flows typically during flood events and hurricanes. Scouring has a negative effect on the foundation support of the bridge abutments and piers, and can ultimately lead to bridge failure risking public safety. Bridges are essential for the evacuation and relief efforts during a flood or hurricane event; thus, scour induced bridge failure causes losses beyond those associated with the replacement bridge. Scour monitoring using portable or permanently installed instrumentation is listed as a viable option for prevention of scour induced bridge failure in the Federal Highway Administration's Highway Engineering Circular #23 [1]. The primary advantage of scour monitoring over other countermeasures (such as riprap and/or armor installation as well as channel and bank control measures) is its low cost and rapid deployment potential.

Since early 1990s, significant efforts have been dedicated to the development and investigation of scour monitoring systems. Some prominent examples of existing monitoring systems include sonar fathometers, the time domain reflectometry (TDR) and magnetic sliding collar. These monitoring techniques, however, are sensitive to environmental and/or flow conditions, such as temperature, salinity, turbidity, air entrainment and debris. Such sensitivity hinders the effectiveness of these existing monitoring sensors and makes the development of a sensing technique that exhibits robustness against such environmental and flow conditions necessary.

The concept behind our novel approach is simple: several dynamic sensors mounted on thin, flexible plates (referred to as VTPs) distributed along the length of a pier or abutment reveal the difference in the vibration levels of sensors that are in the river and subjected to the natural turbulence of the river flow and those that are in the sediment and not exposed to the turbulence [2]. A number of VTPs placed in a sealed pipe that is driven into the riverbed near the abutment or pier of interest measure the time history of the vibration levels. The proposed scour monitoring method yields point measurements, hence, the steel pipe must be buried near the bridge pier or abutment at the location where the riverbed measurements are desired. Once the vibration levels are monitored in the time domain, various metrics (such as root mean square) can be used to reduce the high dimensionality of the collected time-history data and extract information regarding the changes in

S. Atamturktur (✉) • A. Khan
Glenn Department of Civil Engineering, Clemson University, South Carolina, USA
e-mail: sez@clemson.edu

© The Society for Experimental Mechanics, Inc. 2015

C. Niezrecki (ed.), *Structural Health Monitoring and Damage Detection, Volume 7*, Conference Proceedings of the Society for Experimental Mechanics Series, DOI 10.1007/978-3-319-15230-1_13

the vibration levels of VTPs. Because these VTPs are distributed throughout the depth of the bridge pier, changes in the vibration levels can be straightforwardly correlated to the changes in the bed level, i.e., scour formation as well as refill.

The concept behind this approach for scour detection is fundamentally insensitive to many of the environmental or flow conditions that are reported to hinder the abilities of existing scour monitoring devices [3]. VTP sensors are excited by the turbidity and hence yield even more reliable outcomes as turbidity increases, a condition which would hinder the performance of sonar fathometers. Also, VTP method detects the water/sediment interface in a manner that is not affected by debris (which tends to cause false echoes in sonar fathometers) in the channel. Salinity and temperature, which negatively affect the riverbed measurements of TDR method, has *practically* no influence on the VTP mechanism. In this paper, we discuss the development of the VTP sensors, present the experimental data obtained in the laboratory conditions and demonstrate preliminary results from the field deployment.

## 13.2 Sensor Development: General Concepts

The sensor system is constructed with eight neoprene VTPs of 0.02 mm radius mounted on a 0.10 m diameter steel pipe. The VTPs are spaced approximately 0.10 m apart, on center (Fig. 13.3). The sensor is assembled using a compression pipe coupling mounted on the steel pipe with a toroid disk sandwiched in between. The flexible plate is fixed to the toroid disk, as shown in Fig. 13.1. Schematic view of different components of VTP device is shown in Fig. 13.2.

The laboratory experiments are conducted in a 1.2 × 1.2 × 18 m flume available at the Clemson Hydraulic Laboratory. The pipe is buried in the sediment keeping a portion of sensors exposed to the flow. The pipe is supported as shown in Fig. 13.3. The riverbed is represented in the flume with quartz sand, which practically represents a worst case evaluation regarding the performance of the VTP method, as sand is expected to provide less stiff support compared to clay, and lead in higher vibration levels in the sediment. The flow rates in the flume are controlled to vary in between 0.028 and 0.14 cubic meters per second.

Commercially available uniaxial accelerometers (B&K 4507 B 006) are used to measure the acceleration at the center of each flexible plate. These accelerometers are connected to a data acquisition system (B&K LAN-XI 3050A-060) through cables. During the acquisition of the time domain acceleration measurements, sampling frequency is set to 25.6 kHz.

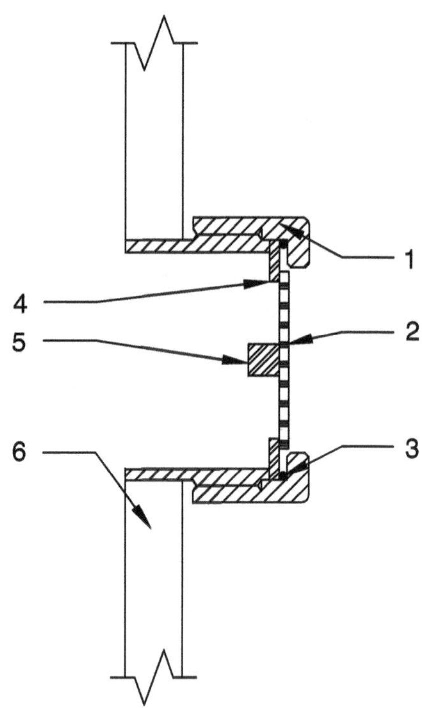


**Fig. 13.1** Prototype VTPs configuration. Components include: (*1*) the compression fitting, (*2*) the vibrating membrane, (*3*) a washer, (*4*) toroid disk, (*5*) accelerometer, and (*6*) the support pipe (Reproduced with permission from [3])

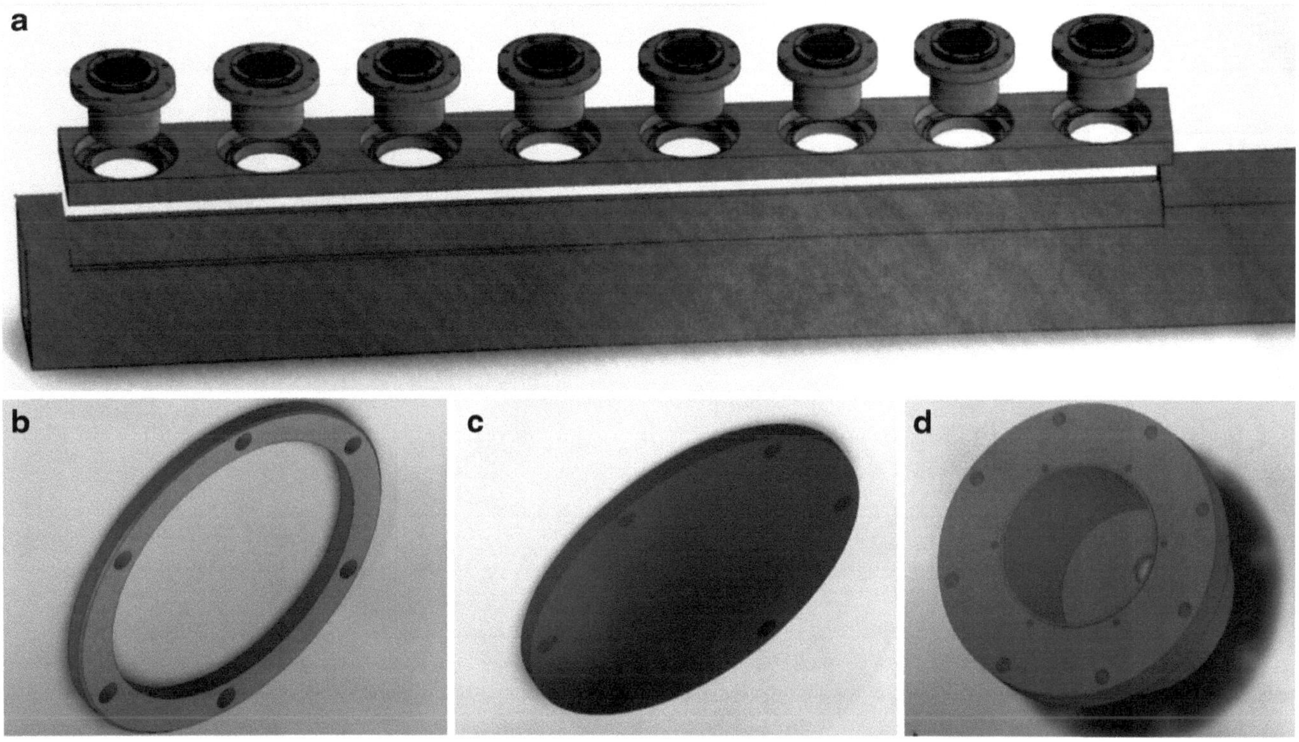


**Fig. 13.2** Schematic view and different components of VTP device include: (**a**) whole VTP device with sensor inserts, (**b**) washer ring, (**c**) vibrating membrane, (**d**) sensor housing insert

**Fig. 13.3** Prototype VTP array installed in flume bed. VTPs 1–4 are shown (*right/left*) above the sand bed, housed in support pipe (*middle*) connected to the flume frame (Printed from Fisher 2013b)

## 13.3 Laboratory Testing

During laboratory testing, a prototype VTP system with eight sensors located in a PVC pipe is used. Four VTPs are submerged in the flume: two buried in the sediment and two positioned in the flow, and the remaining four are located above the water free surface. VTP#8 is situated at the lowest position (0.16 m below the sediment bed) while VTP #5 is situated at the highest position (0.14 m above the sediment bed). The results of the tests are shown in Fig. 13.4, where mean-square acceleration response is plotted against the distance from the sediment interface.

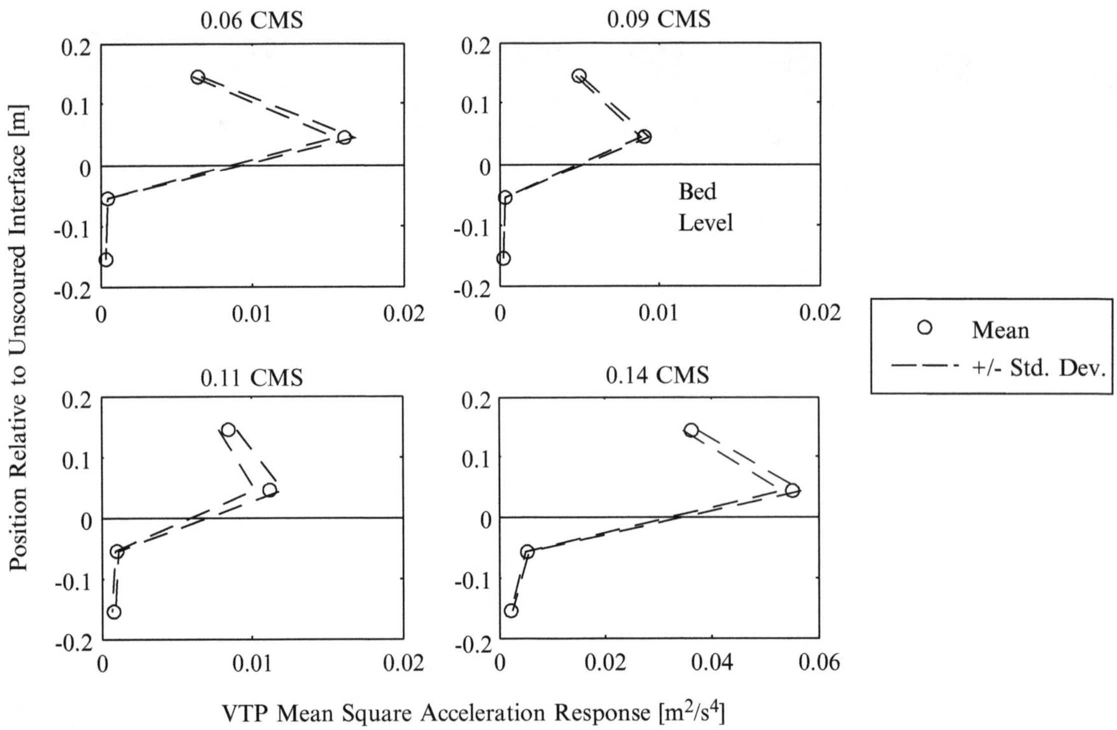


**Fig. 13.4** MSA response of prototype VTPs as a function of distance from the water/sediment interface. Flow rates varied from 0.060 to 0.14 cms. Mean values plotted including ± 1 standard deviation, as well as the unscoured bed level

In Fig. 13.4, it can be seen that the VTP that is closest to the bed measures the highest response. The response level decreases for the sensors that are located closer to the water surface. This observation reflects the flow turbulence profile that is expected along the depth of the channel. An important observation that can be garnered from the measurements is that the measured VTP response in the sediment is significantly lower than that measured in the flow. In this figure, not only the mean values, but also the standard deviations are shown to demonstrate that the detection of the riverbed is not influenced by the natural variability of the flow.

An important factor that is worth investigation is the role of channel flow rate in the accurate detection of the riverbed. In Fig. 13.5, the measured response level of VTPs located in the sediment is compared against those that are located in the flow for a range of flow rates. Data is collected over a 10 s duration and the process is repeated ten times to average the mean-square acceleration response. In this figure, it can be clearly seen that the response of the VTPs in the sediment is observed to be at least an order of magnitude less than those in the flow. As scouring typically occurs during high flow events, it is particularly important that the proposed method functions for varying levels of flow rate and thus, flow velocity. It can be observed in Fig. 13.5 that the distinction between the sensors in the sediment and flow become more noticeable for higher flow velocities.

## 13.4 Field Deployment: Preliminary Results

We have evaluated the performance of the VTP system in comparison to the TDR device during field deployment. Devices are installed at the bridge site located on Highway 76 over Eighteen Mile Creek, Pendleton, SC (USGS Station No. 02186702) as shown in Fig. 13.6. A high pressure pump was used to create two jets. These jets were used to lower the instrument in the bed. The VTP system has been measuring bed elevation since March 2014.

Difference in vibration response between the VTP sensors in sediment and in flow is used to locate the bed as described in the preceding section (Fig. 13.7). VTP sensor response under normal flow conditions are shown in Fig. 13.7a, where bed level is detected between sensors 4 and 5. VTP sensor responses under flood condition are shown in Fig. 13.7b, where bed level is detected between sensor 2 and 3. Comparing these figures, the distinct change in the bed level during normal and flood conditions becomes evident.

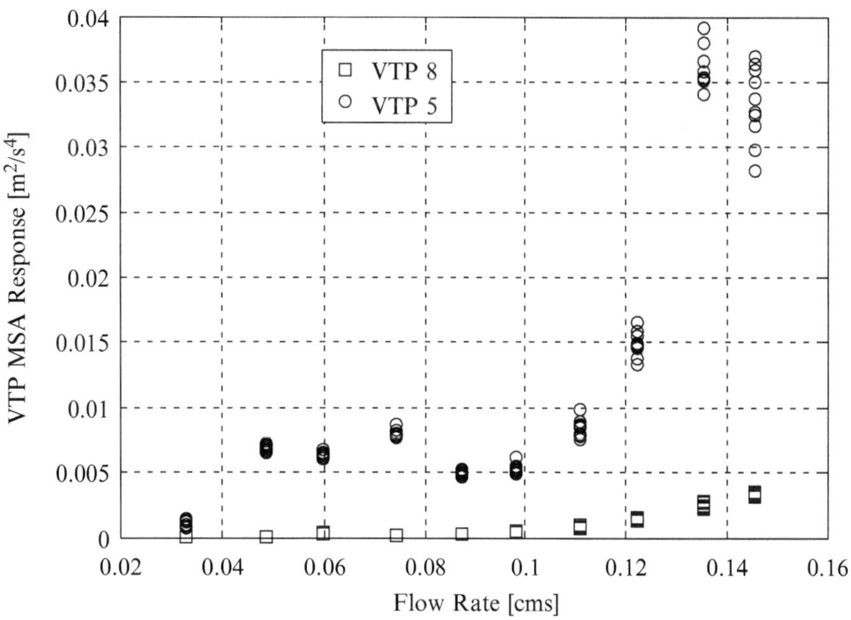


**Fig. 13.5** VTP MSA response of prototype VTPs versus flume flow rates. The slight drop in the mean square acceleration can be attributed to the 10 s measurement time, which may not be sufficiently long to capture the large eddies in the flow. Hence, as the flow rate increases, one should increase the measurement duration

**Fig. 13.6** Installed instruments at the field site

Figure 13.8 compares the bed elevation history collected by the VTP system against the water surface elevations recorded by the USGS gauge station to confirm the consistency and the sensitivity of our instrument in recording the scour resulting from flood events. It can be seen in Fig. 13.8 that an increase in water surface level leads to a decrease in the VTP identified bed level.

Next, the bed elevation histories of TDR and VTP for the Eighteen Mile Creek bridge site are compared in Fig. 13.9. Bed elevations from each instrument are measured with respect to the distance from a common datum (the zero elevation in the plots), which is taken as the top of the concrete pad of the pier (recall Fig. 13.6). Figure 13.9 shows a flood event that occurred around April 7–9, 2014, during which the water level increased up to 305 cm above the datum. The TDR and VTP are both responding to the change in bed level due to the flood event. However, there are differences in the recorded bed elevation partly because the instruments are placed about 46 cm apart resulting in different bed elevations being recorded by TDR and VTP. Both instruments can only record up to their probe length. In our application, the length of TDR probe

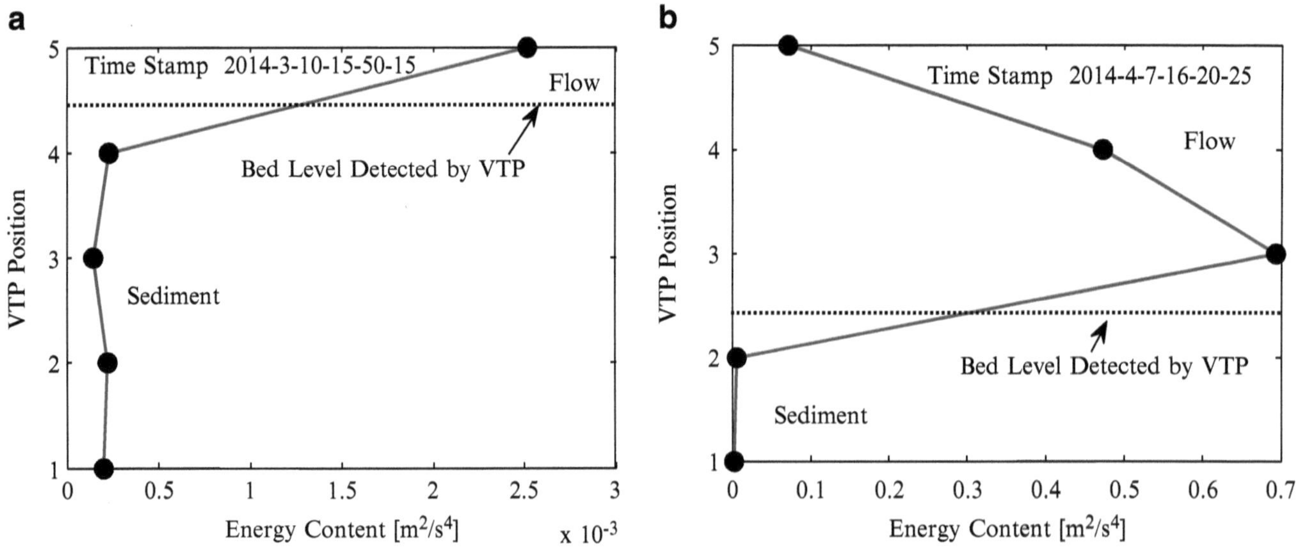


**Fig. 13.7** Typical vibration responses from VTP sensors. Positions of the sensors are numbered from between 1 and 5 (Starting from bottom of the instrument)

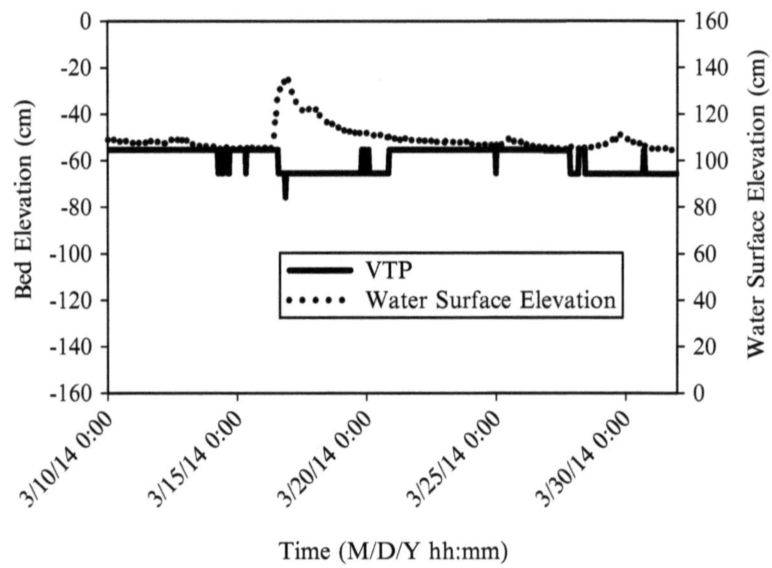


**Fig. 13.8** Bed elevation history recorded by VTP

was longer than that of the VTP and hence, the TDR was able to register scour beyond the limit of the VTP probe. For this reason, the TDR gives greater maximum scour depth than the VTP for this flood event.

Figure 13.10 shows the response of the instruments to the consecutive flood events during April 13–23, 2014. Again, both TDR and VTP are recording bed level consistently. The maximum scour depth recorded by the instruments is similar since the scour depths reached during these events were within the probe length of the two instruments.

## 13.5 Conclusions

Many of the bridge failures within the U.S. have been attributed to scouring damage to bridge piers and abutments and hence, it is critical that a robust real-time monitoring system that can detect the development and infill of scour be developed. A review of related literature has shown that available methods for scour monitoring exhibit sensitivity to the environmental conditions within rivers such as water temperature, salinity, and debris in the channel. It is therefore imperative that this new scour monitoring system is developed keeping the environmental conditions in mind.

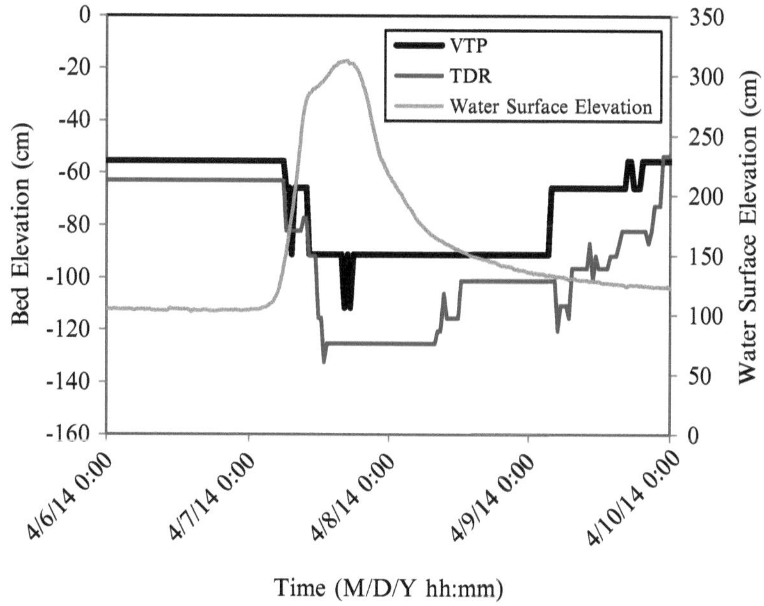


**Fig. 13.9** Performance comparison of VTP and TDR devices in a flood event

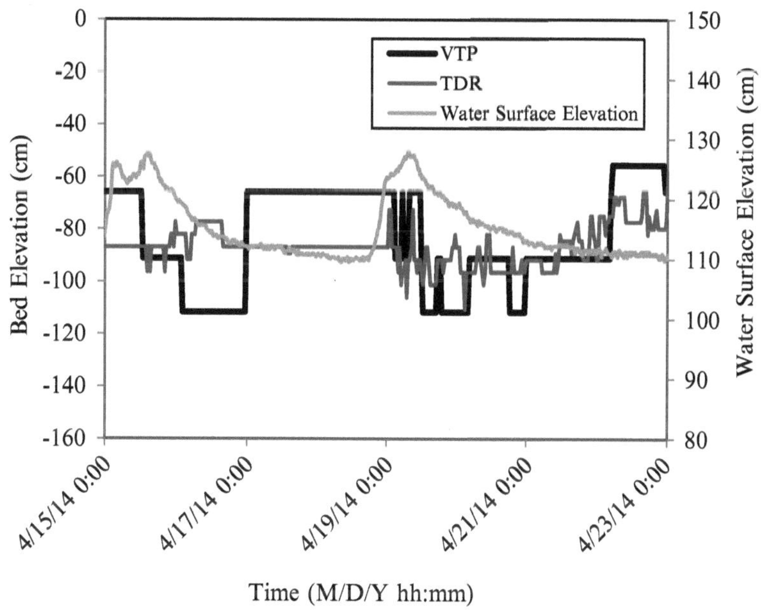


**Fig. 13.10** Performance comparison of VTP and TDR devices in consecutive flood events

This study presented a new instrumentation system that can determine scour formation and refill in real time while remaining insensitive to many of the conditions that negatively affect the available scour monitoring systems. In the proposed methodology, a support pipe is furnished with uniformly distributed vibration-based turbulent pressure sensors, referred to as VTPs, along its length. The pipe is then buried in the channel bed. The VTPs contain accelerometers allowing the time history acceleration response of the plate to be measured to determine if the material surrounding the VTP is water or sediment. Measured accelerations are significantly higher for the VTPs that are located in the flow compared to those that are located in the sediment due to the natural excitations resulting from the dynamic pressure in the flow associated with turbulent fluctuations. Hence, measuring the profile of the acceleration response for VTPs distributed along a bridge pier makes it possible to determine the location of the water/sediment interface.

The experimental measurements conducted in the laboratory demonstrated that the mean-square response of the VTPs located in the channel flow is one to two orders of magnitude higher than that of the VTPs located in the sediment. Furthermore, the formation of scour is observed to improve the distinction between the sensors in the sediment and flow as the turbulence in the scour hole is higher than that in the flow channel.

Performance of VTP device is also evaluated through field investigation. Results reveal that VTP can be used to measure bed elevation accurately. VTP device responds to the bed change due to flood event and can record both scour and refill processes. Although the VTP device is insensitive to the environmental conditions of the channel, it seems to be affected by the floating and accumulated debris in front of the sensor. Debris obstructs the impact of flow turbulence on the vibrating membrane, which reduces the response level of the sensors in the flow. Thus, it can become difficult to accurately distinguish the sensors in the sediment and in the flow. This problem is expected to be more pronounced in shallow depth channels whereas in the case of moderately deep channels, obstruction by debris is expected to be minimal. Rotating the pipe 90° however is observed to alleviate the problems associated with debris accumulation.

**Acknowledgements** The authors would like to thank the South Carolina Department of Transportation for supporting this work through grant number 1,417. The authors thank Murray Fisher and Md. Nasimul Chowdhury for their contributions in obtaining the experimental results for this study.

## References

1. Lagasse PF, Clopper PE, Pagán-Oriz JE, Zevenbergen LW, Arneson LA, Schall JD, Girard LG (2009) Bride scour and stream instability countermeasures: experience, selection and design guidance. Hydraulic engineering circular no.23, 3rd edn. Publication No. FHWA-NHI-09-111, U.S. Department of Transportation, Federal Highway Administration
2. Fisher M, Atamturktur S, Khan A (2013) A novel vibration-based monitoring technique for bridge pier and abutment scour. J Struct Health Monitor 12(2):114–125
3. Fisher M, Chowdhury MN, Khan AA, Atamturktur S (2013) An evaluation of scour measurement devices. Flow Meas Instrum 33:55–67

# Chapter 14
# Monitoring Fatigue Life Expenditure & Detecting Crack Initiation

**John H. Jensen**

**Abstract** The relationship between dynamic plus static loading and work hardening and fatigue life expenditure is well known. This work confirms the analysis that work hardening due to vibration causes a decrease in modal damping. The frequency and damping of higher frequency "local" modes are the most sensitive to work hardening and other changes in structural and material properties. A low frequency mode of a C-141 engine cowling section and a honeycomb laminate control surface section were vibrated over a 100 h excitation time. The frequency and damping of higher order modes were extracted periodically over the excitation time and trend lines plotted. After the 100 h the a stimulated crack was introduced in the cowling and a separation in the Honey comb composite test articles. The behavior of the honeycomb laminate's response to exciting a 762 Hz mode was not predictable. However the extracted frequency and damping values were expected shift over time. The small separation between the honeycomb core and the aluminum skin produced small detectable changes in modal frequency and damping.

Three trends in the data were revealed: (1) Shaking a lower order mode caused the modal damping to decrease due to increased stiffness and work hardening. (2) The higher frequency modes exhibited the greatest changes. (3) Crack initiation and laminate skin separation caused step changes in the frequency and damping trend lines.

The investigation was funded by Air Force Materiel Command at Warner-Robins AFB, through the prime contractor Optical Air Data Systems. The air frame test sections were supplied by the Materiel Command Lab [1].

**Keywords** Work hardening • Crack initiation • Fatigue Cracking • Crack propagation • Honeycomb laminates • Fiber/resin composites • Steam Turbine Generator • Torsional Modes

## 14.1 Introduction

The Air Force Materiel Command at Warner-Robins AFB, [1] funded this investigation. The principle investigator was Philip Rodgers of Optical Air Data Systems with the experimental work produced by John Jensen of Hewlett Packard Systems Engineering Organization. The primary driving force behind this investigation was the desire of the Air Force Materiel Command to find a way to reduce the high cost of re-certifying the flight worthiness of aging airframes. The inspection process for military airframes is very costly and labor and skill intensive. The process starts with paint removal, followed by inspections by skilled technicians. Current art requires several inspection technologies and skilled technicians to make measurements and interpret results. The state of the art inspection techniques are visual, ultrasound, X-ray, and florescent die. The techniques and equipment can be bulky and difficult to use in tight spaces.

The goal of the study was to determine if a more cost effective method could be found to localize areas where more detailed inspections would be warranted and to minimize inspection of areas where no significant changes were detected. The end result would be a system to do a fast structural survey to isolate and identify areas that should be more closely inspected.

The investigation proposed to measure the frequency and damping of higher frequency local modes, as our analysis showed would be much more sensitive to local stiffness increases due to **fatigue work hardening**. The dynamic and static stresses are concentrated at the nodes and results in "localized" increases in stiffness. The higher frequency mode shape

J.H. Jensen (✉)
Society for Experimental Mechanics, 11140 meadow Brook Dr., Auburn, CA 95602, USA
e-mail: jensenjh@sbcglobal.net

© The Society for Experimental Mechanics, Inc. 2015

C. Niezrecki (ed.), *Structural Health Monitoring and Damage Detection, Volume 7*, Conference Proceedings of the Society for Experimental Mechanics Series, DOI 10.1007/978-3-319-15230-1_14

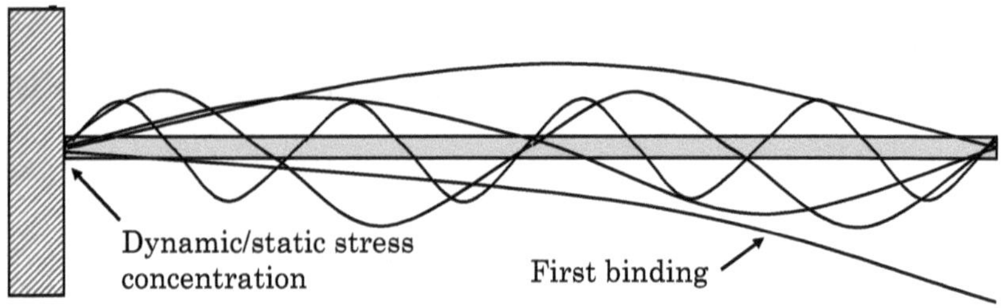


**Fig. 14.1** Higher frequency modes tend to have the steepest slope at the boundary where the static and dynamic stresses are highest and most likely point of failure

trajectory approaches the nodes at the highest slope. The first bending mode may be a major contributor to fatigue life expenditure but its frequency and damping values would be the least affected by the localized work hardening at the node points Fig. 14.1.

The test specimens were supplied by AF Materiel Command Lab and included a honeycomb laminate composite trim control surface trailing edge, and a 2 ft by 2 ft section of a ribbed stiffened engine cowling from a C-141. The cowling test section was fabricated from a forged aluminum billet on a skin mill. The most common failure process for these cowlings is fatigue cracking. The control surface consists of an aluminum skin bonded to a fiber/resin honeycomb core. The failure mechanism of greatest concern is the separation of the skin and the core. Laminate failures are very difficult to predict or detect and can be catastrophic in flight.

## 14.2 Airframe Structural Failure Mechanisms

Fatigue failure mechanisms have been thoroughly studied, understood and well documented. The fatigue driving mechanism for airframes results from high static loading with moderate cyclic loading (flight) with periodic transients of higher impulse and aerodynamic loading (take off and landing). The static plus dynamic loading cause initial work hardening and generally localized in the areas of maximum stress. This hardening process continues as fatigue life is expended. The material becomes brittle and then goes into the next phase in the fatigue failure process, crack initiation.

The **crack initiation** phase is characterized by the breakdown of the metal crystalline interfaces and begins to be visible on the surface. The stress can be amplified by internal material flaws or joint fabrication defects which tend to concentrate the stresses. These cracks usually start on the surface of the structure and begin to propagate in number and in length. These crack sites are usually located at joints, seams, bends, rivets, welds, and other stress concentration areas. The **crack propagation** phase usually continues until crack length reaches what is called "**critical length**". Critical length is usually defined as the length where the crack length grows under normal static and/or dynamic loading. At this juncture failure of the member is imminent.

Other areas of concern in air frames inspection are glued laminates, **corrosion,** fiber resin composites and hydrogen imbrittlement. Fiber resin composites structural components have failure modes that are similar to metal fatigue in that cracking and separations between material interfaces propagate and result in failure. However these failure mechanisms are seldom visible on the surface and are not well understood.

The **fatigue failure mechanisms** in composites are not well defined. The two types of composites in common use in airframes are **fiber/resin composites** and honeycomb core laminates. The primary failure mechanism in **honeycomb laminate composites** has been identified as localized separation between the surface and the honeycomb core. The separation between the core and face sheeting was expected to affect the structural stiffness and would increase damping as delamination propagates. The damping number of the higher order modes was expected to be the most sensitive to this failure mechanism.

**Fiber/resin composites** are becoming widely used in airframes because of their high strength to weight ratio. The primary failure mechanism in fiber resin components is attributed to delamination due to weak bonding between layers. The delamination seldom appears on the surface. The sudden catastrophic failures of composite laminates and **fiber/resin composites** have occurred on air craft and are of ongoing concern. It has been observed and reported that "identical" fiber/resin composite structural members can exhibit wide variability in structural behavior. The fabrication process has been the prime suspect in the variability and repeatability of these structural elements. However because fiber resin failures are often explosive it has been very difficult to find the smoking gun. The weakness between layers would behave like **constrained layer damping**. Again periodic measurement of damping values could provide a window into the structural integrity of these components.

**Corrosion** is usually found in seams and joints where moisture can accumulate and support electrochemical corrosion. **Corrosion** erodes material at the structural interface causing pitting and changes the dynamics at the structural interface. **Corrosion** should increase damping and produce small mass changes due to metal loss and pitting.

## 14.3 Analytical Model

The Modal mass, stiffness, and damping model depicted below and the equations of motion will be used to show the relationships between modal Mass, stiffness, damping, and work hardening.

$$\mathrm{F}(t) = M\frac{d^2x(t)}{dt^2} + C\frac{dx(t)}{dt} + Kx(t)$$

F = force, M = Mass, $C$ = damping, and $K$ = spring constant.

The natural frequency $\omega_n = \sqrt{K/M}$

For damping factor $\zeta < 0.01$

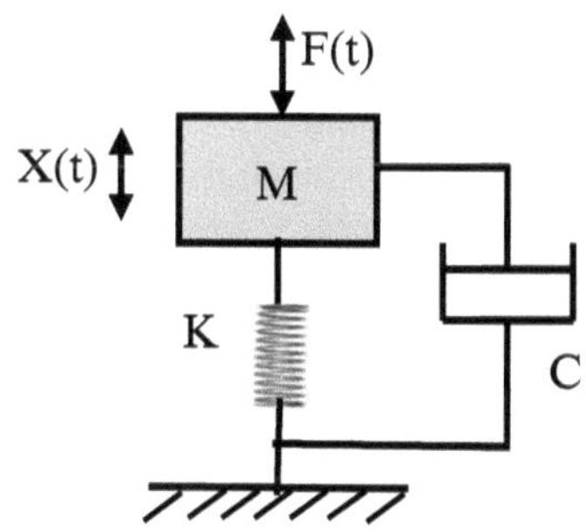


For the modal fatigue model the Mass, $M$, of the local mode is constant, while the spring constant, $K$, or stiffness would increases due to work hardening. The damping, $C$, would decrease with work hardening. For higher frequency modes $K$ is large and M small. Work hardening will produce an increase in the stiffness $K$ and $\omega_n$ will increase by the square root of a small change in $K$. The relationship between damping factor $\zeta$ and the spring constant $K$ is:

$$\zeta = \frac{C}{\sqrt{2KM}}$$

Work hardening causes $C$, damping, to decrease and $K$, the stiffness to increase while $M$ the modal mass remains fixed. Thus the damping factor $\zeta$ is directly proportional to the damping $C$ and inversely related to the square root of $K$. Note $K$ is a large number while $C$ is a small number. Damping, $C$, is a function of velocity so the higher the modes frequency the larger the value of damping. Thus the damping of the higher frequency modes will be affected less by stiffness increases than lower frequency modes.

The relationship between modal natural frequency $\omega_n$ the modal mass, $M$ and stiffness, $K$ for higher order modes is:

$$\omega_n = \sqrt{K/M}$$

Assuming the modal mass is fixed, and the stiffness $K$ is a large number and increases due to work hardening, one can see that the natural frequency will increase but not very much. The square root of a large number equates to a much smaller number and a small change in the value of $K$ produces a small change in the natural frequency, $\Delta\omega_n$. Therefore the damping of the higher frequency local modes will be affected much more than the frequency by work hardening and fatigue life expenditure.

As work hardening continues, inter crystalline separation occurs on a microscopic level. This is the crack initiation phase. The crystal boundaries interact dynamically and the damping C increases. As the damping increases and the stiffness may increase as the cracks grow. These two material properties appear to reinforce each other and amplify the increase in damping factor. In the transition through crack propagation, stiffness decreases and damping increases. Knowing what happens to the damping values as the material approaches and goes through the crack initiation and propagation phases is the key to heading off failures.

## 14.4 Failure Detection via Frequency Response

The use of frequency response as a tool to detect metal defects has been in use for centuries and is still in use today. The inspector taps the structure with a hammer and listens to the response. As the QA person builds up experience they can relate a given acoustic response to the location and type of defect in the material. There are a few issues with this approach. First the data base is locked up between the inspector's ears. Second your data base may quite, be fired, retire, or die. Third the inspectors cannot put a quantitative value on what they hear.

The use of frequency response measurements to detect structural flaws has been investigated for some time. The interpretation of measurements was very difficult and investigators could not discriminate between different types of flaws. However, most attempts focused on relating frequency shifts to structural defects. These investigations did observe shifts in modal frequency responses due to fatigue cracking but the frequency shifts failed to give sufficient lead time on failures or reveal the type of defect and location.

Using frequency shifts as an indicator of structural failure has been plagued by poor repeatability, complexity, and poor sensitivity issues. Initial attempts focused on lower frequency global modes. These global modes are often major contributor to fatigue life expenditure. However these low frequency modes tend to be insensitive to localized defects and work hardening localized at stress concentration points. These global modes are also sensitive to structural configuration (mass) changes, wet wing, temperature and loading. The lower frequency modes have been observed to produce only very small frequency shifts due to crack initiation. The best results have been achieved in the lab with "designed in" stress concentration areas so finding the crack initiation sight was not a problem [2]. The approach of looking at frequency shifts continues to be plagued by insensitivity, repeatability, and inability to pinpointing crack problems. These issues have made it difficult to move frequency response measurements out of the lab to field deployment.

## 14.5 New Approach

The new approach investigated in this study focused on higher frequency local modes in general and damping values specifically. Analysis indicated damping of higher order modes would be more sensitive to the localized stiffness increases at the modal nodes and result in decreases in damping. The slopes of the higher order modes at the joints and connection points are much greater than that of the lower frequency modes. Thus the higher order modes would exhibit larger shifts in frequency and damping due to these localized areas of work hardening.

In this study the procedure was to first measure the natural frequency and damping and mode shapes. The investigator was not interested in measuring "nice looking" mode shapes which would require many more measurement points and time. Then a lower frequency mode was selected and excited over a 100 h time frame. The frequency and damping of the higher order modes were measured and logged periodically over the 100 h that the lower order modes were excited. The mode shapes were measured again after the excitation interval and the defects were installed. The expectation was that changes in the mode shapes would help in localizing where to look for damage.

## 14.6 Test Methodology

The fatigue damage study was done on two airframe sections supplied by the Air Force. One test article was a section of a C-141 engine cowling approximately 2 ft. by 2 ft. The cowling was fabricated by machining a billet of aluminum on a skin mill and was curved with machined in rib stiffeners. The other section was a honeycomb laminate composite C-141 control surface trailing edge. The center of mass was located as the driving point. The test sections were mounted to two electro-dynamic shakers and a low order bending mode of the structures were excited. The test set-up is shown in Fig. 14.2. The sections were shaken over 100 h. The frequency responses were measured periodically and the modal frequencies and damping were extracted using a single degree of freedom curve fitter provided in the Agilent 35670A Dynamic Signal Analyzer. The Agilent DSA provided the sinusoidal excitation to one shaker's power amplifiers. The analyzer also measured the modal frequency response functions via impact hammer excitation. The analyzer also provided hard copy plots. Animated mode shapes were extracted before excitation and after the faults were installed to see if changes in the mode shapes would localize the defect sight. The frequency and damping of the modes of interest were extracted along with the modal data from several higher frequency modes [4]. Time constrains required that the number of modal test points for the higher frequency

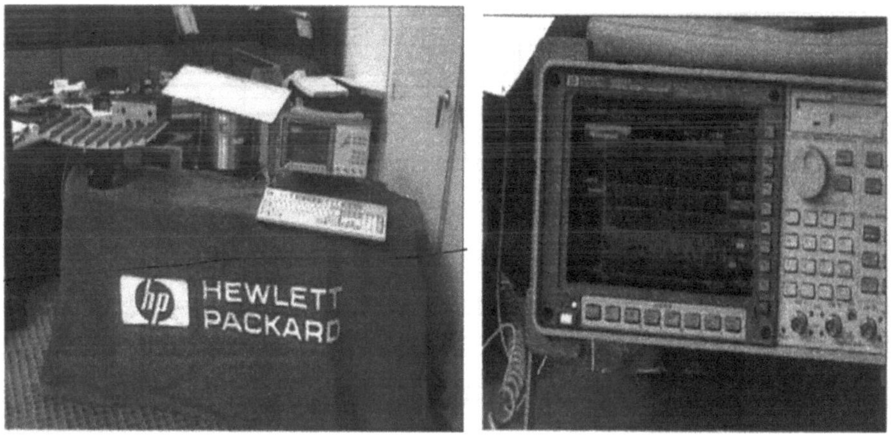


**Fig. 14.2** The test setup for the two airframe assemblies are shown above. The two shakers were driven by their power amplifiers and the HP/Agilent 35670A dynamic signal analyzer and an oscillator

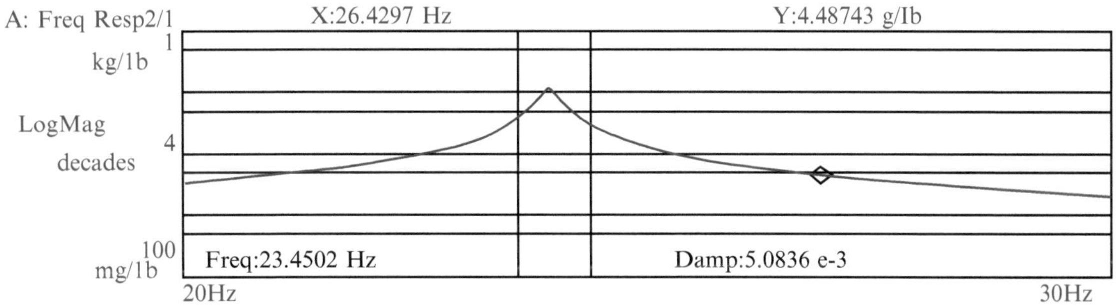


**Fig. 14.3** The plot is the initial frequency response measurement on the C-141 engine cowling section. The frequency and damping values for the 23 Hz mode is shown

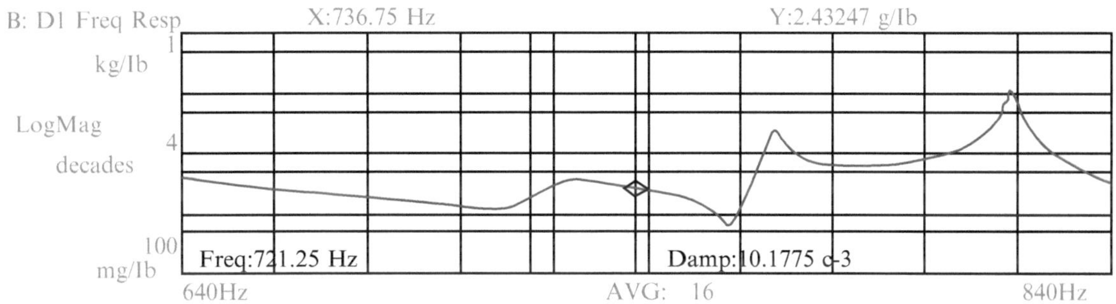


**Fig. 14.4** The plot show the frequency response measurement of the higher frequency modes of the C-141 engine cowling section. The measurement was between the driving point, pt#18, at the center of mass and pt #7

modes be minimized. The point spacing was too sparse to define nice smooth mode shapes. As it turned out more detailed mode shapes was of little consequence. See appendix.

The plot in Fig. 14.3 is the frequency response for the engine cowling. The 23.4 Hz first bending mode is bracketed by the analyzer's markers and a single degree of freedom curve fitter extracted the modal frequency and damping values. The first bending modes damping value was 5.0836 $e^{-3}$. The damping estimate for the 721 Hz mode is 10.177$e^{-3}$ and is shown in Fig. 14.4. Over the 100 h the structure was excited the frequencies and damping values of the modes were logged. In Figs. 14.4 and 14.5 the curve fit values for the three modes tracked. The frequency and damping values were plotted verses excitation time in the plots in Fig. 14.6 below.

We see the damping of each of the four modal frequencies measured on the rib stiffened cowling section. The damping versus hours (cycles) was predicted to decrease as the panel work hardened. The downward slope of the damping values of the 723 Hz mode is clearly visible. The other two higher frequency modes also exhibit negative slope in damping although not as pronounced. Note that the first bending mode at 23 Hz exhibits a scattering of damping values with no apparent slope. The increase in damping for the 723 Hz mode after the "crack" was installed was 44 %. The 767 Hz mode exhibited a 8.8 %

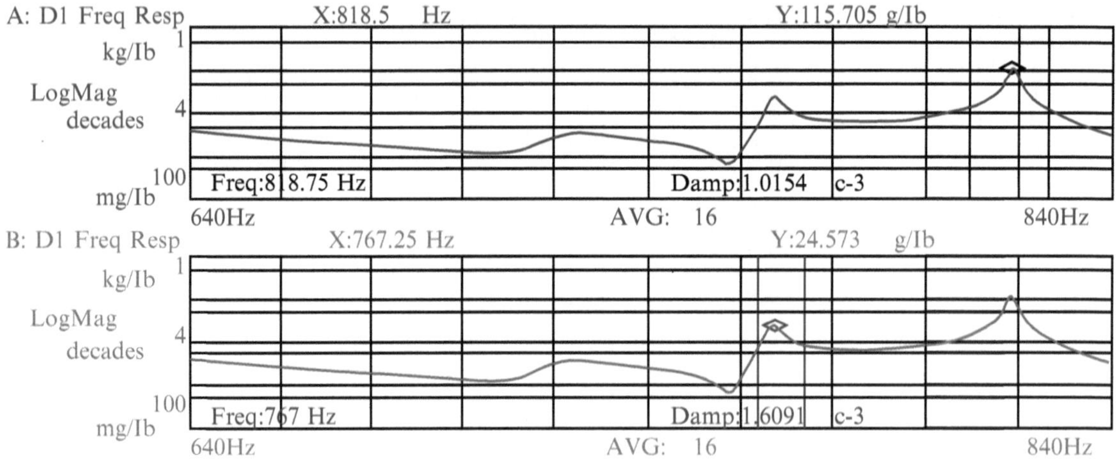


**Fig. 14.5** The frequency response measurement of the cowling used to extract the frequency and damping of the modes after the crack was installed

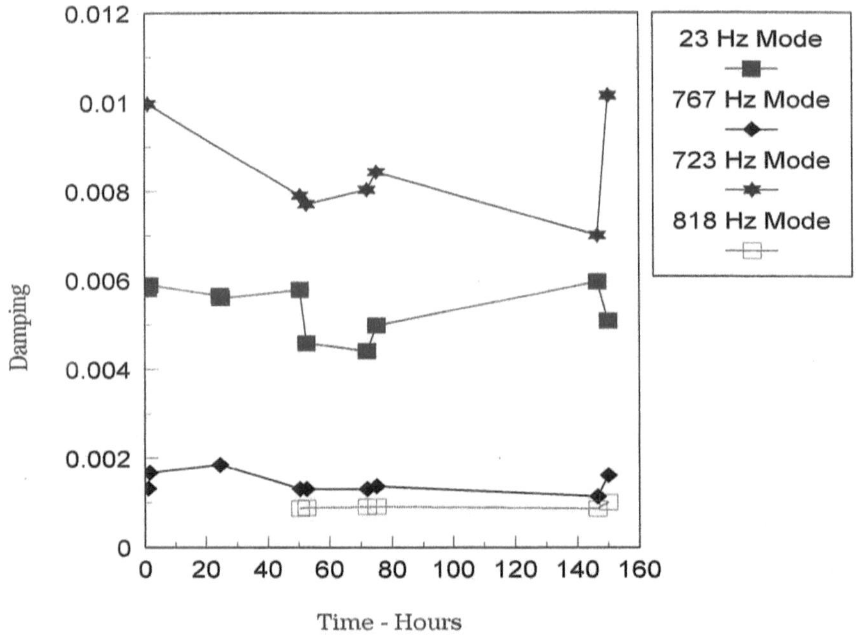


**Fig. 14.6** Damping values for the four cowling modes verses excitation time

increase in damping value after the fault was installed. The predicted damping increase after **crack initiation** is clearly visible as a step increase in damping.

There is a small step increase in the three higher order modal frequencies after the damage is done and a decrease in the modal frequency of the first bending mode Fig. 14.7. Shaking the first bending mode through the crack initiation point would reveal the transitional changes in both frequency and damping of the higher order modes. This transitional data would be the key to detecting crack initiation.

## 14.7 Honeycomb Laminate Control Surface

The **honeycomb laminate composite** control surface section was not expected to behave in the same manner as the all metal cowling section. The C-141 trailing edge control surface was a rectangular shape with a triangular cross section. The structure consisted of an aluminum skin bonded to a honeycomb core. The modal frequencies were widely spread with heaver damping so it was only practical to monitor two modes in one frequency span. Two sets of damping values were extracted

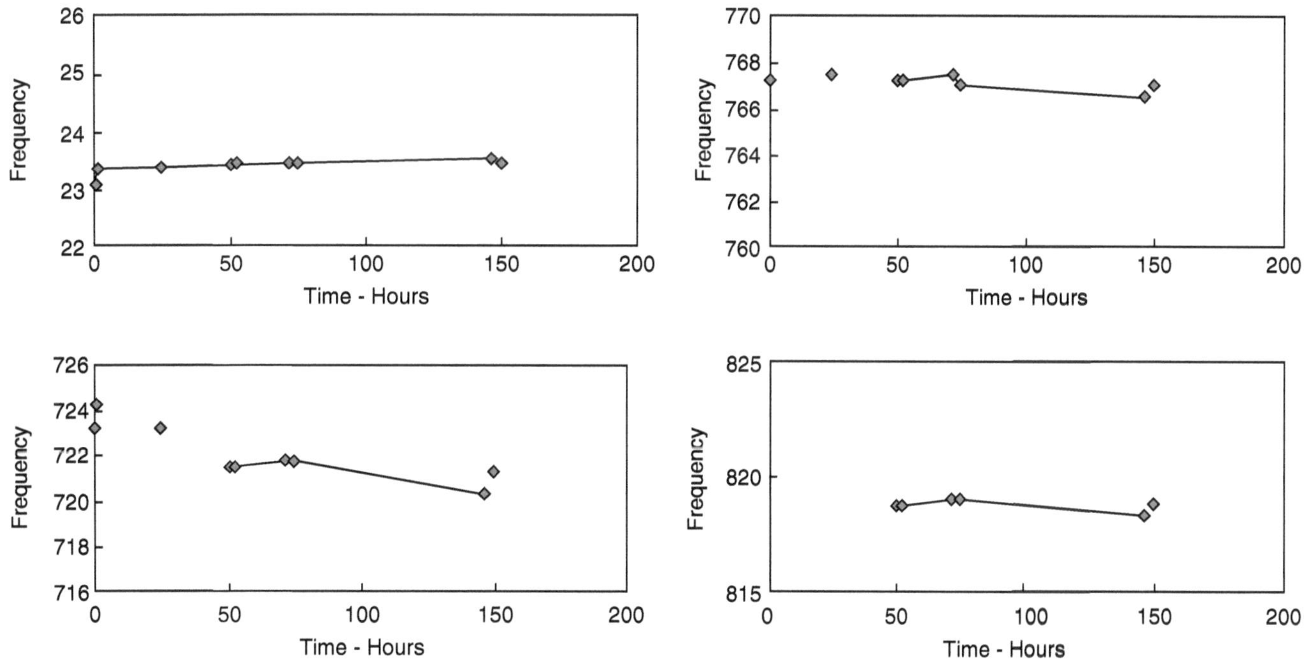


**Fig. 14.7** The four frequencies monitored for the cowling test article are plotted above. Note the slightly downward slope for the 723 Hz mode and small step increase in all three higher frequency modes

**Fig. 14.8** The frequency response measurement of the honeycomb control surface is shown at left. The 760 Hz control surface mode was excited for 100 h

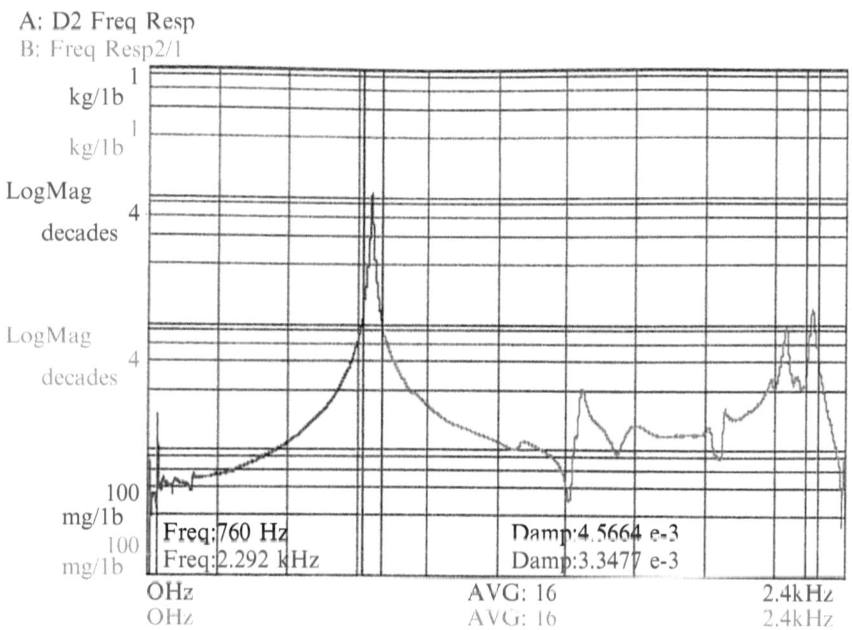


before and after fatigue cycling and damage. The test section was returned to the Air Force lab to install a small separation between the aluminum skin and the honeycomb core to simulate a separation failure.

In Fig. 14.8 the frequency response measurement between the driving point (pt 23) and the response point (42) is shown with the curve fitter on the 760 Hz mode. The 760 Hz mode was excited rather than the first bending mode. The damping value is 4.5664 $e^{-3}$. The second set of markers bracket a mode at 2.292 Hz mode with a damping estimate of 3.3477 $e^{-3}$. Figure 14.9 below is the identical measurement made after a small skin separation was installed. Note the changes in the damping values for both modes and no significant shifts in the frequency of the modes.

The plot of damping versus excitation hours is plotted in Fig. 14.10 for the honeycomb control surface trailing edge. The driven 760 Hz mode's damping increase 2.1 %. This small increase is within the scatter of the other damping values. The

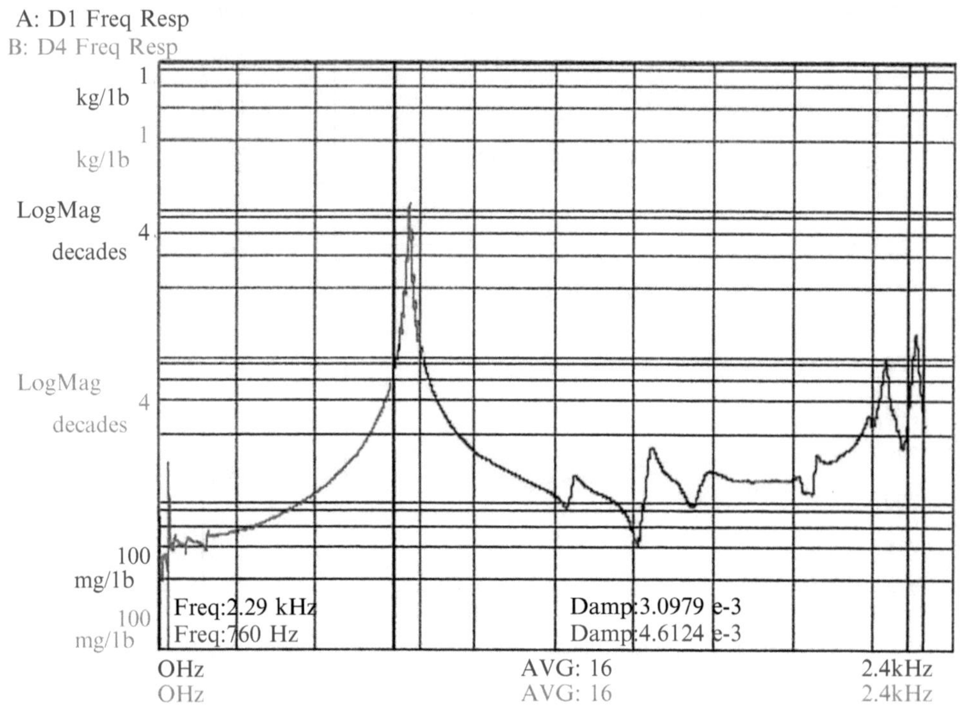


**Fig. 14.9** The plot of the frequency response of the control surface after the small separation was installed. Note the increase in damping values verses Fig. 14.8

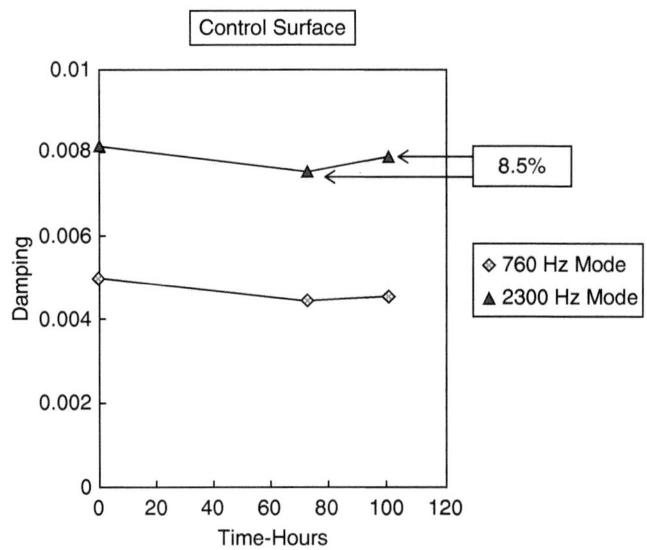


**Fig. 14.10** The modes of the honeycomb laminate control surface exhibited a small step increase in damping

2,290 Hz mode showed an 8.5 % increase in damping after the skin separation. Although the change was small it is a visible step increase from the trend line. A shaker with more force may have produced a larger shift.

Again the higher frequency modes show more sensitive to fatigue life expenditure for the composite section. It can also be observed in Fig. 14.11a that there is step decrease in the frequency trend of the 2,300 Hz mode of only 2 Hz. unlike the all metal cowling section, the honeycomb section exhibited observable changes in modal frequency, as well as in damping, as a function of work hardening and fatigue life expenditure. Again if time allowed it would have been interesting to measure what happens for "real" faults and crack initiation. Would damping values continued down the slope for higher frequency mode and produce a larger step increase after the separation fault? It was also observed that the amplitude of the mode increased for the measurement points nearest the separation. See appendix

In Fig. 14.11b the 760 Hz mode exhibited a straight line decrease in frequency. Where the 2,300 Hz mode shows a 2 Hz step decrease in frequency after the separation fault.

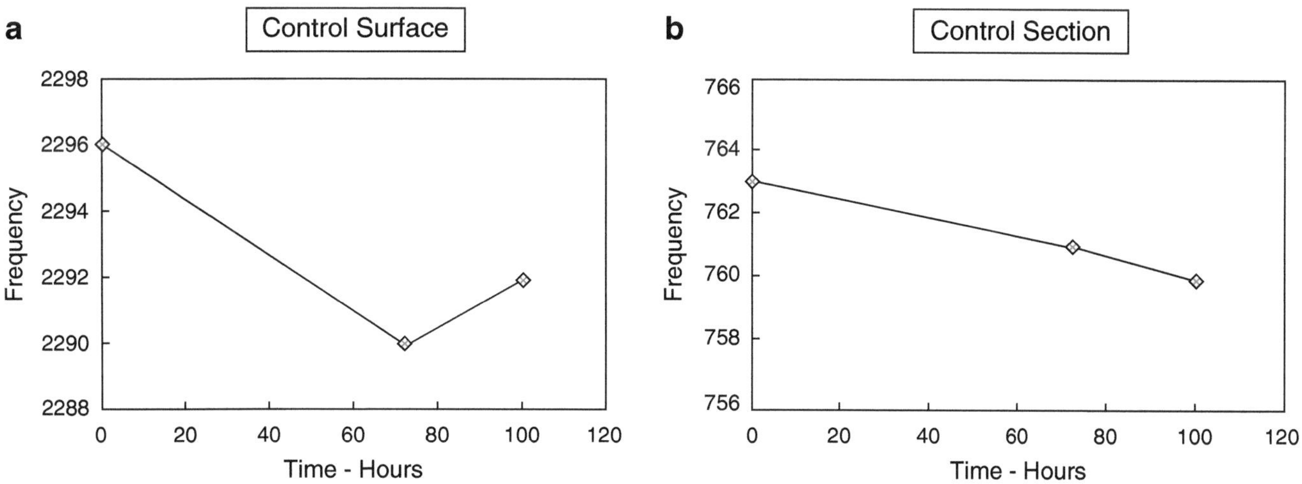


**Fig. 14.11** (**a**) The honeycomb laminate control surface habited small step decrease in the 2,300 Hz mode of 2 Hz (**b**) No step shift in the frequency of the 763 Hz mode is exhibited

## 14.8 Conclusions

The periodic logging of frequency and damping of higher order modes can provide a window into fatigue life expenditure of materials and changes in structural properties. Monitoring damping would benefit a broad range of industries that need an ability to track fatigue life expenditure to predict and prevent catastrophic structural failures. A small field portable data acquisition and measurement system could handle most structures. A Dynamic Signal Analyzer with inputs for an impact hammer/force sensor and accelerometers would be appropriate for small structures and production QA. A field portable hand held system would be suitable for monitoring critical piping systems, bridge I bars (eyebars) and other critical structural members.

**Air Frame Testing**: For larger structure such as airframes an X, Y, Z coordinate positioning gantry system would position the measurement head to make Frequency response measurements around an impact force point. An array of non-contact laser vibrometers mounted on the positioning head would allow automated measurements. **Steam Turbine Generator**: Damage to turbine shafts due to transients such as loss of load or step increases in load can cause crack initiation. Measuring frequency and damping of the shaft's torsional modes can provide a window into the health of the shaft. The frequency and damping can be measured while the turbine generator remains in operation. If a step increase in damping in the torsional modes is seen after a transient event then crack initiation should be suspected. If no step increase in damping is seen then no damage occurred. However the transient event may have used up some fatigue life. Ongoing monitoring of damping values can provide the lead time to head off catastrophic failures and unnecessary down time.

**I-Bars**: Eyebars are non-redundant structural elements that hold up bridges. Eyebars (I-bars) usually consist of several bars stacked side by side and pinned. They are subject to **corrosion** and fatigue cracking. The inner most elements in an eyebar (I-bar) stack are at best very difficult to inspect. Seeing **corrosion** pitting and fatigue cracks on the inner I-bars is very difficult. When I-bar faults are detected their fatigue life can be restored through welding, heat treatment, and shot peening.

**Piping:** Catastrophic piping failures due to fatigue, thermal stress, vibration, hydrogen imbrittlement and **corrosion** can generate headlines and can be very costly in life and collateral damage. The key to preventing catastrophic failures is advanced warning and periodic damping measurements could provide that warning.

**Fiber/resin composites:** Production testing of fiber/resin would allow another level QA over the tap and lessen approach. Frequency and damping values would allow associating numbers to different types of failure mechanisms.

## 14.9 Opportunities for Further Investigation

The time constraints and limited shaker force did not allow a more detailed investigation. The results identified many questions and opportunities for further investigation.

What happens to the damping of test components driven through crack initiation, crack propagation, and failure? More detailed sampling of frequency and damping data in the transition from crack initiation through critical crack length is needed. Does the slope in the damping values verses shake time continue in a straight line slope.

How large of a step decrease in damping occurs at crack initiation?

Can the damping values be related to stress strain and hardiness measurements?

Does the damping data differ for test specimens under high static loads and light dynamic cycling verses no static loading with high levels of dynamic cycling?

What is the damping characteristics of **fiber/resin composites** with built in flaws such as entrapped air and weak inter layer bonding?

How does **corrosion pitting** affect the frequency and damping values?

How do flaws in welds affect the frequency and damping data?

How do missing or loose rivets affect the frequency and damping data?

How does hydrogen imbrittlement affect the frequency and damping?

Can "learning algorithms" be applied to frequency response measurements to detect very subtle flaws not visible to the eye?

## 14.10 Experimental Issues

The first bending mode of the cowling was very lightly damped at 0.5 % and required manual tuning to stay on resonance. Using a shaker/power amplifier that can be locked onto and track the first bending mode would have greatly reduce the time to produce crack initiation. Summing in random noise into the power amplifier would allow continuous measurement and monitoring of the modes. Care must be taken in the measurement of frequency and damping. One needs at least 15 frequency lines in the resonant bandwidth to avoid bias errors in the damping estimates.

## Appendix: Mode Shapes

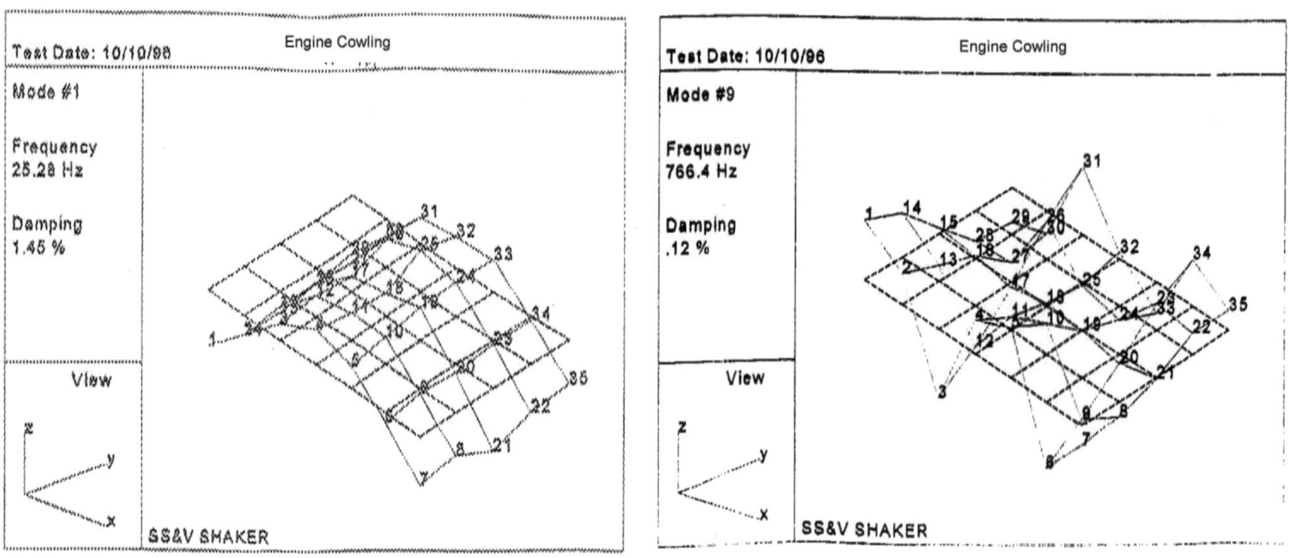


Engine cowling mode shapes for 25 Hz and 766 Hz modes.

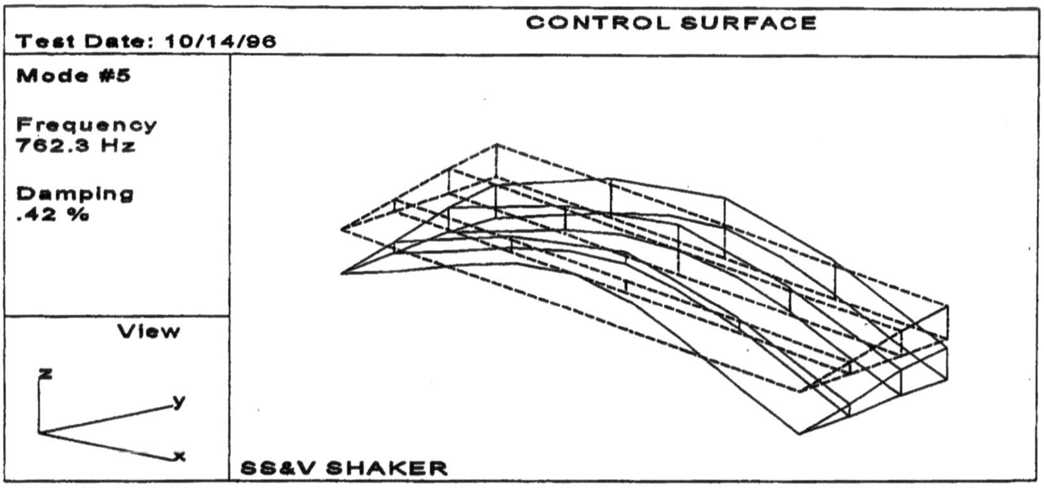


Control surface mode shape.

## References

1. Laser Vibrometer Early Warning Aircraft Damage Detection System. Philip L. Rodgers Principal Investigator, Optical Air Data Systems. John Jensen Hewlett-Packard. Prepared for: Air Force Materiel Command WR-ALC/TIEDM. Robins Air Force Base, GA 31098-1640 Contract: F09650-96-C-0380. Final Technical Report February 1997.
2. Evaluation of Bridge Eyebars Using Dynamic Testing. Principal Investigators: Frank Reed P.E. Senior Materials and Research Engineer Madhwesh Raghavendrachar PhD., P.E. ID Number 638041-33157, June 30, 1994. State of California, Department of Transportation, Division of New Technology, Materials and Research.
3. Application Note 243-3 The Fundamentals of Modal Testing. Keysight Tech/Agilent Technologies
4. Hammer – 3D and Shaker – 3D Seattle Sound and Vibration
5. Better, Safer Power Generation Measurements in Less Time. By John Demcko, Arizona Public Service and John Jensen, Hewlett Packard. Realtime Update 1995–1996

### Publications

- Authored and Co-authored articles and application notes on the use of Fourier and Dynamic Signal Analyzers for physical measurements.
- A partial list follows.
- Servo Mechanism Open Loop Gain Measurements and Stability Analysis
- Measuring Ultra Sonic Wire Bonding
- Power Generation Measurements with Dynamic Signal Analyzers
- Publication Number 5964-6045E
- Rotor Dynamics Measurement Techniques Product Note: HP 35670A-1
- Publication Number 5966-0519E
- Disk Spindle Bearing Runout Measurement Techniques

# Chapter 15
# Characterization and Prognosis of Multirotor Failures

**Joseph M. Brown, Jesse A. Coffey, Dustin Harvey, and Jordan M. Thayer**

**Abstract** Multirotor (MR) unmanned aviation systems are becoming more prevalent in the commercial, philanthropic, and military communities. Because of these public environment applications, hardware malfunctions pose serious safety concerns. Propeller, motor, and structural damage can cause substantial failure of the MR vehicle and endanger surrounding people and structures; thus, early identification and prognosis of these failure modes is necessary to mitigate harm. An embedded structural health monitoring (SHM) system is optimal for identification and diagnosis of these failure modes in time to alter or abort the mission. To achieve autonomous SHM, statistical data must be accrued from a series of sensor measurements. This information is utilized in the development of appropriate damage metrics for failure modes of interest, which determine the real-time state of hardware elements. A comprehensive sensor network was successfully designed and implemented on an MR vehicle to determine which instruments provide valuable information. Utilizing this relevant data, a compatible set of tools was developed for signal processing, and the resulting SHM system is capable of classifying propeller, motor, and structural hardware failures.

**Keywords** Unmanned Aerial Vehicle • UAV • Multirotor Vehicle • Damage Prognosis • Structural Health Monitoring

## 15.1 Introduction

### *15.1.1 Motivation*

Multirotor (MR) Unmanned Aerial Vehicles (UAVs) are becoming increasingly prevalent in modern society, as they have a wide array of applications in a variety of industries. The military uses UAVs to save soldiers' lives in otherwise hazardous situations [1]. Commercial corporations such as Amazon and UPS are researching the use of UAVs for speedy delivery of packages [2]. Philanthropic and medical groups have investigated UAVs extensively in order to quickly deliver medical supplies to areas with limited road access [3]. Some predict that, in the future, UAVs could be used for common transportation in major cities. Currently, the United States Federal Aviation Administration (FAA) has restrictions on the types of vehicles permitted to fly; however, changes in favor of unmanned aerial vehicle use are anticipated [4]. Because commercial use of UAVs is likely, steps must be taken to ensure their safe operation.

Unmanned aircrafts used in uncontrolled areas pose a risk to the general public. As devices with several complex moving components, UAVs are susceptible to a variety of mechanical failures. For example, high-speed motors and propellers induce vibrations that weaken the airframe or fatigue joints over time. The lithium-polymer batteries used to power the aircraft pose a fire risk if the vehicle crashes, especially in a dry area. Failure of an aircraft's navigation system or sensors could misdirect the flight of a UAV and result in component failures leading to a crash [5]. In order to address these safety concerns, operators

J.M. Brown
Department of Mechanical Engineering, Worcester Polytechnic Institute, Worcester, MA 01609, USA

J.A. Coffey III
Department of Aeronautical and Astronautical Engineering, Stanford University, Stanford, CA 94305, USA

D. Harvey (✉)
Department of Structural Engineering, University of California San Diego, San Diego, CA 92093, USA
e-mail: harveydy@lanl.gov

J.M. Thayer
Department of Mechanical Engineering, University of Southern California, Los Angeles, CA 90007, USA

© The Society for Experimental Mechanics, Inc. 2015

C. Niezrecki (ed.), *Structural Health Monitoring and Damage Detection, Volume 7*, Conference Proceedings of the Society for Experimental Mechanics Series, DOI 10.1007/978-3-319-15230-1_15

must be aware of the causes of failure and how to mitigate them. Ideally, UAVs will be equipped with an onboard monitoring system that autonomously assesses damage as it occurs and adjust its flight accordingly.

### 15.1.2 *Background*

Autonomous monitoring of MR UAV failures requires utilization of Structural Health Monitoring (SHM) and Damage Prognosis (DP) principles. SHM is the detection and characterization of damage through statistical pattern recognition. Coupled with operational and environmental data from past and current states, known as usage monitoring, SHM can be expanded to include expected future states of operation and environmental data. Such predictions allow for the assessment of a component's remaining useful life and failure intensity, which is referred to as damage prognosis. Most damage prognosis is dependent on data-based models due to its probabilistic nature [6].

To achieve successful damage prognosis, some basic SHM concepts are important for design considerations. The inherent presence of natural defects in all materials requires the development of a range of acceptable failure rather than an absolute threshold for damage. Analysis of damage can be performed in either supervised or unsupervised learning modes. The former refers to damage diagnosis performed with knowledge of damaged and undamaged states, and can achieve continuous classification of failure intensity through regression analysis. Unsupervised learning only uses information from the undamaged state; the resulting system is exclusively capable of detecting the presence of damage through outlier detection or other such metrics [7]. Because of their low cost, simple reparability, and relatively short times to failure, MR UAVs provide an opportunity to perform SHM in the supervised mode. As a result, complete damage prognosis of these vehicles is plausible.

SHM is considered a prerequisite to complete damage prognosis, so it has been the main focus for analysis of dynamic systems to date. With the exception of a few isolated applications, including rotating machinery and seismic probabilistic risk assessment, the challenges associated with successful damage detection and characterization have prevented progress in full prognosis [6]. As defined above, damage prognosis depends on predictive models of future loading and environmental conditions, which are not readily available for many dynamic systems. For single rotor machinery though, supervised learning has allowed extensive data sets of damage types, locations, and conditions to be accumulated.

The helicopter industry has demonstrated great success in the area of single rotor prognosis. Vibration analysis has been utilized to predict maintenance requirements and increase component life by 15 % in some cases [6, 8]. Health and usage monitoring systems (HUMS) have been added to gearbox and main rotor components to increase performance and lifetime. HUMS have received such notable successes that they are endorsed by the Federal Aviation Administration and are commercially used to prevent failures in large rotorcraft. Success with helicopter damage prognosis can be attributed to a few key flight conditions: rotor speed is "maintained typically within 2 % of nominal for all flight regimes" and a single load path exists without redundancy [6]. A combination of these characteristics creates steady vibrations in the rotor during normal conditions, which are easily distinguishable from behavior during failure; thus, prognosis of helicopters is rather simplified [6]. Unfortunately, MR vehicles do not operate under the same flight conditions, as multiple load paths exist and variable speeds are required to maintain stability. To exacerbate the problem, few data records on failure types and locations currently exist for multirotor systems. In the case of MR vehicle flight prognosis, a more advanced data-based predictive model must be developed.

In part because of the above challenges associated with MR vehicle prognosis, very little research has been conducted about structural health monitoring of multirotor systems. A qualitative study of multirotor UAV failures by Olson and Atkins yielded significant conclusions. Failure modes were identified throughout the vehicle, such as problems with the navigation program, ground communication, propellers, motors, and mounting of components. [5]. An understanding of these qualitative failures is invaluable in developing a robust damage prognosis system for MR vehicles. Previous work with usage monitoring, SHM, and prognosis in single rotor vehicles serves as a starting point for the associated quantitative analysis of MR failure modes. Especially with the increase of MR UAV applications, such a damage prognosis sensor network is needed to mitigate safety concerns.

### 15.1.3 *Project Scope*

Intimate knowledge of the structural health of an MR vehicle is necessary for safe and effective mission planning. An MR vehicle is an intricate system that depends on hardware and software components working in tandem, and vehicle failure can be attributed to either element. Ideally, an onboard damage prognosis system will analyze the current state of an MR vehicle's components, assess the potential for failure, and predict the remaining useful life. Most research in this field has

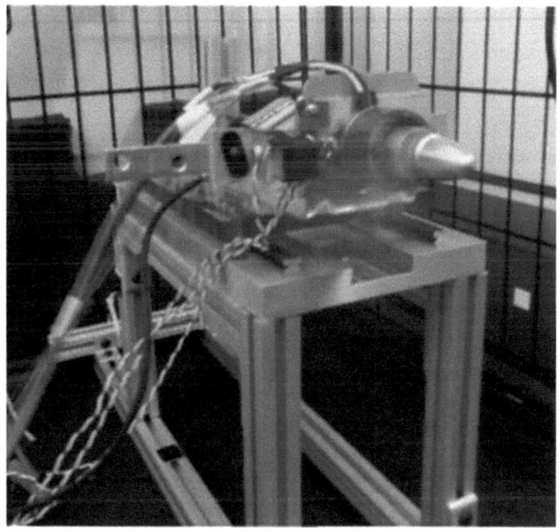

**Fig. 15.1** Single propeller structure constructed for preliminary tests and controlled component testing

been focused on structural health monitoring. Working toward the goal of damage prognosis, common failure modes of UAV hardware were characterized and quantitative indications of damage were identified. For simplicity and because hardware failures are of particular interest, software issues were neglected. The trends discovered can be applied to future work in full damage prognosis of robust, high-quality multirotor UAVs. The following sections describe the testing platforms, sensors and procedures used, and present an analysis of the obtained results.

## 15.2 Instrumentation and Experimentation

### *15.2.1 Testing Platforms*

Two testing platforms were developed to fully characterize the behavior of healthy and damaged MR UAV hardware. Structural health and damage prognosis of single rotor systems have been studied extensively, and their behavior is generally understood. A single propeller structure (shown in Fig. 15.1) was used to test the sensors and intended experimental procedures, determine qualitative trends of damage data, and provide a controlled environment to test the components of the MR system individually. Duplicate components on the single and multirotor platforms allow for comparative analysis. The structure was constructed from aluminum framing to house a three-phase brushless DC motor with an attached propeller. An electronic speed controller (ESC) provides the motor signal given a pulse width modulation (PWM) input signal, and a lithium polymer battery supplies the power. Motor speed is determined by the pulse width of the PWM signal rather than the frequency. A high quality motor and ESC, rated at 400 W and 100A respectively, were used for a robust, high performance testing apparatus.

The same hardware was applied to a quad-rotor testing apparatus. The motor, propeller, and ESC configuration was repeated on each of the four arms of the quad-copter, and a single battery was utilized for power. A central control hub was implemented on the quad-rotor with infrared proximity sensor, radio transmission, and an autopilot—all linked to commercial Mission Planner software. A handheld remote controller was also used to operate the vehicle manually. The autopilot utilizes its internal gyroscope to autonomously adjust the rotational speed of each motor and achieve stability control. A depiction of the vehicle is shown in Fig. 15.2; this apparatus is designed to be substantially similar to commercial unmanned aerial vehicles.

## 15.3 Measureable Failure Modes

Although sensors cannot directly detect damage, the presence of damage can be measured from abnormalities in parameters such as acceleration, power consumption, impedance, and temperature. This diagnosis requires a database of these parameters in healthy conditions to recognize damage of motors, propellers, and ESCs with confidence. Redundant measurements were desired in most cases to develop a more robust sensor network. All measurements were acquired using National Instruments Data Acquisition Hardware at a sampling rate of 10,000 Hz, and pre-processed through LabView. Table 15.1 summarizes the component failures and their measureable parameters, indicated damage, and sensors utilized at the end of the section.

**Fig. 15.2** Quad-rotor testing apparatus

**Table 15.1** Summary of failure modes, measurable parameters and sensors

| Component | Failure mode | Measureable parameter | Sensor |
|---|---|---|---|
| Motor | Motor damage, unbalanced structure | Angular velocity | Infrared phototransistor/LED (IR PT) |
| | Communication, unbalanced structure | Control signal (input to ESC) | Voltmeter |
| Propeller | Vibrations in system | Acceleration | PCB Piezotronics 3-axis accelerometer |
| | Broken or fractured propellers | Lift | S-beam load cell |
| | Unbalanced structure from damaged propellers | Angular velocity | Infrared phototransistor/LED (IR PT) |
| Electronic speed controller | ESC/Motor failure or overuse | Input current | Hall effect sensor |
| | ESC overheating | Temperature | Type K thermocouple |
| Battery | Weak/dying battery | Voltage | Voltmeter |

### *15.3.1 Motor*

If a motor is damaged from overuse or a collision, it typically spins significantly slower than usual; in the case of the quad-rotor, the remaining motors compensate with faster rotations. Continuous measurement of motor speed allows for observation of this phenomenon. Further, acquiring the motor control signal data from the onboard autopilot ensures that such a change in the motors is due to hardware failure rather than a communication or control error.

An infrared phototransistor and LED were used in combination to create a tachometer for motor RPM measurements. Unlike one of its competitive counterparts, the Inductive Proximity Sensor (IPS), the LED and phototransistor combination does not require precise positioning and close proximity (<1 mm) sensing. Instead, unobtrusive reflective tape on the motor housing allows a phototransistor positioned next to an infrared LED to register each rotation in the time domain, as shown in Fig. 15.3. The generated output signal is a pulse for each rotation, from which the RPMs of the motor are calculated. Motor control measurements were taken between the autopilot and each ESC in order to monitor the PWM signal sent to each motor, as shown in Fig. 15.4.

### *15.3.2 Propeller*

Propeller damage on UAVs is often caused by impact collisions, and creates a structural imbalance of the vehicle that can be detected with various metrics. Dramatic changes in acceleration of the structure indicate excess vibrations in the system, which are generally induced by propeller failures. Also, the lift of the vehicle decreases if portions of the propeller are sheared. Three-axis accelerometers were used to measure vibrations in the structure at each motor mount. For both testing platforms, in-plane and axial acceleration with respect to the motor and propeller system were measured. An in-line S-beam load cell was utilized in the motor mount of the single propeller structure to measure lift forces; however, this sensor was not implemented on the quad-rotor as it would have inhibited the movement of the aircraft.

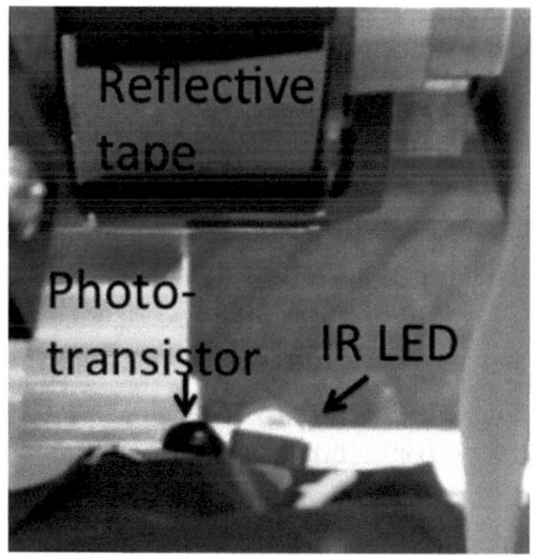


**Fig. 15.3** Infrared LED and phototransistor used as tachometer

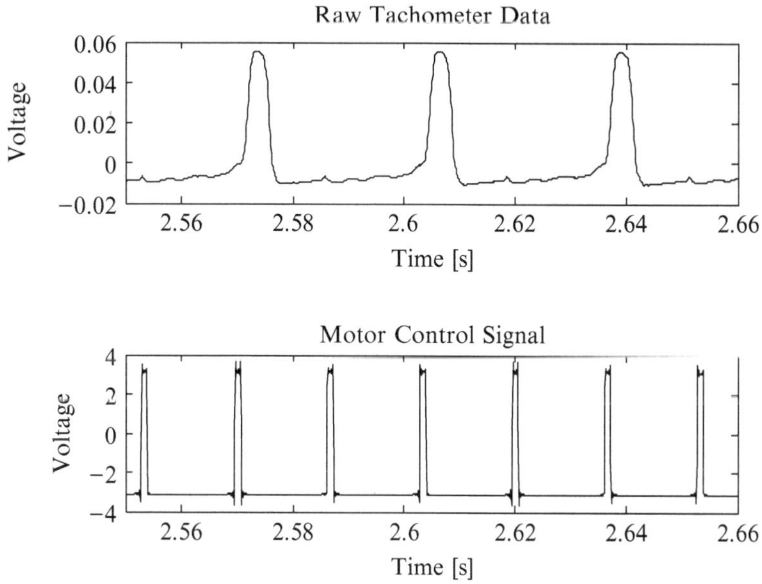


**Fig. 15.4** Sample phototransistor data with corresponding motor control signal

### *15.3.3 Electronic Speed Controller (ESC)*

Electronic Speed Controllers (ESCs) draw current from the battery and control the signal conversion from PWM to the appropriate three-phase motor signal. The duty cycle of the PWM input controls the three-phase motor output. Because ESCs are susceptible to overheating, measurements of their internal temperature can indicate such potential failures. A type K thermocouple was wedged into each ESC and held in place by electrical tape. Hall effect sensors were also installed between the battery and each ESC to measure current draw in order to determine whether changes in current could indicate damage to either the ESC or the motor.

### *15.3.4 Battery*

In some cases, battery failures are incorrectly categorized as motor, propeller, or ESC damage, as low battery can cause similar response in current, lift, and speed measurements. To eliminate uncertainty in the cause of failure, the battery output voltage was measured and recorded.

## 15.4 Testing Procedure

### 15.4.1 Single Rotor Testing Procedure

To enable supervised learning, data from both healthy and damaged components was collected. The above instrumentation was applied to the single rotor structure as illustrated in Fig. 15.5, and a general testing procedure was developed.

A single trial consisted of 10, 5-s measurements at each of five different PWM duty cycles, which control the motor speed. These five duty cycles were kept constant within each trial, and were evenly spaced in ascending and descending order. For each measurement period, data from the sensor network was acquired and recorded. This trial process was repeated five times for each test. In total, the single propeller apparatus was tested with five different motors, three ESCs, and eight $35.56 \times 13.97$ cm propellers. Two of these propellers were healthy, and six were in various states of controlled damage. Propeller condition was classified as either symmetric or asymmetric with heavy or light damage. Figure 15.6 displays the varying damage states of the six propellers compared to one of the healthy propellers.

Propeller 4 is an example of symmetric heavy damage, while propeller 7 is classified as asymmetric light damage. The three ESCs were tested with no induced damage. A range of motor conditions was used during experimentation: three of the motors were new and undamaged, one was operated extensively but remained undamaged, and the final motor had confirmed damage but was operable. For each motor, all three ESCs were tested with two healthy propellers. In addition, a single ESC was used to test each of the six damaged propellers. In total, 55 tests were performed. As a general procedure, each test was run until the five trials had elapsed, or until the motor stopped unexpectedly.

To observe failure during operation and establish a greater understanding of component behavior before, during, and after damage, the system was run at a constant speed until motor or ESC failure was induced. Impact damage to a propeller in

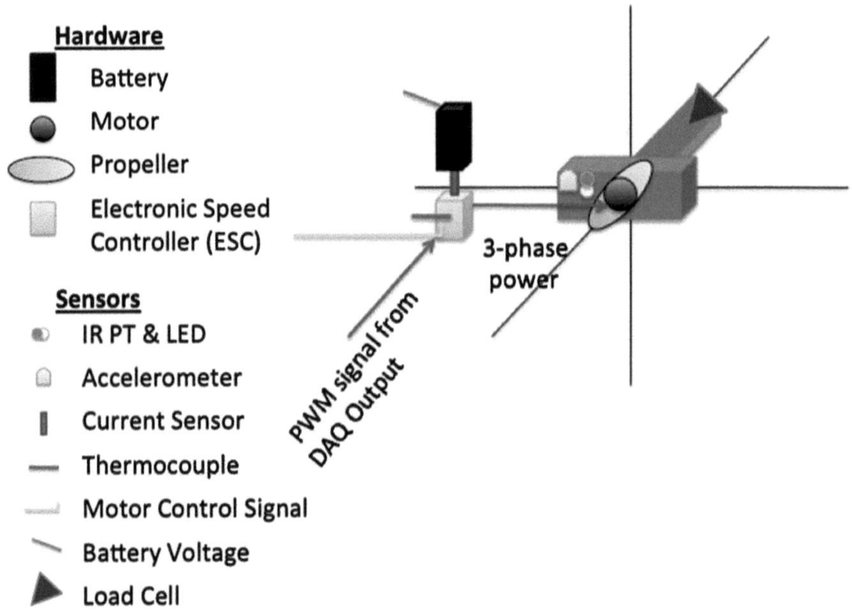


**Fig. 15.5** Diagram of single rotor structure sensor placement

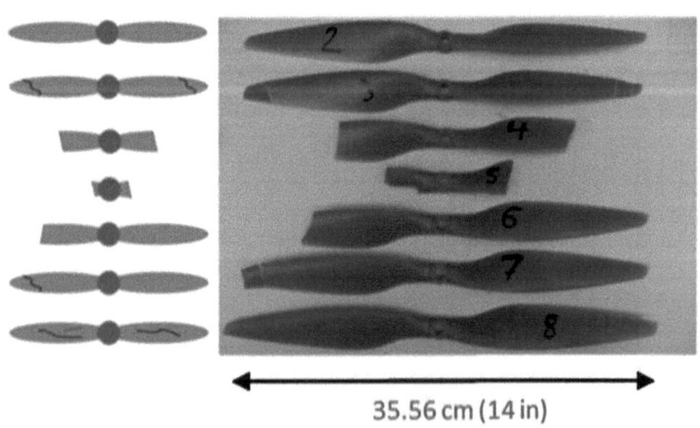


**Fig. 15.6** Range of damaged propellers compared to a healthy propeller

Fig. 15.7 Impact simulation the single rotor by dropping a weight on the propeller while spinning

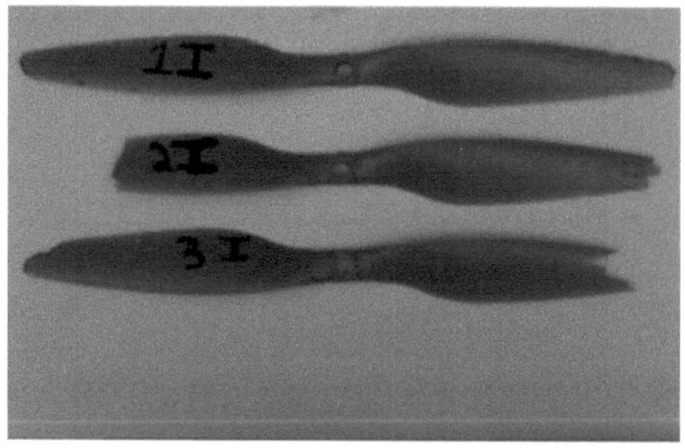

Fig. 15.8 Broken propellers due to impact simulation

flight was also simulated in three separate trials by dropping a weight onto a healthy propeller as it was spinning at a constant speed, as shown in Fig. 15.7. In two cases, the motor stalled as a result of the impact. Figure 15.8 shows how each of the three propellers sustained substantial damage as a result of this test.

Throughout failure testing on the single-rotor apparatus, three of the five motors experienced catastrophic failure from overuse, but none of the ESCs reached such terminal damage. Extensive logs of healthy and damaged behavior were produced. Such information was invaluable in determining which sensors are most useful for each damage state and what type of sensitivity they provide. An understanding of this behavior proved necessary for analysis of the more complex quad-rotor system.

### 15.4.2 *Multirotor Testing Procedure*

As with the single propeller structure, the quad-rotor was instrumented with the selected sensors; a diagram of sensor placement on the quad-rotor is shown in Fig. 15.9. A phototransistor, accelerometer, Hall effect sensor, motor control measurement, and thermocouple was mounted on each of the four arms of the aircraft, which totaled 22 sensors on the vehicle, including one battery voltage measurement and an altitude measurement.

Much like the single propeller test procedure, the MR vehicle was first tested with four healthy motors, ESCs, and propellers. The apparatus was elevated to a specified height and autonomously maintained at a constant altitude during each testing period of 1 min. After recording each sensor's data from the healthy vehicle, damaged component behavior was observed. The ESC, motor, and propeller were each systematically replaced with damaged components in separate trials to simulate various states of flight damage to the vehicle. As above, five damaged propellers, varying conditions of motors, and different ESCs were utilized; the damage cases, including asymmetric and symmetric cases, are shown in Fig. 15.10. Loosening of bolts on the motor housings was observed during testing of the quad-rotor system as well. To determine damage metrics for this structural damage, bolts were intentionally loosened prior to testing.

The quad-rotor system was then tested to achieve continuous classification of failure. Beginning with a healthy structure, hovering of the quad-rotor was observed until catastrophic failure was reached. Analysis of this sort allowed for the development of metrics for a range of damaged states, including healthy, noticeable failure without hover performance loss, and catastrophic failure of the vehicle. In most cases, particularly because of the high performance motors used, battery drain was observed before failure of the components.

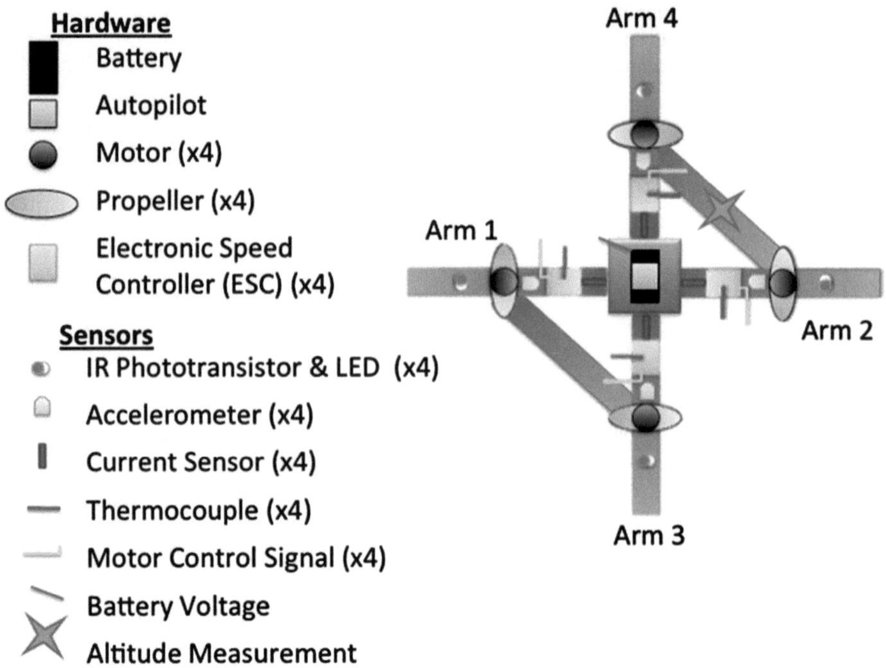


**Fig. 15.9** Illustration of sensor type and placement on the quad-rotor

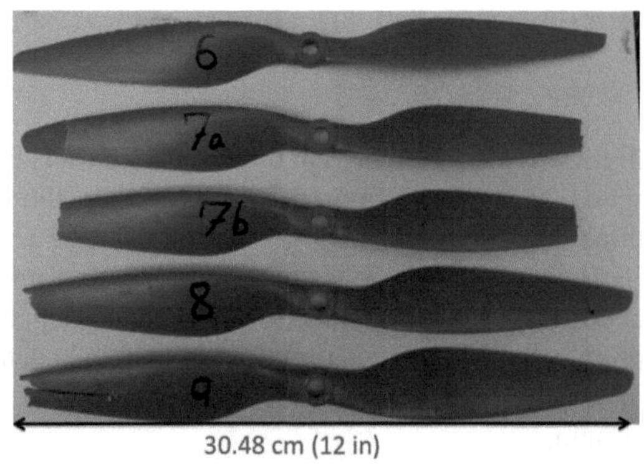


**Fig. 15.10** Range of damaged propellers for the quad-rotor damage testing

### *15.4.3 Data Processing*

Raw signals were pre-processed to extract motor angular speed from the phototransistor data, temperature from the thermocouple, force from the load cell voltage, current from the Hall effect sensors, and battery power from the voltage and current measurements. Efficiency of the system was determined by the amount of lift force generated per unit of power drawn. A sample set of this converted data from one trial of the single rotor structure with healthy components is shown in Fig. 15.11. Data was plotted similarly for each arm of the quad-rotor. Waterfall plots like the one in Fig. 15.11 were valuable in determining interactions between different measurements and which metrics were valuable for classification.

Accelerometer data was processed and analyzed separately because of its pre-processing requirements. Due to differences in motor speeds on the single rotor and the erratic nature of the quad-rotor stability control, the accelerometer data was normalized by motor speed to compare across different test cases and different arms of the quad-rotor platform. The accelerometer raw data was segmented by the synchronous signal from the tachometer and resampled from the time domain to the motor's rotational position [9].

## 15.5 Results

Given the time series behavior of the single rotor and quad-rotor systems, a few measurements were selected for further analysis. In addition to developing successful damage classification methods, the comparison of single rotor and quad-rotor

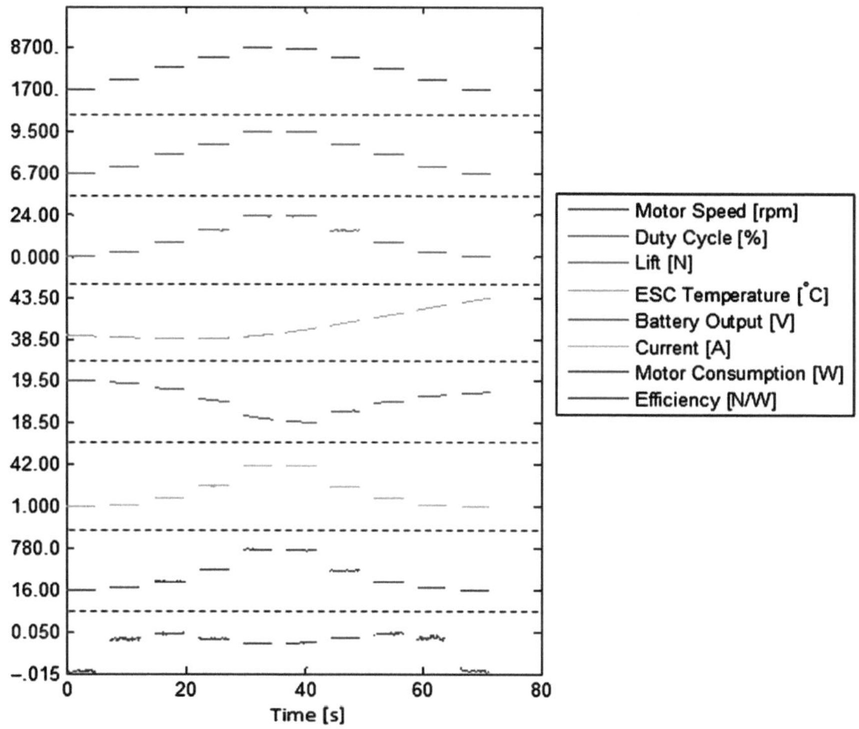


**Fig. 15.11** Sample data set for single rotor structure displayed in a waterfall plot

failure identification was of particular interest. Some of the single rotor damage metrics were directly transferrable to the quad-rotor apparatus. However, because of complexities in the MR vehicle, the remaining metrics were determined to require modifications before their application.

## *15.5.1 Lift/Load Cell*

### 15.5.1.1 Single Rotor

As mentioned in Sect. 15.3.2, lift was only measured on the single rotor apparatus due to structural and instrumental constraints on the quad-rotor. This data proved useful in determining heavy propeller damage. As the graph in Fig. 15.12 shows, the lift behaves as expected and significantly decreases at a given motor speed for heavily damaged propellers. In order to utilize this trend for damage prognosis on a multirotor vehicle and determine the location and extent of heavy propeller damage, smaller load cells could be placed underneath each motor. However, the results may vary from that of a single rotor due to the interactions between motors on a multirotor vehicle.

## *15.5.2 Temperature/Thermocouple*

The intended purpose of the thermocouples was to measure the temperature of the ESCs and classify their damage accordingly. Unlike their predicted behavior, the ESCs on the single rotor system did not generally experience overheating to the point of failure; thus, no significant correlation between ESC temperature and failure was observed. However, by utilizing the waterfall plots to determine noteworthy interactions between components, an increased ESC heating rate was observed in trials with eventual motor damage. To test the significance of this observation, the slope of temperature increase was computed for each 5 s interval of testing. A kernel density estimator was applied to these heating rates for both the healthy and damaged motor cases to produce their probability density functions. A total of 200 damaged and 1,400 healthy rates were utilized in the analysis. A comparison of these probability densities is shown in Fig. 15.13.

In the motor damaged case, the peak value of the distribution is shifted by approximately 0.1 °C/s from the healthy data. More importantly, the positive tail end of the damaged case covers a larger area than that of the healthy tests. Thus, if the

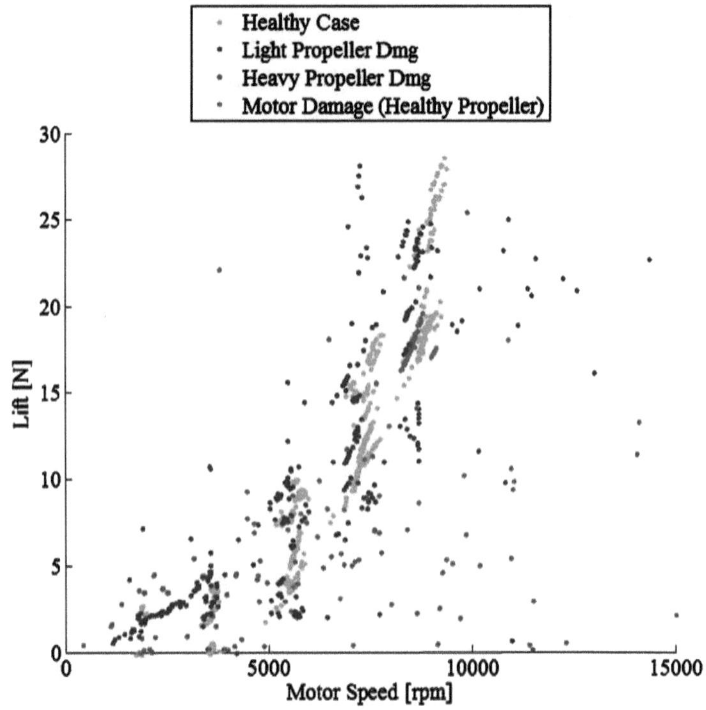


**Fig. 15.12** Lift of the single rotor plotted against motor speed

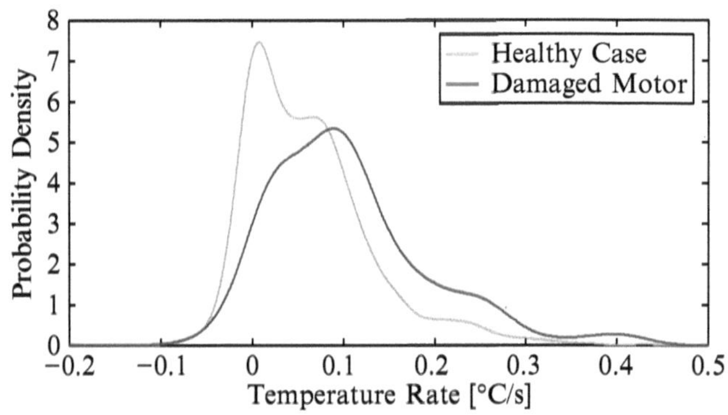


**Fig. 15.13** Probability density functions of healthy and damaged motor temperature rates

slope of temperature increase is measured to be larger than 0.1 °C/s, it is more likely to be damaged than healthy. While these are valuable damage trends for the single propeller structure, temperature measurements at the ESCs do not produce a distinct threshold for motor damage. Direct measurement of the motor temperature, using a different mounting procedure for the thermocouples, is preferred to improve classification and extend the metric for use on the MR vehicle. The quad-rotor platform also offers the opportunity for comparative temperature analysis between motors at any given time. For healthy motors of similar angular speed, temperature should be approximately the same. According to qualitative observations during testing, motor damage may be more likely if one motor's temperature varies significantly.

## *15.5.3 Current or Power/Hall Effect Sensor*

### 15.5.3.1 Single Rotor

For the single rotor platform, the duty cycle of the ESC control signal was set for each trial. As a result, the power draw of the motor was dictated by the duty cycle and was relatively constant for each test. It was found that change in motor torque, due to either aerodynamic drag or mechanical resistance, is exhibited by motor speed rather than the power drawn. In Fig. 15.14, electrical power draw of the motor is compared to the resulting motor speed for the single rotor platform.

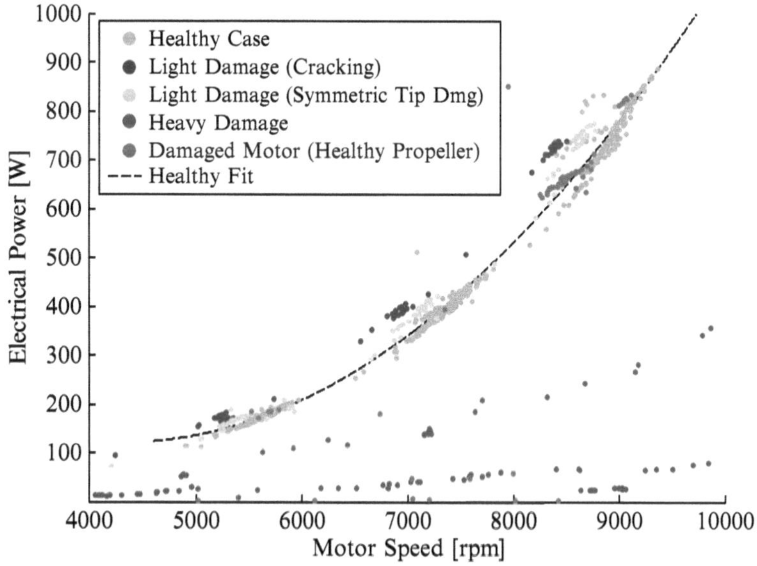


**Fig. 15.14** Electrical power draw of the motor plotted against motor speed

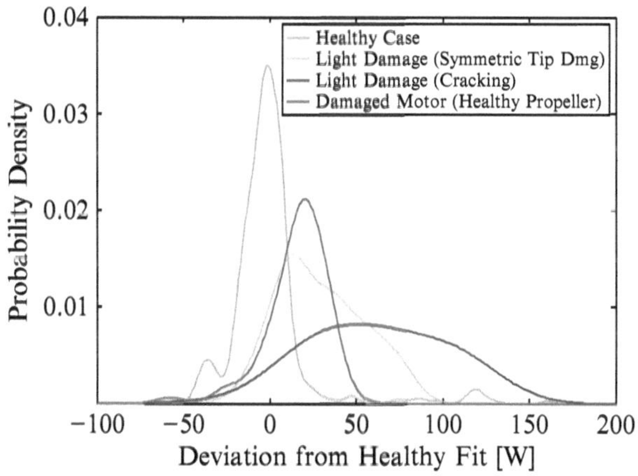


**Fig. 15.15** Distribution of the deviation of each damaged case from the healthy fit, generated by a kernel density estimator

The healthy case shows a trend line that is distinctly different from the damaged cases and is related to the second order drag relation for velocity ($v$) shown in the equation below.

$$F_D = \frac{1}{2} C v^2$$

Where $F_D$ is force due to drag and $C$ is a constant that depends on the geometry and material of the airfoil, as well as the density of air. Because heavily damaged propellers are characterized by full shearing of blades, they provide little aerodynamic drag or motor torque and can be easily distinguished from healthy cases. For lightly damaged cases, the flow over the airfoil is disrupted and causes slightly increased drag. The reduction in motor speed from increased torque causes a trend line that is offset from the healthy case. In order to quantify this offset, a second order polynomial was fit to the healthy data in accordance with the drag relation. The distribution of the deviation from the healthy fit, generated by a kernel density estimator, is shown in Fig. 15.15.

Both symmetric light propeller damage cases, which are undetectable with any other method, are able to be distinguished by their offset from the healthy distribution. Motor damage also shows a distinct offset, most likely due to increased internal mechanical resistance.

#### 15.5.3.2 Quad-Rotor

On the MR test platform, the autopilot has a feedback control system through the internal gyroscope. The power draw is dictated by this control system to produce the required lift from each motor and achieve stable flight. The data collected on

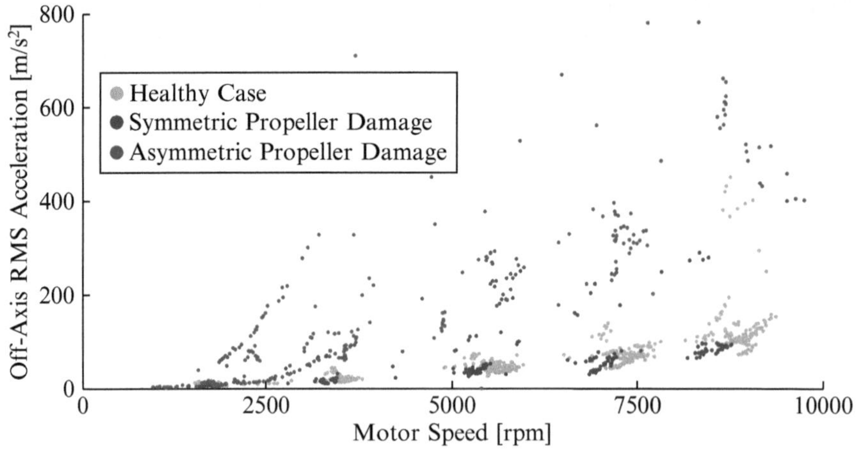


**Fig. 15.16** Comparison of damaged and healthy propeller cases for RMS acceleration versus motor speed

the quad-rotor platform does not show as defined of a trend line, most likely due to inherently erratic nature of the motor speed from the stability control and insufficient battery output to meet the demand of the motors. While detection of damage by this method was not possible on the quad-rotor test platform, further investigation is warranted with a well-designed power system and more precise stability control.

## *15.5.4 Vibrations/Accelerometer*

Analysis of vibrations in the single rotor and quad-rotor platforms proved to be the most valuable metric for damage. Asymmetric propeller damage and structural failures were detectable with amplitude and frequency analysis. In all cases, only the in-plane acceleration was useful; thus, replacement of the three-axis accelerometers with single direction, in-plane accelerometers is recommended for sensor reduction.

### 15.5.4.1 Magnitude of Vibrations

According to S. Braun, the Root-Mean-Square (RMS) acceleration of a rotating structure is valuable in detecting asymmetry [11]. Analysis was performed for the single rotor structure to determine if damage detection is possible with RMS acceleration. A plot of off-axis RMS acceleration versus angular speed of the motor is shown in Fig. 15.16 for the single propeller apparatus. Each data point corresponds to the RMS acceleration computed for each 5 s increment of testing.

As expected, the RMS acceleration increases as a function of motor angular speed, and exhibits a larger positive slope for asymmetrically damaged propellers. While the symmetric propeller damage cases are generally indistinguishable from the healthy data, a clear distinction between asymmetric damaged and healthy cases exists. Given the separation of asymmetry from healthy cases, a linear threshold could be developed depending on the desired cost function or percentage of false positives or negatives.

Because of its successes in damage classification on the single rotor system, RMS acceleration was also analyzed on the quad-rotor apparatus. A plot of off-axis RMS acceleration versus motor angular speed is shown in Fig. 15.17 for arm 3 of the quad-rotor platform. A range of damage cases, for both symmetric and asymmetric cases, are shown. Much like the single propeller trends, a clear distinction between the asymmetric propeller damage and healthy or symmetric propeller damage exists. Again, a linear threshold for damage classification could be developed according to operational specifications. A similar analysis of arm 1 on the quad-rotor was performed; a comparison of the damage levels for arm 1 is also shown in Fig. 15.17.

Unlike the previous cases of RMS measurements, damaged and healthy cases are not distinctly different, as a large amount of healthy data is mixed with the damaged cases. This phenomenon occurs particularly at motor speeds above 9,000 rpm of Fig. 15.17, which correspond to the same test data of a single continuous flight. To determine the reason for such misclassification in this particular test, a comparison of each of the arms of the quad-rotor for off-axis RMS acceleration versus time was produced, and the result is shown in Fig. 15.18. Refer to Fig. 15.9 for orientation of the quad and arm numbering.

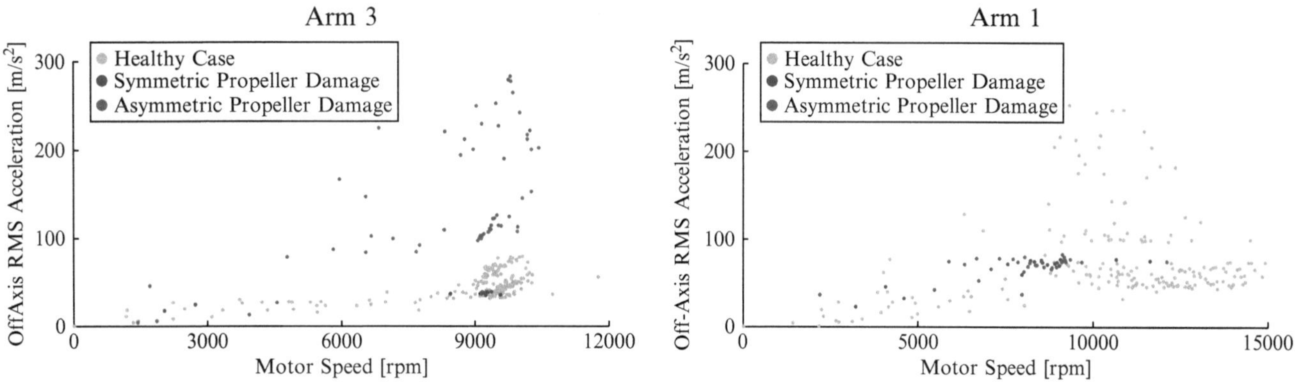


**Fig. 15.17** Comparison of healthy and damaged propellers for off-axis acceleration versus motor speed in arms 3 (*left*) and 1 (*right*) of the quad-rotor. Ranges of asymmetric and symmetric damage were used as in Sect. 15.4.2

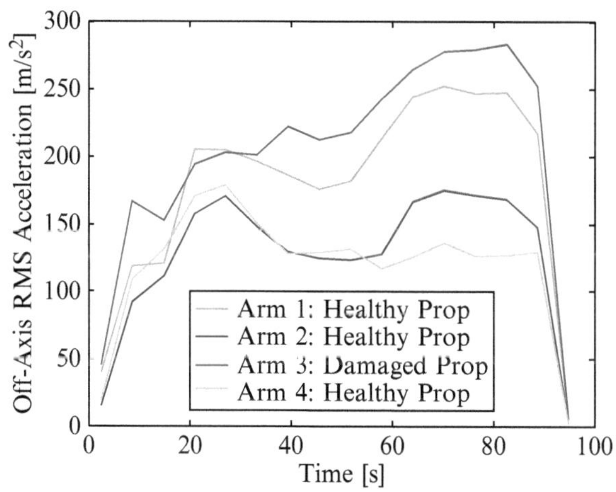


**Fig. 15.18** Comparison of each quad-rotor arm for RMS acceleration versus time

Arm 2 and 4 follow a similar trajectory over the course of the 90 s test, which is expected for two healthy propellers. Much like trends evaluated in Figs. 15.17 and 15.18, the asymmetrically damaged propeller on arm 3 induces larger vibrations and, thus, RMS acceleration on its arm. However, despite its healthy conditions, arm 1 follows the same trajectory of the damaged propeller; this behavior implies mechanical coupling between arms 1 and 3. The presence of crossbeams between arms 1 and 3 (and 2 and 4) is likely the cause of such coupling. Because vibrations in one arm can interact with another arm and induce excess vibrations, unnecessary and accelerated wear of components may occur throughout testing. Implementation of vibration isolation methods, such as dampers between crossbeams, is recommended to mitigate this problem. Even without such modifications, though, damage can be successfully localized to a motor pair, which is a valuable damage metric.

#### 15.5.4.2 Vibrations in the Frequency Domain

Single Rotor

To further characterize vibrations, the Power Spectral Density (PSD) of the angular resampled signal (ARS) acceleration data was analyzed to determine the dominant frequencies of the accelerometer data. The ARS signal is generated by normalizing the frequencies of the measured acceleration to the corresponding motor arm's rotational velocity using the SHMTools software ARS implementation [10]. Because of this, the PSD of the ARS signal is also normalized and the motor frequency is aligned when comparing arms with vastly different motor speeds. The PSD of the ARS signal also shows any dominant frequencies that differ from the harmonics of the motor frequency, which are most likely due to damage [11]. For the single rotor platform, a healthy baseline was established to ensure that behavior was predictable for several different healthy cases. In Fig. 15.19, a single healthy case is compared to the averaged PSD from all healthy data. As shown, the average matches the healthy case well, and does not have any significant off-harmonic frequencies that are not represented in the average. This behavior was also observed in the majority of healthy cases.

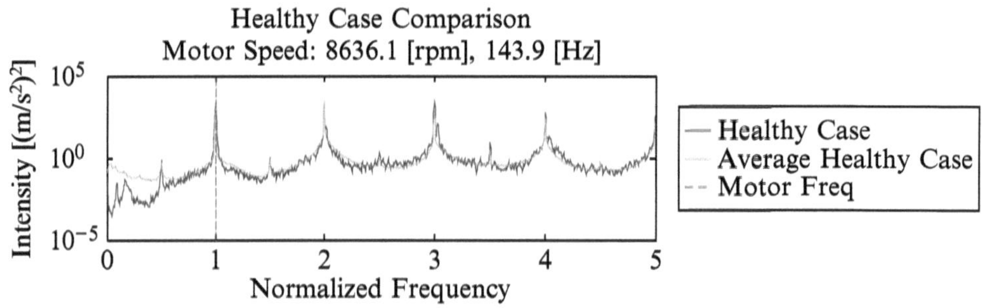


**Fig. 15.19** Comparison of singe healthy case to the average PSD of all healthy data

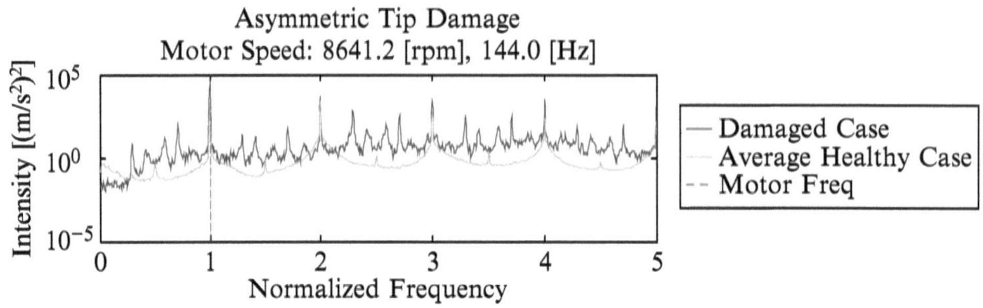


**Fig. 15.20** Comparison of PSD of propeller with asymmetric tip damage to healthy average

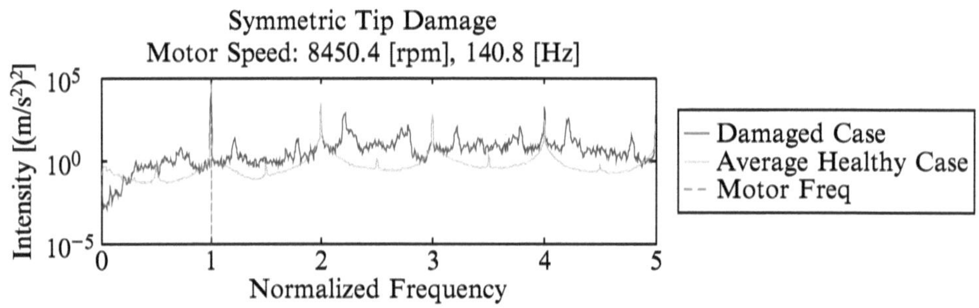


**Fig. 15.21** Comparison of PSD of propeller with symmetric tip damage to healthy average

The average was also compared to several damage cases. In Fig. 15.20, the propeller with asymmetric tip damage shows multiple peaks at off-harmonic frequencies that are easily distinguished from the healthy case. Figure 15.21 also shows the PSD of the ARS acceleration signal for the symmetric tip damage case. This is less easily distinguished than the asymmetric case, but still distinctly different than the healthy data. Both cases show the potential for damage metrics based on the deviation in the frequency domain.

### Quad-Rotor

For the quad-rotor, each damaged test has three systems of healthy components to compare to the damaged component. Although each motor rotates at a different speed for a given time, using the ARS signal normalizes the frequencies with each motor's individual frequency as discussed before. Similar to the single rotor platform, the healthy arms are dominated by the motor frequency and its harmonics. Due to the coupling between arms and the increased complexity of the quad-rotor system, the healthy signals exhibit many other small peaks. Figure 15.22 is a PSD of each of the four arms on the quad-rotor for the same asymmetrically damaged tip case demonstrated on the single rotor platform. Figure 15.23 shows the PSD of one of the structural damage tests performed through loosening of bolts. Both types of damage produce intense low frequency and off-harmonic vibrations that are not present in the arms with healthy components. In addition, the clear detection of structural damage will be extremely useful to alter or abort the flight of an MR vehicle before a motor mount fails catastrophically.

For both the single rotor and quad-rotor platforms, frequency analysis of the vibration data along a single in-plane accelerometer shows distinctly different trends for the damaged case. Both platforms indicate deviation in the frequency could be used in determining damage metrics.

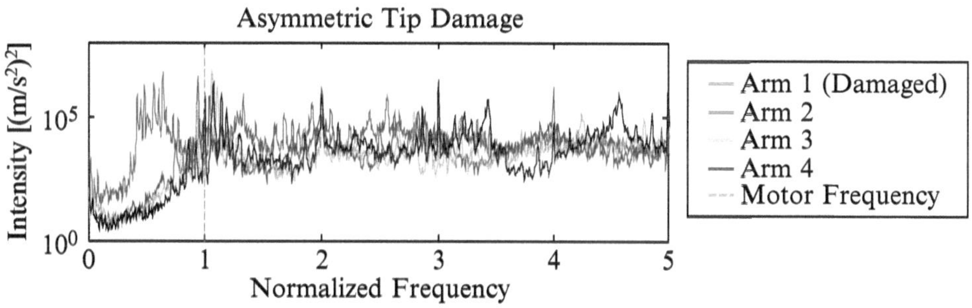


**Fig. 15.22** PSD comparison for asymmetric damage on one arm of the quad-rotor

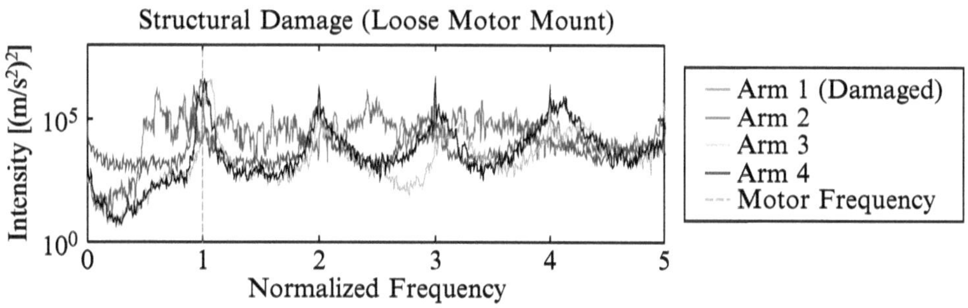


**Fig. 15.23** PSD comparison for symmetric damage on one arm of the quad-rotor

### *15.5.5 Additional Observations*

Beyond the quantitative analysis, certain trends that were not reflected in the data are important to note. These observations were recorded periodically throughout the testing.

#### 15.5.5.1 Battery

Battery overheating was a common failure mode observed, which was recorded to avoid misclassification of battery failure as hardware malfunction. The five-cell lithium polymer batteries used on the test platforms were not rated high enough to simultaneously control the four high performance motors and often overheated, occasionally causing catastrophic failure in the system. The high current sourced to the motors induced the melting of several power connectors, which opened the electrical circuit, cutting off the power supply to the vehicle and causing the aircraft to crash instantly. Such high battery temperatures were indicated by deformation of the plastic coating around the battery and corresponding melting odor. Temperatures of the battery exceeded 160 °C during testing. In future research, it is advisable to include a constant temperature measurement on the battery in addition to the voltage readings.

#### 15.5.5.2 ESC

In one case, an ESC on the quad-rotor vehicle burnt out due to its PWM signal wire becoming unplugged during normal operation. This change presumably caused an unregulated current to continuously enter the ESC for the duration of the test. The thermocouple embedded in this particular ESC failed, likely due to the extreme internal temperature and melting of the ESC casing. A thermal image taken after the test showed the ESC at 105 °C, as shown in Figs. 15.24 and 15.25. Figure 15.26 shows the burnt ESC after it had cooled down and been removed from the system. This test case highlights the need for an improved monitoring system for electronic speed controllers, with either a more sophisticated temperature sensor or improved placement of the thermocouple within the ESC.

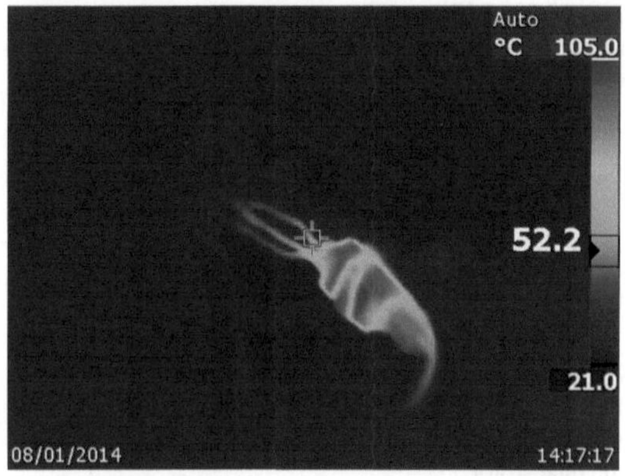


**Fig. 15.24** Thermal image of overheated ESC at a maximum recorded temperature of 105 ° C

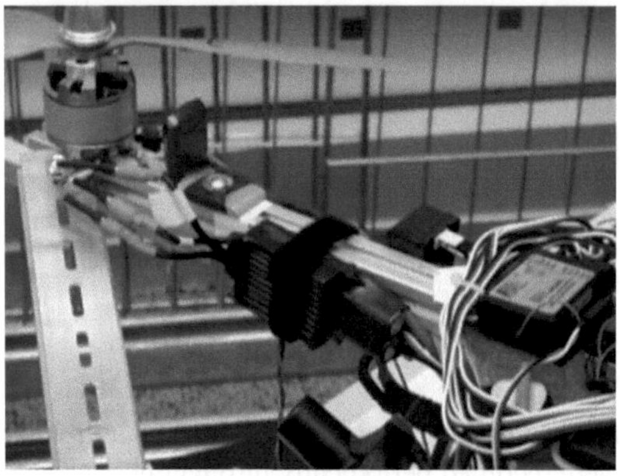

**Fig. 15.25** Photograph of overheated ESC

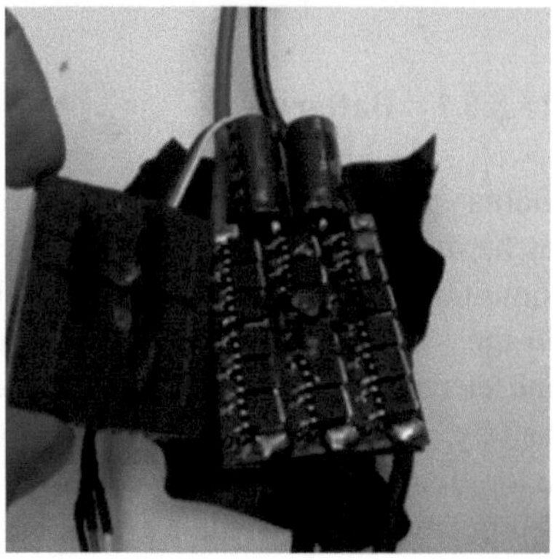

**Fig. 15.26** Closer inspection of burnt ESC

## 15.6 Conclusion

The single rotor and quad-rotor platforms were successfully instrumented with a robust SHM system, and the value of the different sensors as damage classifiers was determined. Accelerometer information about structural vibrations proved to be the most informative measurement for damage, but temperature, power, and lift measurements also provided failure metrics. In some cases, such as vibrations in the frequency domain and temperature, the principles and damage classifiers of the single rotor platform transferred directly to the quad-rotor. Component interactions on the quad-rotor platform introduced coupling and application issues that made analysis of the magnitude of vibrations and power measurements more complex than the single rotor system. In general, data obtained from the sensors exhibited clear trends, which can be used to create statistical models for determining the current health of the vehicle. Complete damage prognosis will require additional collection of data from the SHM system, loading conditions, and environmental factors to build models that characterize a timeline for predicted failure. In the event of damage, these models can successfully estimate the remaining useful life of the vehicle and provide valuable information to actively alter or abort a mission.

## References


1. Zenko M (2012) 10 things you didn't know about drones. Foreign policy. http://www.twc.edu/sites/default/files/assets/academicCourseDocs/4.%20Zenko,%2010%20Things%20You%20Didn't%20Know%20About%20Drones.pdf. 16 June 2014
2. Raptopoulos A (2013) Matternet. http://matternet.us/ted/. 16 June 2014
3. Banker S (2013) Amazon and drones – here is why it will work. Forbes. http://www.forbes.com/sites/stevebanker/2013/12/19/amazon-drones-here-is-why-it-will-work/. 16 June 2014
4. Dent S (2014) What you need to know about commercial drones. Engadget. http://www.engadget.com/2014/06/13/commercial-drone-explainer/. 16 June 2014
5. Olson IJ, Atkins EM (2013) Qualitative failure analysis for a small quadrotor unmanned aircraft system. In: Proceedings of the AIAA guidance, navigation, and control (GNC) conference. Boston, 19–22 Aug 2013
6. Farrar CR, Worden K (2013) Structural health monitoring: a machine learning perspective, 1st edn. Wiley, Chichester, pp 461–477
7. Worden K, Farrar CR, Manson G, Park G (2005) Fundamental axioms of structural health monitoring. In: Proceedings of 5th international workshop on structural health monitoring, Stanford, 12–14 Sept 2005, pp. 26–41
8. Silverman H (2005) T-HUMS AH64 lead the fleet summary and a glimpse at hermes 450 MT-HUMS. In: Proceedings of AIAC HUMS primary conference, Melbourne, March 2005
9. Lebold M, McClintic K, Campbell R, Byington C, Maynard K (2000) Review of vibration analysis methods for gearbox diagnostics and prognostics. In: Proceedings of the 54th meeting of the society for machinery failure prevention technology, Virginia Beach, 1–4 May 2000, pp. 623–634
10. Flynn EB, Kpotufe S, Harvey D, Figueiredo E, Taylor S, Dondi D, Mollov T, Todd MD, Rosing TS, Park G, Farrar C (2010) SHMTools: a new embeddable software package for SHM applications. In: Proceedings of SPIE2010, 7647, San Digeo, 7–11 March 2010
11. Braun S (1986) Mechanical signature analysis: theory and applications. Academic, London. (Print)

# Chapter 16
# Statistical Tools for the Characterization of Environmental and Operational Factors in Vibration-Based SHM

**C. Rainieri, D. Gargaro, and G. Fabbrocino**

**Abstract** In recent years the interest in permanent monitoring of civil structures has raised because of the needs of controlling the ageing of a huge number of existing infrastructures. The recent advances in sensing technologies and data processing have made the installation and operation of permanent monitoring systems more and more attractive. Vibration-based monitoring is based on the analysis of the evolution in time of damage features. A lot of these features are obtained from experimental estimates of the modal parameters. However, these estimates are usually influenced by environmental and operational factors. The variations they induce in the estimates may hide the small changes due to damage, so their influence has to be appropriately considered in practical applications.

Using a large number of experimental data, models relating modal properties and environmental and operational factors can be set. However, the selection of the factors to be measured is typically not straightforward. As an alternative, statistical tools can be used to correct the estimates without the need to measure those factors.

In the present paper, after a review of the available approaches to quantify the influence of environmental and operational factors, the opportunity of applying robust blind source separation techniques in this field is assessed.

**Keywords** Vibration based Structural Health Monitoring • Natural frequencies • Environmental factors • Blind source separation • Second order blind identification

## 16.1 Introduction

The recent increase in the number of applications of modal based damage detection techniques for Structural Health Monitoring (SHM) of civil structures takes advantage of the recent development of several algorithms for automated identification [1] and tracking [2] of modal parameters based on Operational Modal Analysis (OMA) methods. Damage detection techniques based on changes of the modal parameters of the monitored structure over time are well-established [3]. Thus, the continuous monitoring of modal parameters has a large potential in performance and health assessment of civil engineering structures [4]. Applications range from prompt detection of damage and degradation phenomena [5] to post-earthquake health assessment and emergency management [6, 7]. An automated, accurate estimation of modal parameters also plays a primary role in the assessment of the dynamic behavior of complex structural systems such as geotechnical [8, 9] and historical structures [10, 11].

Extensive surveys and dedicated books are available in the literature about vibration-based SHM [4, 12, 13]. The monitoring process consists in the observation of the structure over long periods of time. Appropriate sensors and measurement systems continuously acquire records of the structural response; damage sensitive features are then extracted from the collected data and analyzed to assess the health state of the structure.

From a general point of view, damage is defined as any change of the structure that adversely affects its performance [12]. This change can be in the form of stiffness change (for instance, cracking), mass change, connectivity change (for instance, looseness in a bolted joint) or boundary condition change (for instance, bridge scour). An effective SHM system should be able to automatically detect damage at an early stage [4]. Five damage detection levels have been defined [12]:

- Level 1: identification of damage existence;
- Level 2: localization of damage;

---

C. Rainieri (✉) • D. Gargaro • G. Fabbrocino
DiBT Department, Structural and Geotechnical Dynamics Laboratory StreGa, University of Molise,
Viale Manzoni s.n.c, 86100 Campobasso, Italy
e-mail: carlo.rainieri@unimol.it; danilo.gargaro@unimol.it; giovanni.fabbrocino@unimol.it

© The Society for Experimental Mechanics, Inc. 2015 
C. Niezrecki (ed.), *Structural Health Monitoring and Damage Detection, Volume 7*, Conference Proceedings of the Society for Experimental Mechanics Series, DOI 10.1007/978-3-319-15230-1_16

- Level 3: identification of the type of damage;
- Level 4: quantification of damage severity;
- Level 5: prediction of the remaining service life of the structure (prognosis).

Modal based damage detection starts by recognizing that the modal parameters depend on the physical parameters (mass, stiffness and damping). Assuming that damage yields a change in the physical properties of the structure, this is reflected by a change in the modal properties. Thus, it is theoretically possible to identify damage from the analysis of the variations of the modal parameters. A number of damage sensitive features have been, therefore, defined in terms of modal parameters.

Damage sensitive features can be defined in terms of natural frequencies and mode shapes. Natural frequency variations provide the easiest way to detect the presence of damage, because they can be accurately estimated even in the presence of a few sensors. However, the information they provide is limited to Level 1 damage detection. Thus, other features have been defined in terms of mode shapes and mode shape curvatures, because mode shapes can provide information also for damage location. However, they are typically estimated with lower accuracy with respect to natural frequencies.

One of the main drawbacks of modal based damage detection is related to the sensitivity of natural frequency estimates to environmental and operational conditions that can cause changes of the same order of magnitude of those induced by damage. As a consequence, the estimates have to be depurated from the effects of environmental factors in order to effectively detect damage. Using a large number of experimental data, models relating modal properties and environmental and operational factors can be set. However, the selection of the factors to be measured is typically not straightforward. As an alternative, statistical tools can be used to correct the estimates without the need to measure those factors. Once the damage sensitive features have been filtered by removing the environmental effects, a number of tools can be applied for feature discrimination. They can be broadly classified as supervised and unsupervised learning approaches [13]. The former are applied when data are available for both the undamaged and damaged structure, while the latter are applied when reference data are available only for the structure in healthy state.

In the present paper, after a review of the available approaches to quantify the influence of environmental and operational factors, the opportunity of applying robust blind source separation techniques in this field is assessed.

## 16.2 Removal of the Influence of Environmental Factors from Natural Frequency Estimates

Environmental effects (temperature, humidity, wind, ...) and operational factors influence the natural frequency estimates. Such an influence has to be quantified and removed in order to ensure the ability of the continuous monitoring system to detect slight changes induced by damage. Temperature has a major influence on natural frequency estimates, as demonstrated in several studies [14, 15]. In order to remove its effect on natural frequencies, an attractive approach consists in the definition of models able to represent the physical phenomenon behind the frequency changes. In particular, it is possible to rely on black-box models, whose parameters are tuned by collecting a large number of observations, to establish a relation between the natural frequencies and a set of environmental and operational factors. These have to be measured as well. However, the selection of these factors is not straightforward. Whenever the factors influencing the estimates cannot be clearly identified or they cannot be measured, approaches based on statistical tools can be profitably applied to correct the natural frequency estimates without the need of measuring the environmental and operational factors.

If measurements of the parameters influencing the estimates are available, approaches based on regression analysis can be applied. For instance, the Multiple Linear Regression can be used to identify the relation between a single dependent variable and several independent variables, the so-called predictors. The established model allows the prediction of future values of the dependent variable when only the predictors are known. In the context of SHM, after the quantification of the influence of environmental factors based on a given set of observations, the predicted natural frequencies are compared with the values directly estimated from the collected acceleration time series in order to remove the environmental influence on the estimates. The following equation characterizes the regression model:

$$\{y\} = [Z]\{\beta\} + \{\varepsilon\} \tag{16.1}$$

where the vector $\{y\}$ collects the n observations of the dependent variable, $[Z]$ is the n-by-p matrix collecting the corresponding n values of p selected independent variables, and $\{\beta\}$ is the vector of the unknown coefficients relating the dependent variable to each predictor; finally, $\{\varepsilon\}$ is the vector that takes into account the effect of measurement errors and other variables not explicitly considered in the model. The error vector is assumed to have zero mean and constant variance:

$$E\left[\{\varepsilon\}\right] = \{0\}$$
$$E\left[\{\varepsilon\}\{\varepsilon\}^T\right] = \sigma^2\left[I\right] \tag{16.2}$$

If the same predictors explain the variability of the natural frequency for a number of modes, a multivariate multiple regression analysis can be applied. In this case, it is possible to generalize Eq. 16.1 by adopting the following matrix formulation:

$$[Y] = [Z]\,[B] + [E] \tag{16.3}$$

where [Y] is the n-by-m matrix with the time series of the m identified natural frequencies in each column, [Z] is the n-by-p matrix with the corresponding n values of p selected predictors, [B] is the p-by-m matrix collecting the coefficients of the models to be determined and [E] is the matrix holding the information about the modeling errors associated with each natural frequency.

The main objective of regression analysis is then to estimate the model coefficients that provide a good fit between the observations and the values of the independent variable provided by the model. These are usually obtained through the least squares method, which minimizes the sum of the squared errors. Once a good model has been obtained, it can be used to predict the value of the dependent variable from given values of the predictors, eventually not considered in the definition of the model.

In the framework of modal based SHM, the setting of multiple regression models should rely on data collected during a sufficiently long period of time so that the influence of environmental factors on natural frequencies can be exhaustively characterized; for instance, data from summer as well as winter periods have to be considered in the definition of the model. In fact, regression models are able to predict the values of the natural frequencies associated with given predictors only in the region of the data used to set the model.

When measurements of the parameters influencing the estimates are not available, statistical methods able to eliminate the effects of environmental and operational factors in the absence of exploitable information about them have to be adopted. In this case, some features are extracted from the collected data over a reference period of time through the decomposition of the covariance matrix of the natural frequency time histories. Examples of these methods are the Principal Component Analysis (PCA) [16] and the factor analysis [17].

The PCA is a multivariate statistical tool used to express the covariance structure of a set of variables through a few linear combinations of these variables. In other words, the original variables {y} are transformed into a new set of variables {z} by the following equation:

$$\{z\} = [T]\,\{y\} \tag{16.4}$$

In order to remove the effects of environmental and operational factors on natural frequency estimates continuously collected by a monitoring system, the PCA is applied to the covariance matrix of the natural frequency time series collected during a sufficiently long period of time and holding the information about the full range of variability of environmental and operational factors. Such a period is typically one year associated with a "healthy" state of the structure.

The PCA highlights the effect of relevant factors, such as the temperature, and reduce the dimensionality of the model by the elimination of secondary effects, such as those due to random errors in the identification of the natural frequencies. The differences between the observed and the re-mapped values are used afterwards as features insensitive to the factors modeled by the principal components. As a consequence, these residues can be used to detect abnormal values that might be related to damage.

Factor Analysis (FA) is another multivariate statistical tool that can be used to remove the influence of environmental factors [17]. It can be considered as an extension of PCA because it is based on the decomposition of a covariance matrix as well. In the context of dynamic monitoring, the estimated features (normally natural frequencies) can be expressed as the sum of two components. The relation between m observable features (the natural frequencies) and p unobserved factors is expressed by the following equation:

$$\{y\} = [A]\,\{s\} + \{\varepsilon\} \tag{16.5}$$

where [A] is a m-by-p matrix. This equation is similar to Eq. 16.1 for multivariate linear regression, but in factor analysis the {s} components are not measured. The elements of matrix [A] are called factor loadings, the components of {s} are referred to as common factors and the $\varepsilon_i$ (i = 1,...,m) are defined specific factors. Moreover, the factor model has the following properties:

$$E[\{y\}] = E[\{s\}] = E[\{\varepsilon\}] = \{0\} \tag{16.6}$$

$$E\left[\{s\}\{s\}^T\right] = [I] \tag{16.7}$$

$$E\left[\{\varepsilon\}\{\varepsilon^T\}\right] = [\Psi] = \begin{bmatrix} \psi_1 & \dots & \dots & 0 \\ \vdots & \psi_2 & \dots & 0 \\ \vdots & \vdots & \ddots & \vdots \\ 0 & 0 & \dots & \psi_m \end{bmatrix} \tag{16.8}$$

$$E\left[\{\varepsilon\}\{s\}^T\right] = [0] \tag{16.9}$$

Thus, the common factors are uncorrelated and characterized by a unitary variance, as remarked by the structure of the covariance matrix of {s}. The specific factors are also uncorrelated each other and with the common factors. These properties lead to the following fundamental equation:

$$E\left[\{y\}\{y\}^T\right] = [A][A]^T + [\Psi] \tag{16.10}$$

Thus, the objective of factor analysis is the determination of the matrices [A] and [Ψ] that fit a set of observations {y}. A possible approach is based on an iterative procedure and the computation of the Singular Value Decomposition (SVD) of the covariance matrix estimated from the data. This and other algorithms are extensively described in classical books of multivariate statistical analysis (see, for instance, [18]).

## 16.3 Second Order Blind Identification for Removal of Environmental Effects

The use of blind source separation (BSS) techniques in the context of OMA and SHM has been recently proposed [19, 20]. BSS techniques extract a set of signals, the so-called sources, from observations of their mixtures [21] based on fairly general assumptions about the sources and the mixing process. PCA is an example of BSS technique. In this section attention is focused on the applicability of another BSS technique, the Second Order Blind Identification (SOBI) procedure, for the removal of environmental effects from natural frequency estimates in the absence of data about the environmental and operational factors. The basics of SOBI are herein briefly reported to highlight how the method can be applied to remove the influence of temperature on natural frequency estimates.

The structure of the fundamental equation of SOBI is similar to Eq. 16.5. The $l$ recorded time series are, therefore, modeled as a linear combination of sources {s} (the unknown environmental factors) plus noise {ε}. As a consequence, if there are N active sources, they can be identified only if rank([A]) = N. Since the mixing matrix has dimension $l \times N$, this implies that the number of monitored modes has to be larger than or equal to the number of active sources.

A fundamental assumption in SOBI is that the sources are stationary, uncorrelated, and scaled to have unit variance, so their covariance matrix is the identity matrix. With the appropriate translations, this assumptions leads to an equation that is similar to Eq. 16.7. The additive noise is assumed to be a temporally and spatially white stationary random process, with zero mean and constant variance – see also Eq. 16.2. The added noise is also assumed to be independent of the source signals – see Eq. 16.9 with the appropriate translation of the meaning of symbols -. Moreover, it is possible to show [22] that the covariance matrix of the observed mixture {y} can be expressed as follows:

$$\left[R_{yy}(0)\right] = E\left[\{y\}\{y\}^T\right] = [A][A]^T + \sigma^2[I] \tag{16.11}$$

This points out that the basic assumptions in SOBI are similar to those of factor analysis. The main difference is in the method adopted to estimate the matrices [A] and [Ψ]. The main steps of SOBI are herein outlined; a detailed description of the algorithm can be found elsewhere [22].

The first step consists of whitening the signal part of the observed data {x} = [A]{s}. This is achieved by a linear transformation of {x} such that the whitened data {z} are uncorrelated and have unit variance:

$$\{z\} = [W]\{x\} \Rightarrow [R_{zz}(0)] = E\left[\{z\}\{z\}^T\right] = [I] \tag{16.12}$$

The matrix [W] defining this transformation is referred to as the whitening matrix. From a practical point of view, once the measured data have been centralized by removal of the mean value from each component of {y}, whitening is obtained as follows. First of all, the eigenvalue decomposition of $[R_{yy}(0)]$ is computed:

$$\left[R_{yy}(0)\right] = E\left[\{y\}\{y\}^T\right] = [V][D][V]^T \tag{16.13}$$

where [V] is the matrix of eigenvectors and [D] is the diagonal matrix of eigenvalues. If only the N largest eigenvalues $d_1,\ldots,d_N$ and the corresponding eigenvectors are retained, the average of the remaining *l*-N eigenvalues yields an estimate $\sigma^2$ of the noise variance, under the assumption of white noise [23]. The whitened signals are then computed from the largest eigenvalues and the corresponding eigenvectors as:

$$\{z\} = \left([D_N] - \sigma^2[I_N]\right)^{-1/2}[V_N]^T\{y\} = [W]\{y\} \tag{16.14}$$

where $[D_N]$ is the submatrix of [D] holding only the N largest eigenvalues, $[V_N]$ is the submatrix of [V] collecting the eigenvectors corresponding to the N largest eigenvalues of $[R_{yy}(0)]$, and the whitening matrix is given by:

$$[W] = \left([D_N] - \sigma^2[I_N]\right)^{-1/2}[V_N]^T \tag{16.15}$$

On the analogy with PCA, matrix [D] can be partitioned into two diagonal matrices, where the second matrix holds the smallest *l*-N singular values, which are not relevant to explain the variability of the components of {y}. Thus, the number of active sources can be estimated by looking for a gap in the plot of all the singular values arranged in descending order.

In SOBI the noise variance is assumed to be the same for all channels – see also Eq. 16.2 -. The white noise assumption is needed to get an estimate $\sigma^2$ of the noise variance as the average of the smallest *l*-N eigenvalues of $[R_{yy}(0)]$. Once the whitened signals have been obtained, the following *p* time-shifted covariance matrices have to be computed:

$$[R_{zz}(\tau_k)], k = 1,\ldots,p \tag{16.16}$$

In order to estimate the sources and the mixing matrix, SOBI carries out an approximate joint diagonalization of those *p* time-shifted covariance matrices according to the joint approximate diagonalization (JAD) technique [23].

The objective of JAD is to find the unitary matrix [Φ] that approximately diagonalizes the time-shifted covariance matrices. An optimization problem is defined with respect to the matrix [Φ] that minimizes the sum of all off-diagonal terms of $[\Phi]^T[R_{zz}(\tau_k)][\Phi]$ ($k = 1,\ldots,p$) for the *p* time-shifted covariance matrices. The solution to the minimization problem is found by means of a numerical algorithm based on the Jacobi rotation technique [23]. Once the matrix [Φ] has been obtained, the demixing matrix [U] and the mixing matrix [A] can be computed:

$$[U] = [\Phi]^T[W] \tag{16.17}$$

$$[A] = [W]^+[\Phi] \tag{16.18}$$

where the superscript $^+$ denotes pseudoinverse. The resulting sources are shift-uncorrelated because the matrices $[R_{ss}(\tau_k)]$ are nearly diagonal. The sources are obtained as follows:

$$\{s\} = [U]\{y\} \tag{16.19}$$

In order to apply SOBI for the removal of environmental influence on natural frequency estimates, sufficiently long time series of the natural frequencies of interest have to be collected with reference to the health state of the structure. These estimates are used to identify a reference mixing matrix $[A_{healthy}]$. The number of active sources N is determined from the inspection of the sequence of singular values of $[R_{yy}(0)]$ as previously discussed. The additional data collected after the estimation of the matrix $[A_{healthy}]$ are also decomposed by SOBI and the corresponding sources are estimated. These are recombined by the previously identified $[A_{healthy}]$ matrix:

$$\{\widehat{y}\} = \left[A_{healthy}\right]\{s\} \tag{16.20}$$

and the resulting values are subtracted from the original ones in order to compute the residue:

$$\{\varepsilon\} = \{y\} - \{\widehat{y}\} \tag{16.21}$$

which is independent of the environmental factors and, therefore, can effectively be used as a feature in the context of damage detection procedures.

## 16.4 Proof of Concept

The proposed approach is applied to data collected by the permanent dynamic monitoring system developed to monitor the tensile load in the cable of a sample arch in the steel roof (Fig. 16.1) of the University of Molise Sports Hall in Campobasso, Italy [24].

The monitoring system takes advantage of an effective algorithm for automated output-only modal identification, which is based on the combination of different OMA techniques in order to make easier the analysis and interpretation of the stabilization diagram. More details about the monitoring system and the automated OMA procedure can be found elsewhere [24, 25]. The monitoring system started operating at the end of 2013, allowing the collection of a database of the fundamental modal properties of the cable.

In order identify the reference mixing matrix, the time series of the first three natural frequencies collected on March 2014 (298 estimates per mode) have been processed by SOBI. Only one parameter has been assumed to determine the variability of the estimates (N = 1). A comparison between the measured data and the reconstructed time series is shown in Figs. 16.2, 16.3 and 16.4 for the cable modes I, II and III, respectively, demonstrating a very good agreement. SOBI has been applied afterwards to the time series collected in the period September-October 2014.

**Fig. 16.1** Layout of the monitoring system

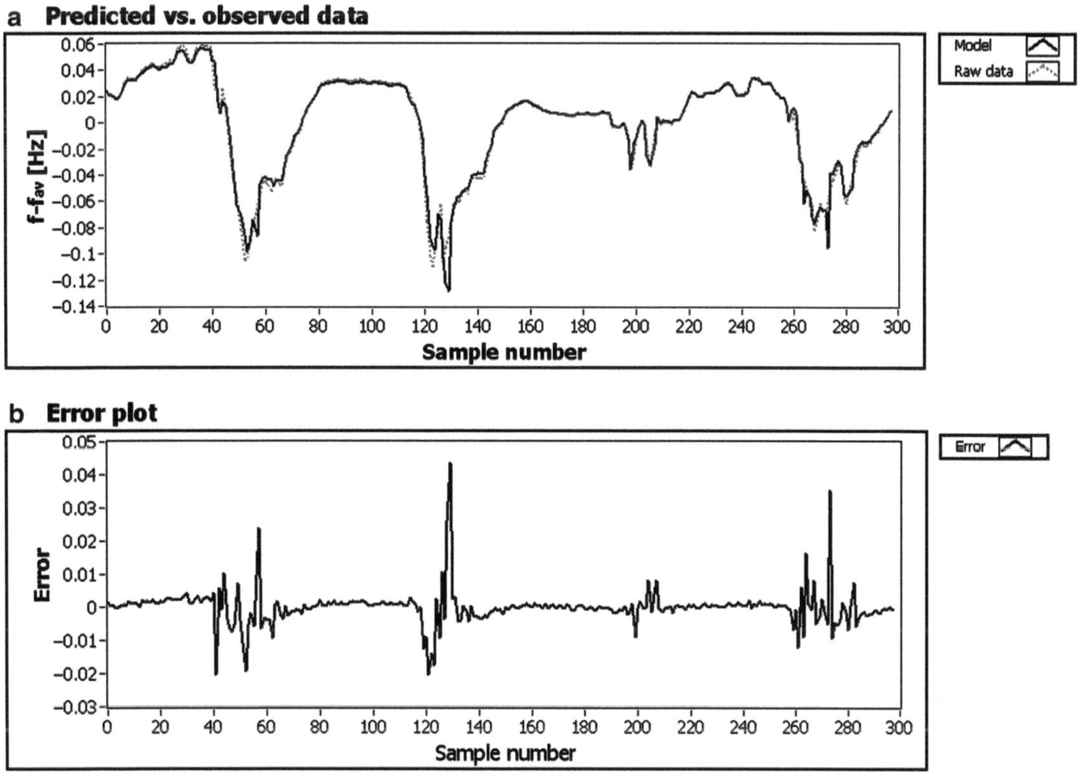


**Fig. 16.2** Predicted vs. observed data (**a**) and residue (**b**) for mode I of the cable

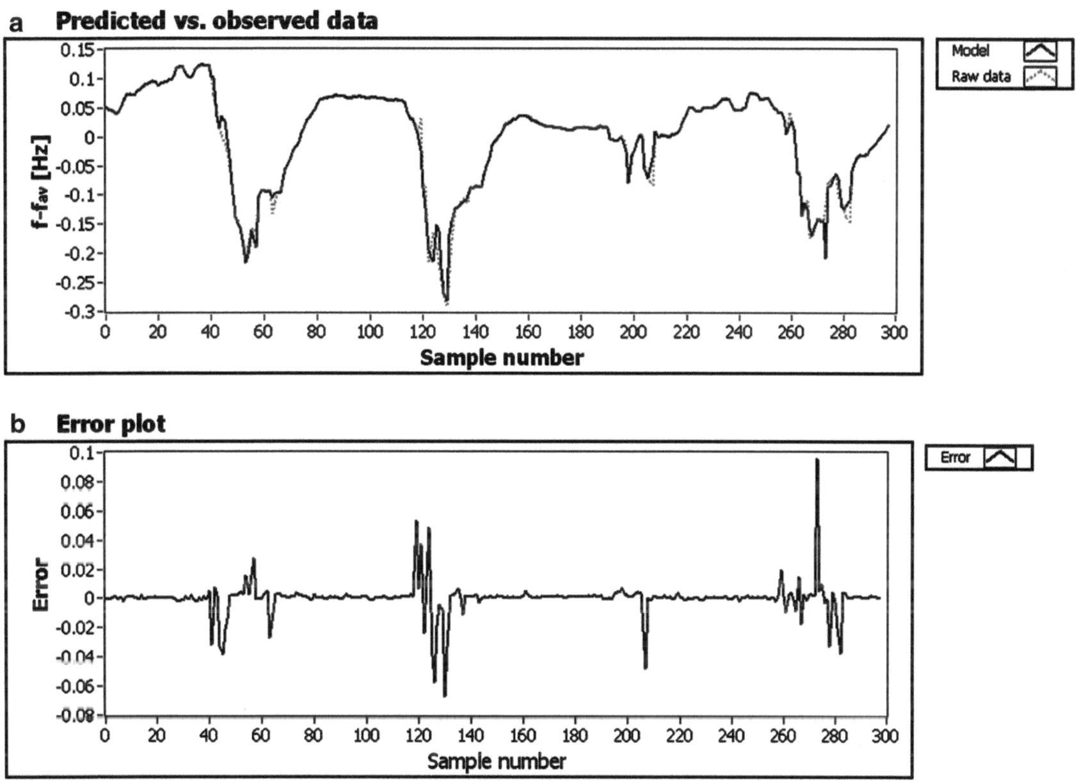


**Fig. 16.3** Predicted vs. observed data (**a**) and residue (**b**) for mode II of the cable

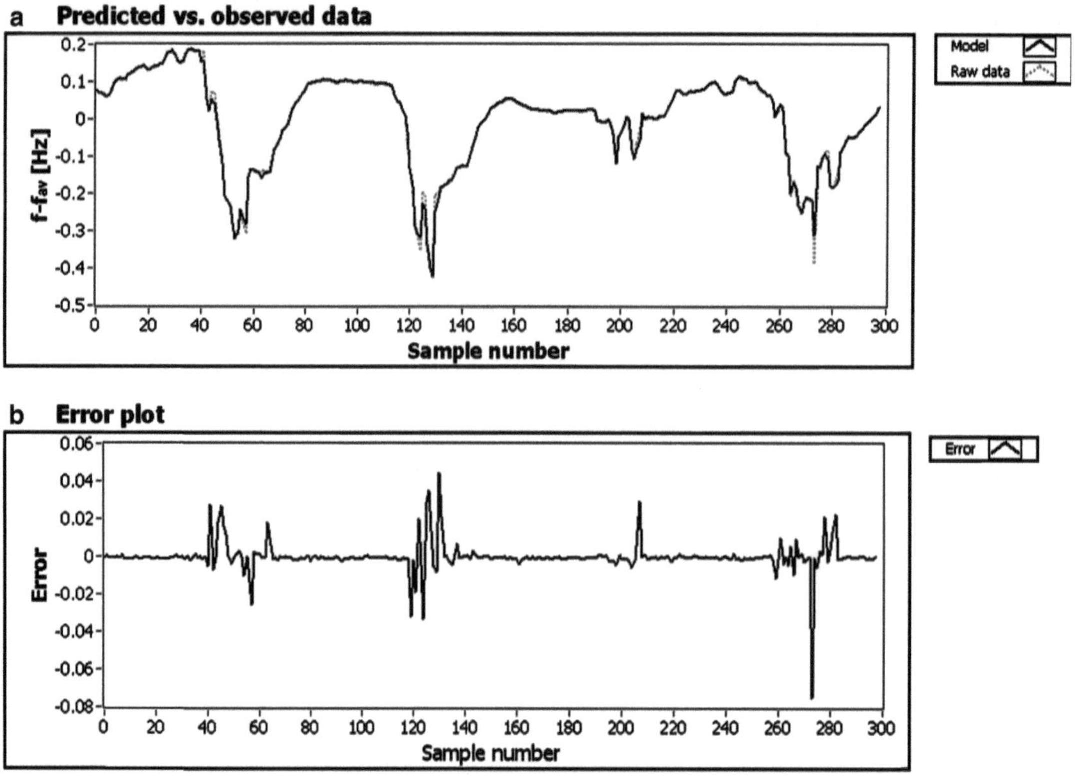


Fig. 16.4 Predicted vs. observed data (a) and residue (b) for mode III of the cable

Fig. 16.5 Correlation between the first natural frequency of the cable and the temperature (March 2014)

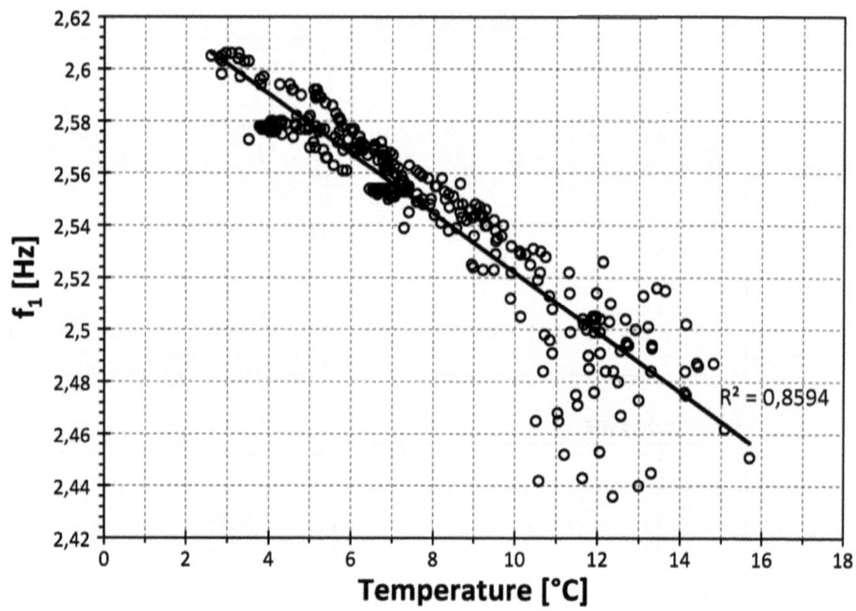


In order to characterize the performance of the method from a quantitative point of view, the reconstructed signals obtained by combining the previously mentioned reference mixing matrix with the source directly identified from the time series collected in the period September-October 2014 have been compared to the raw data and the residue has been computed. Moreover, since records of the temperature were also available for these periods and the natural frequency estimates are well correlated to the temperature (Fig. 16.5), a regression model has been developed based on the time series recorded on March and used to predict the values of the natural frequencies in the period September-October 2014 from the recorded temperature values. The residue between original and reconstructed signals has been computed also in this case (Fig. 16.6b). The time series have been centralized by removing the mean when SOBI and linear regression have been applied.

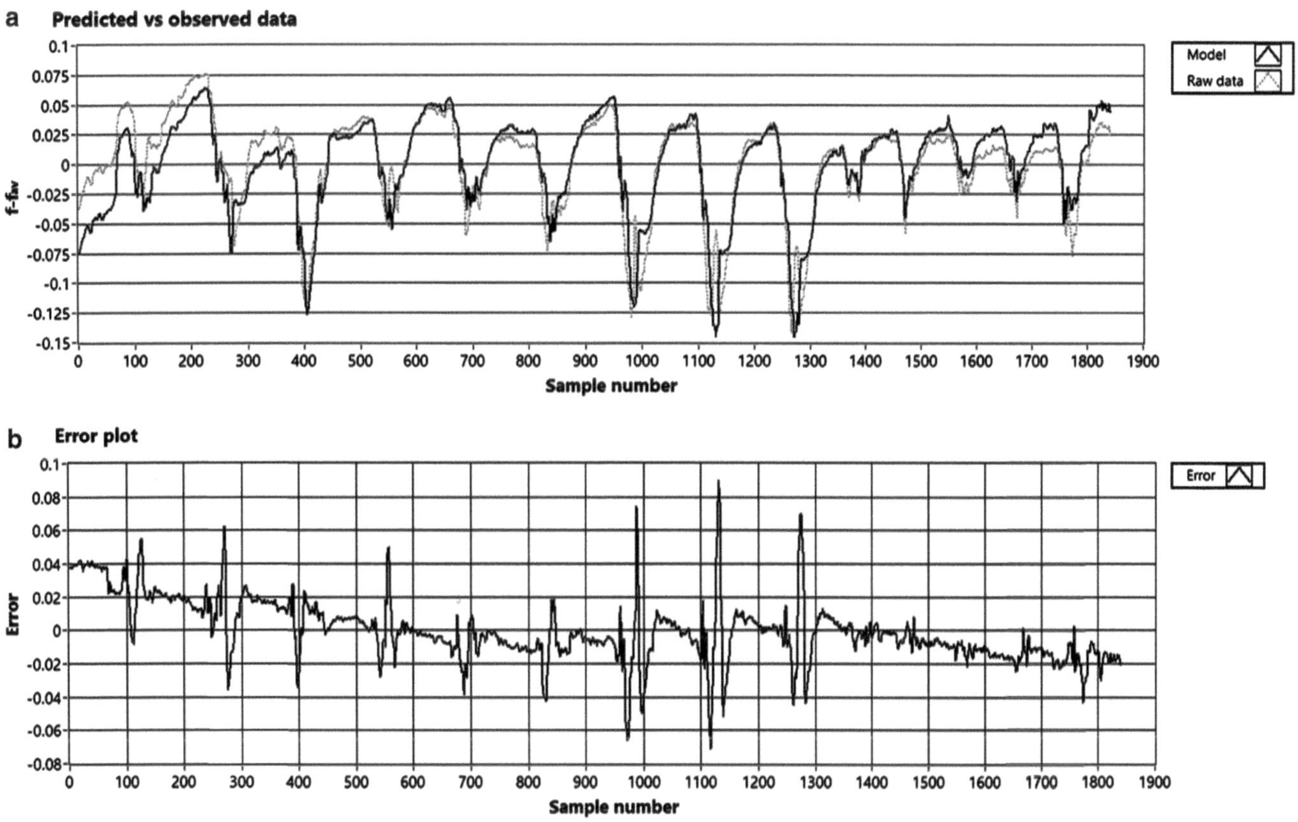


Fig. 16.6 Predicted vs. observed data (a) and residue (b) from the regression model (September-October 2014, mode I)

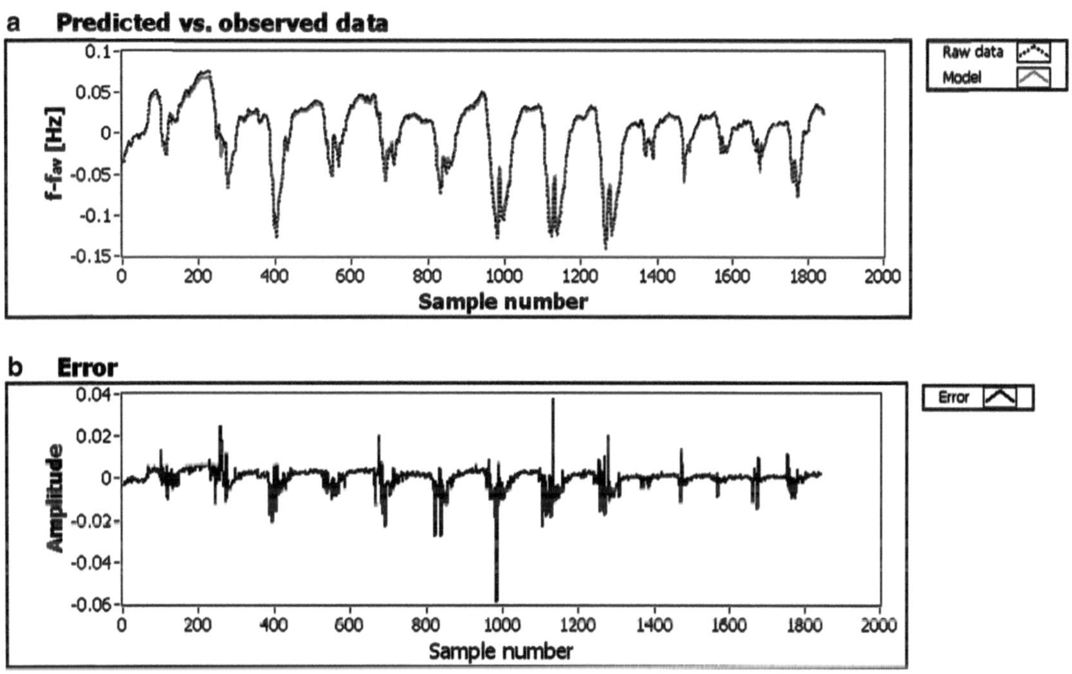


Fig. 16.7 Predicted vs. observed data (a) and residue (b) from SOBI (September-October 2014, mode I)

Inspection and comparison of Figs. 16.6a and 16.7a show that the mixing model provided by SOBI is able to accurately predict the evolution of the natural frequencies over time. Moreover, lower residues are obtained with respect to the regression model (Figs. 16.6b and 16.7b). Even if further investigations are needed, these preliminary results are definitely encouraging, remarking the promising applicative perspectives of SOBI for removal of the influence of environmental and operational factors without measuring them.

## 16.5 Conclusions

The opportunities provided by SOBI in the characterization of the influence of environmental and operational factors on natural frequency estimates in the absence of measures of those factors have been investigated in this paper. After a review of common approaches for removal of environmental influence on estimates, where the similarities of the basic assumptions of SOBI with those of other available methods have been remarked, quantitative comparisons between SOBI and linear regression have been made to validate the proposed approach. Time series of the natural frequencies of a continuously monitored cable have been processed, obtaining encouraging results and useful indications for future developments.

**Acknowledgments** The present work has been supported by the ReLuis-DPC Executive Project 2014-2016, Special Project "Monitoring", whose contribution is gratefully acknowledged.

## References

1. Rainieri C, Fabbrocino G (2010) Automated output-only dynamic identification of civil engineering structures. Mech Syst Signal Process 24(3):678–695
2. Rainieri C, Fabbrocino G, Cosenza E (2011) Near real-time tracking of dynamic properties for standalone structural health monitoring systems. Mech Syst Signal Process 25(8):3010–3026
3. Doebling SW, Farrar CR, Prime MB, Shevitz DW (1996) Damage identification and health monitoring of structural and mechanical systems from changes in their vibration characteristics: a literature review, technical report LA-13070-MS, UC-900, Los Alamos National Laboratory, New Mexico 87545
4. Rainieri C, Fabbrocino G, Cosenza E (2008) Integrated systems for structural health monitoring: worldwide applications and perspectives. In: Proceedings of the fourth European workshop on structural health monitoring, Cracow
5. Magalhaes F, Cunha A, Caetano E (2012) Vibration based structural health monitoring of an arch bridge: from automated OMA to damage detection. Mech Syst Signal Process 28:212–228
6. Rainieri C, Fabbrocino G, Manfredi G, Dolce M (2012) Robust output-only modal identification and monitoring of buildings in the presence of dynamic interactions for rapid post-earthquake emergency management. Eng Struct 34:436–446
7. Rainieri C, Fabbrocino G, Cosenza E (2010) Integrated seismic early warning and structural health monitoring of critical civil infrastructures in seismically prone areas. Struct Health Monit Int J 10(3):291–308
8. Fabbrocino G, Laorenza C, Rainieri C, Santucci De Magistris F (2009) Seismic monitoring of structural and geotechnical integrated systems. Mater Forum 33:404–419
9. Rainieri C, Fabbrocino G, Santucci de Magistris F (2013) An integrated seismic monitoring system for a full-scale embedded retaining Wall. Geotech Test J 36(1):1–14
10. Rainieri C, Fabbrocino G (2011) Operational modal analysis for the characterization of heritage structures. Geofizika 28:127–143
11. Conte C, Rainieri C, Aiello MA, Fabbrocino G (2011) On-site assessment of masonry vaults: dynamic tests and numerical analysis. Geofizika 28:109–126
12. Sohn H, Farrar CR, Hemez FM, Shunk DD, Stinemates DW, Nadler BR (2003) A review of structural health monitoring literature: 1996–2001. Technical report LA-13976-MS, UC-900, Los Alamos National Laboratory, New Mexico 87545
13. Farrar CR, Worden K (2012) Structural health monitoring: a machine learning perspective. Wiley, Chichester, p 631
14. Farrar CR, Doebling SW, Cornwell PJ, Straser EG (1997) Variability of modal parameters measured on the Alamosa Canyon Bridge. In: Proceedings of IMAC 15, international modal analysis conference, Orlando
15. Peeters B, De Roeck G (2001) One-year monitoring of the Z24-Bridge: environmental effects versus damage events. Earthq Eng Struct Dyn 30:149–171
16. Yan A-M, Kerschen G, De Boe P, Golinval J-C (2005) Structural damage diagnosis under varying environmental conditions – Part I: a linear analysis. Mech Syst Signal Process 19:847–864
17. Deraemaeker A, Reynders E, De Roeck G, Kullaa J (2008) Vibration-based structural health monitoring using output-only measurements under changing environment. Mech Syst Signal Process 22:34–56
18. Johnson RA, Wichern DW (1992) Applied multivariate statistical analysis. Prentice Hall, Upper Saddle River
19. Zang C, Friswell MI, Imregun M (2004) Structural damage detection using independent component analysis. Struct Health Monit 3:69–83
20. Kerschen G, Poncelet F, Golinval JC (2007) Physical interpretation of independent component analysis in structural dynamics. Mech Syst Signal Process 21:1561–1575
21. Ans B, Herault J, Jutten C (1985) Adaptive neural architectures: detection of primitives. In: Proceedings of COGNITIVA '85, Paris, pp 593–597
22. Rainieri C (2014) Perspectives of second-order blind identification for operational modal analysis of civil structures. Shock Vib, Article ID 845106, p 9
23. Belouchrani A, Abed-Meraim K, Cardoso JF, Moulines E (1997) A blind source separation technique using second-order statistics. IEEE Trans Signal Process 45:434–444
24. Rainieri C, Gargaro D, Cieri L, Fabbrocino G (2014) Vibration-based continuous monitoring of tensile loads in cables and rods: system development and application. In: Proceedings of IMAC XXXII, Orlando
25. Rainieri C, Fabbrocino G (2013) Accurate damping estimation by automated OMA procedures. Conf Proc Soc Exp Mech Ser 39(4):1–9

# Chapter 17
# An Experimental Investigation of Feature Availability in Nominally Identical Structures for Population-Based SHM

**Evangelos Papatheou, Robert J. Barthorpe, and Keith Worden**

**Abstract** It is perhaps well known that the uncertainty in realistic structures may complicate most efforts for modelling and damage identification. In a population of structures which are considered identical, as in a wind farm for example, it is very often that the accurate modelling of one structure will be inadequate for the robust monitoring of the rest in an SHM approach. This paper presents an exploration of the common features which can be found in nominally identical structures and which can be used for damage identification with the ultimate purpose of population-based SHM. The concept of a population-based approach means that any additional new structures to the population will not need to be fully modelled in order to be monitored. Two different variants of the tail wing of a Piper PA-28 aircraft are used to create two pairs of nominally identical structures by separating the tail wings in half. The new population of four structures thus contains two pairs of them which are similar, but they have different length and different weight. A full modal test is performed in all of the structures and an exploration of possible common features is also done. The results show that common damage-sensitive features exist across the structures, a key requirement if population-based SHM is to be successfull.

**Keywords** Feature selection • Population-based • SHM • Vibration-based • Modal testing

## 17.1 Introduction

The idea of population-based monitoring has been implemented in the medical community as in the disease surveillance field [1], where it has been successful, and several systems are currently operational. The extension of the disease surveillance into structural health monitoring (SHM) has been presented before in [2]. The concept of a population-based approach means that in a population of nominally identical structures, such as in a wind farm, any additional new structures will not need to be fully modelled in order to be monitored. This can prove not only advantageous, but also a cost effective solution to large parts of modern infrastructure. In reality, the uncertainty that even nominally identical structures may present will complicate any such efforts of population-based monitoring. In [3], two different variants of the tail wing of a Piper PA-28 aircraft were used to create two pairs of nominally identical structures by separating them in half. Then the structures were tested and the existence of common patterns among them was explored in their natural frequencies and mode shapes. The study confirmed that while there were differences in the frequencies and modes, as expected, there were also common patterns in their behaviour. The purpose of this paper is to extend this comparison to all four structures and to further explore the potential of common features which can be used for the health monitoring of all of them.

The whole concept of population-based SHM can be illustrated in Fig. 17.1. The top part of the diagram, highlighted in green, displays the general framework of a pattern recognition approach to SHM as it is generally defined in [4] for an arbitrary structure. The process eventually defines a mapping between the normal (N1 in the diagram of Fig. 17.1) and a damaged (D1) condition through a feature selection/classification approach. This process is generally adapted to specific structures and in the case of significant variations it will most likely have to be repeated. However, if the structures are related or common patterns can be identified, then a separate mapping among the normal conditions (highlighted in blue in the diagram) can lead to a mapping among the obtained damage indication functionals (highlighted in yellow), which will in turn allow for the identification of unknown damage conditions in new structures (highlighted in brown in Fig. 17.1). The functional which describes the mapping from a normal to a damaged condition will be based on features, and a feature in this concept is a set of data measured or derived from measured data which can be used to individually identify a structure and

E. Papatheou (✉) • R.J. Barthorpe • K. Worden
Department of Mechanical Engineering, Dynamics Research Group, University of Sheffield,
Sir Frederick Mappin Building, Mappin Street, Sheffield, South Yorkshire S1 3JD, UK
e-mail: e.papatheou@sheffield.ac.uk

© The Society for Experimental Mechanics, Inc. 2015

C. Niezrecki (ed.), *Structural Health Monitoring and Damage Detection, Volume 7*, Conference Proceedings of the Society for Experimental Mechanics Series, DOI 10.1007/978-3-319-15230-1_17

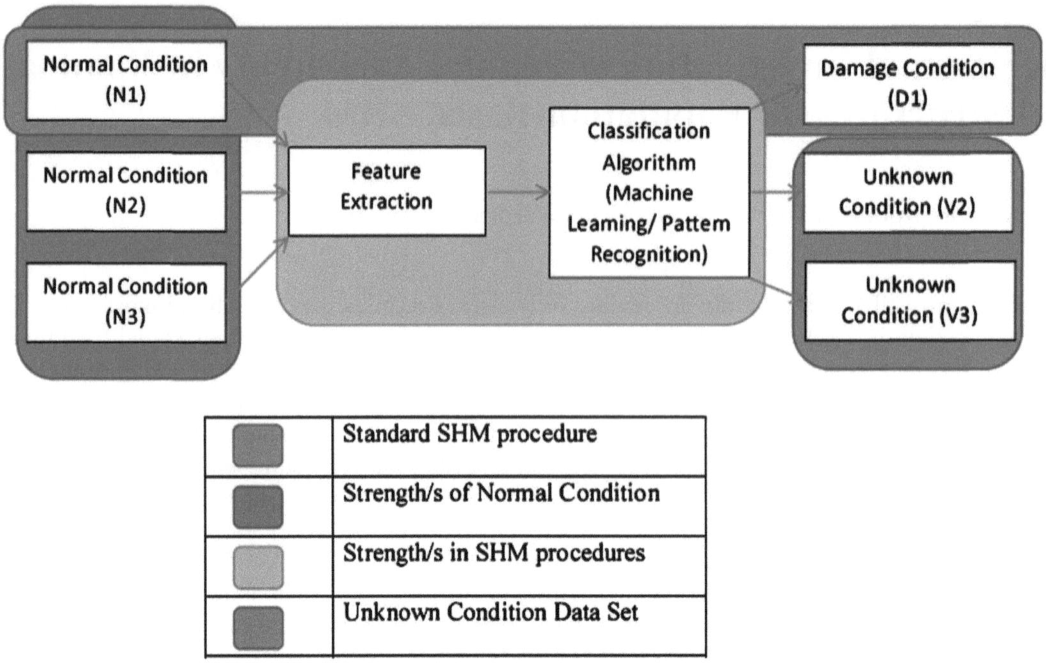


**Fig. 17.1** General framework for population based SHM

its varying states. Essentially, anything recorded or derived from an experimental test can be used. In order to accomplish the feature exploration in the population based SHM concept, a full modal testing was performed in all of the structures in their normal state, as well as in a 'damaged' state—when an added mass was applied. Features are selected by visually scanning through the FRFs and by using the effect of the added mass as a guidance. The suitability of the features is then tested with the help of outlier analysis.

The layout of this paper is as follows, the next section briefly describes the structures which were tested and the test setup. The following section describes the results of the comparison of the structures, firstly in the mode shapes and secondly in the FRFs with the help of outlier analysis. Finally the paper is rounded off with some overall conclusions and some discussion about future work.

## 17.2 Test Structures and Data Acquisition

The experimental process was described in detail in [3]. The main points are repeated here for the convenience of the reader. The structures of interest are the horizontal tail wing sections of a Piper PA-28 'Arrow' and of a Piper PA-28 'Cherokee'. The two wings were cut in half in order to create two pairs of nominally identical structures. The two new sections originating from the 'Cherokee' were named as A1, A2 and the other two sections created from the 'Arrow' as B1 and B2. All of the structures can be seen in Figs. 17.2 and 17.3 and a closer look can reveal that the 'Arrow' (B1–B2) sections are a subset of the 'Cherokee' wing (A1 and A2) as both originate from a variant of the same aircraft, the PA-28. Both wing sections have the same airfoil NACA0012 and the same chord (0.76 m), but different span: tail A is 3.62 m long where tail B is 2.74 m. So, sections B1 and B2 are in theory similar to A1 and A2, but shorter by 0.44 m each.

A full modal test was performed for each of the structures. The acquisition system used throughout the tests was a DIFA Scadas III controlled by LMS software running on a DELL desktop PC, and the sensors used were PCB piezoelectric uniaxial accelerometers. In total, there were 65 FRFs recorded for sections B1 and B2, and 75 FRFs for sections A1 and A2. The sensors were rotated in rows as was described in [3]. The locations of the sensors can be seen in Fig. 17.3, where one row of accelerometers is visible. The sensor mapping between the nominally similar structures is identical, but mirrored. Since B1 and B2 are subsets of A1 and A2, there is also an identical subset of sensor locations among all of them with their number being nominally 55, but because of an inconsistency to the test of A2 there are 44 which can be used for comparison.

The wings were suspended by springs in order to approximate free-free boundary conditions, and they were excited using a Data Physics (dp) electrodynamic shaker attached directly to their lower surface. A narrowband Gaussian excitation

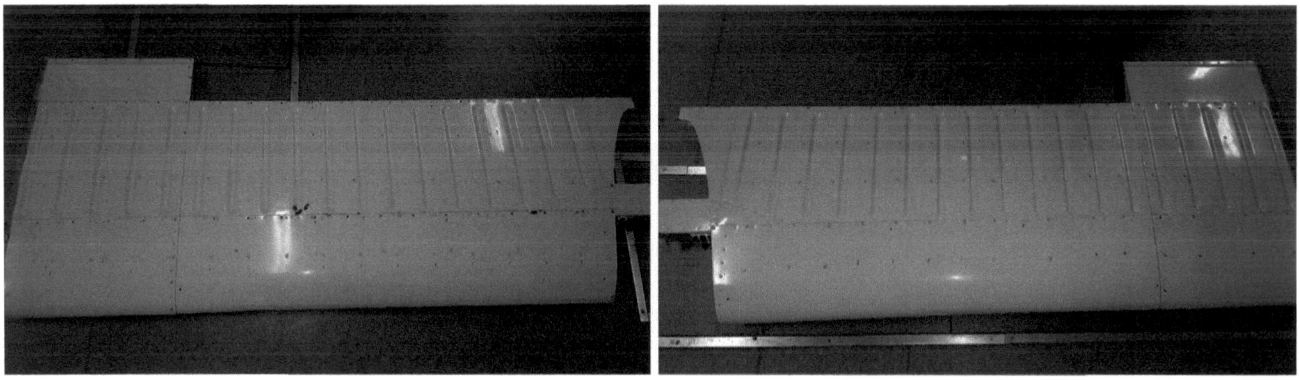

**Fig. 17.2** The two structures created from the separation of the PA-28 'Cherokee' tailplane, A1 (*left*) and A2 (*right*)

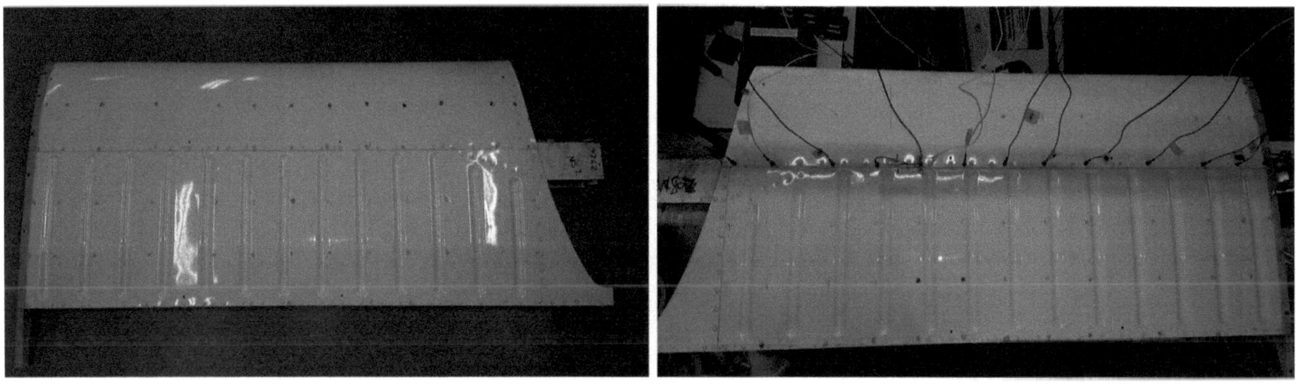

**Fig. 17.3** The two structures created from the separation of the PA-28 'Arrow' tailplane, B1 (*left*) and B2 (*right*). The marks for the sensor locations can also be seen. The sensors on B2 are located on the main stiffener

was generated within the acquisition system and amplified using a Data Physics power amplifier. The excitation itself was measured using a standard PCB force transducer, and care was taken to ensure that the location of the excitation was identical in all structures. All measurements used in the modal analysis were obtained within a frequency bandwidth of 0–512 Hz at a frequency resolution of 0.125 Hz. The Hv FRF estimator [5] was used throughout all the tests and the number of averages recorded for all measurements was 40. After the completion of the modal testing in the normal condition, the procedure was repeated for each structure with an added mass of 104 g. The mass was applied to the same location each time. Because of the difference in weight between the A tail and the B, there was also an overall different percentage of mass increase: 1.5 % in A1 and A2, and 2.2 % in B1 and B2.

## 17.3 Feature Comparison of the Structures

### *17.3.1 Natural Frequencies and Mode Shapes*

In [3] the natural frequencies and the mode shapes of the three structures A2, B1 and B2 were compared, and it was shown that the uncertainty caused by manufacturing differences such as the thickness of the skin or the condition of the test structures (e.g. worn skin) affect the natural frequencies and of course the FRFs of the identical structures, but common patterns were available, especially in the mode shapes and in the effect of the added mass on the FRFs. The work here extends the previous comparison to the fourth structure A1.

The estimation of modal parameters was accomplished with the help of the Polyreference Least Squares Complex Frequency (LSCF) domain algorithm [6], also known as PolyMAX. The rotation of the sensors complicated the modal identification, since it introduced a certain degree of variability on the test setup. It was already found from [3] that there was a slight difference in the thickness between structures A1 and A2, and this was thought to be the main cause for the deviation in their natural frequencies. Figure 17.4 shows FRFs recorded from the same location, the main stiffener, from A1 and A2. It is clear that the difference in the frequencies is stronger in the higher modes. Figure 17.5 displays the comparison

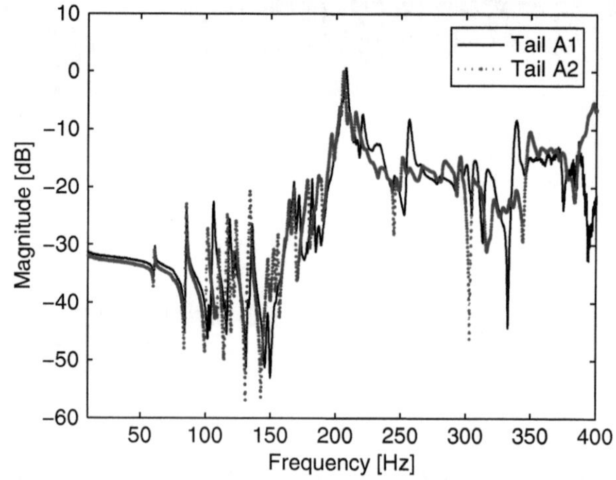


**Fig. 17.4** FRFs recorded from the same location on tail A1 and A2

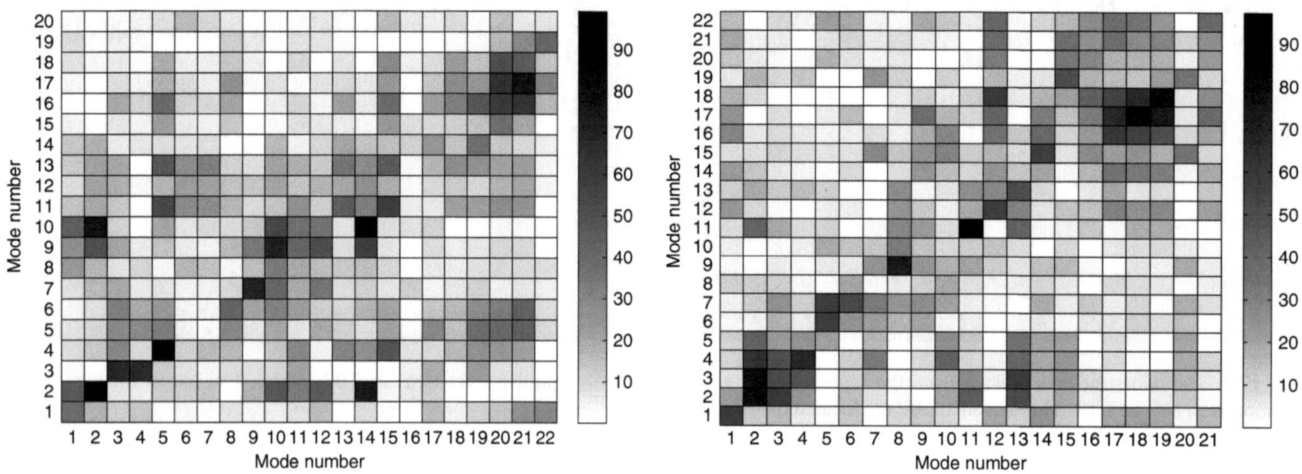


**Fig. 17.5** MAC between structures A1, A2 (*left*) and B1, B2 (*right*) for all identified modes

of the mode shapes between A1 and A2 with the help of the Modal Assurance Criterion (MAC), and also the comparison between B1 and B2 (which has been shown in [3]) for reference. All the modal identification was performed in a frequency bandwidth of 0–220 Hz, and the sensor locations used for the MAC comparison were located on the main stiffeners of the structures.

The MAC comparison indicates that although there may be close natural frequencies between the nominally identical structures, there are still enough deviations which are highlighted by low MAC numbers. It is interesting, that the modes which seem to correlate more between the nominally identical structures are similar (approximately in the same frequency areas) in both types of them. Overall, these results confirm that despite the variations, there are common patterns between the similar structures. It should be noted that the measurement of mode shapes can be challenging in practice, as was the case here, and the sometimes unavoidable measurement error, presented may be a reason for low MAC values. In addition, the MAC values tend to be sensitive to 'large' differences, which could be due to badly identified modes or inconsistency in the test setup (as was in this case with the rotation of the sensors), but also the MAC has been shown unreliable when it is used with few points [7]. For these reasons, and also because the MAC has not proved robust in SHM applications, the MAC comparisons between A1,A2 and B1,B2 which was shown in [3] is not repeated here.

### *17.3.2 FRF Features*

Since one of the main motivations for this study was its potential use in a population based scheme, the introduction of damage in the structures was seen as natural. An added mass of 104 g was added to the structures at the location. The use of an added mass as a non-destructive way of introducing damage to a structure has been extensively explored and validated in [8, 9].

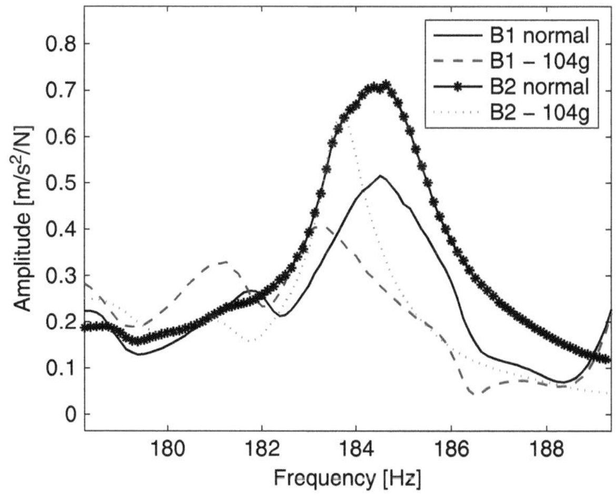


**Fig. 17.6** FRFs of tails B1 and B2. The effect of the added mass of 104 g on peak 184 Hz is shown

There were in total 75 FRFs recorded (from 75 locations) for the A1, A2, and 65 FRFs for the B1 and B2 structures. From all those FRF locations, there is a subset of 44 which are identical for all structures. A feature in this work is defined as a set of data measured or derived from measured data which can be used to distinguish the different states of the structure. By different states, one may consider 'healthy' and 'damaged'. Figure 17.6 shows an example of a similar pattern appearing in both B1 and B2 structures: when a mass is added, the peak at 184 Hz drops both in frequency and amplitude. Essentially, this common pattern is a potential feature, and its suitability can be tested with the help of am appropriate classifier. Any approach would be acceptable, but here outlier analysis is the chosen method. The approach has been described in detail in [10] and it involves the fusion of a multivariate feature into a single quantity called the Mahalanobis squared distance, which is then compared against a threshold. The threshold is calculated through a Monte Carlo approach, and can be inclusive or exclusive if the potential outlier is used in its calculation or not. The Mahalanobis squared distance is calculated by Eq. (17.1).

$$D_{\zeta} = (\mathbf{x}_{\zeta} - \overline{\mathbf{x}})^{\mathrm{T}}\mathbf{S}^{-1}(\mathbf{x}_{\zeta} - \overline{\mathbf{x}}) \tag{17.1}$$

As only single high average FRFs were measured for each state, the potential feature (the frequency lines of interest) was copied 1,000 times and then polluted with normally distributed noise of a standard deviation equal to the 4 % of the maximum value of the original feature. In this particular feature the size of the dimensions used is 50 (50 points around the peak). The FRFs originating from the intact structures are used as the normal data, and their mean and standard deviation is calculated to be used with Eq. (17.1). Then, the Mahalanobis distance from the features created from the FRFs with the added mass are tested against the threshold in order to declare whether there is significant deviation from normality, therefore damage or not. Figure 17.7 shows the Mahalanobis distance created from the feature which was extracted from the FRFs (located at point 65) of structures B1 and B2 around the mode 184 Hz. The threshold in this case was calculated at 112.67. It is clear that this feature could be used for both structures in order to indicate the presence of added mass. What is important, is that this feature can be selected just in one of the two nominally identical structures (B1 or B2) and it will be able to indicate the presence of damage (added mass here) in the other.

The location of the FRF which is shown in Fig. 17.6 is not in the subset of the common sensor positions among all of the structures, therefore a direct comparison of how well this particular feature performs in tails A1 and A2 is not possible. However, the frequency lines (which correspond to the feature) can be tested with outlier analysis in all of the rest of the sensor locations, both for A1 and A2. In order to check easily the performance of the feature, the Mahalanobis squared distance can be normalised by the threshold (which always depends on the dimension). In this way, every feature which produces a mean normalised discordancy value greater than 2 will be considered adequate and successful. Out of the 60 FRFs that there were tested for structures A1 and A2, 47 produced adequate features for both the structures. In the case of B1 and B2 this number was a bit higher at 57. Due to some problems presented during the experimental tests of the structure A2, the FRFs numbering 60–75 were not considered reliable and therefore they are excluded from this study here. Out of the 47 FRFs which produced adequate features for A1 and A2, 33 also work for B1 and B2. This result mainly means that a feature which was 'blindly' selected to indicate the added mass in structures B1 and B2 also works for both A1 and A2.

Since the aforementioned feature was selected by visually checking both the FRFs from the B1 and B2, it may be interesting to explore what would happen if one selects features simply on only one of the four structures. To investigate this, structure A1 was arbitrarily chosen, and in the same spirit, the FRF from sensor location 20 was also arbitrarily chosen.

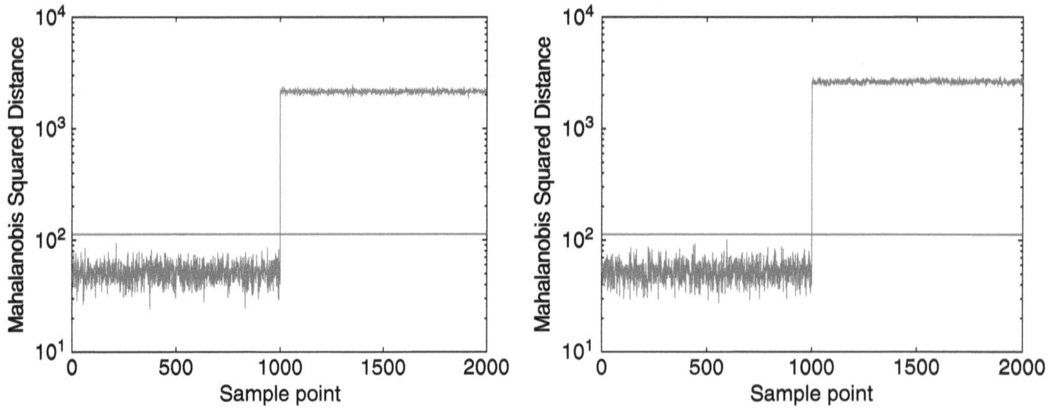


**Fig. 17.7** Outlier statistics for the feature created from the FRFs of structures B1 (*left*) and B2 (*right*) around the mode 184 Hz, the first 1,000 points tested correspond to healthy (no added mass) data

**Table 17.1** Comparison of the feature suitability for features selected from structure A1 (FRF point 20) when tested across all structures

| FRF lines | Number of successful FRFs for A1 and A2 | Number of successful FRFs for B1 and B2 | Number of FRFs for all structures | Successful for the FRF which was originally selected ? |
|---|---|---|---|---|
| 1,317–1,370 | 56 | 56 | 40 | Yes |
| 1,509–1,722 | 60 | 58 | 44 | Yes |
| 1,739–1,926 | 57 | 16 | 11 | No |
| 2,027–2136 | 55 | 25 | 20 | Yes |
| 2,949–3,035 | 47 | 51 | 28 | No |
| 3,115–3,259 | 55 | 53 | 37 | Yes |

By scanning this FRF when plotted against the FRF with the added mass, feature selection is possible. From this procedure six features were finally selected and tested across all four structures. Table 17.1 shows the results for all features when tested across all structures.

It is clear that all six features are able to produce adequate novelty detectors, meaning that they can indicate the presence of the added mass in all of the structures. Of course, no feature works for all locations, but most features are working for a good number of the common locations, with the exception of the third feature (11 out of 44 common sensor locations). Four out of the six selected features are successful across all structures for the same FRF location they were originally chosen. It can also be seen that in five of the features there are more locations which can indicate the presence of the mass in A1 and A2 than in B1 and B2, something which should be expected since they were originally selected in one of the two A structures, the A1. An in-depth exploration of features should make use of all 44 common FRF locations for all possible features, but this was not possible due to time restrictions. However, a limited study has been done by automatically separating all FRFs in parts of dimension 50 (with no overlap) and checking 'blindly' whether they work well as features. This analysis has shown that in the region of 200–400 Hz there are many potential features which will work for all structures. Overall, this short study has shown that it is possible to select a feature in one structure, and then use it to monitor the other three structures, which is certainly a first step towards population-based SHM, and this was consistent in all of the features tested—there were always FRFs which would produce good novelty detectors.

## 17.4 Conclusions

This paper presented a short study on the exploration of common features among nominally identical structures and how these may be used in an SHM concept. The work extended a preliminary study which was reported elsewhere and attempted to explore possible features which can be used in a population-based monitoring scheme. Such a scheme will make use of common patterns across structures in order to monitor new untested structures without the need of a full modelling or a complete knowledge of their characteristics. In order to do this, two wingtail sections of a Piper PA-28 'Cherokee' and of one of its variants, the PA-28 'Arrow' were acquired, and subsequently separated in half, creating thus two pairs of nominally identical structures. The 'Arrow' sections are subsets of the 'Cherokee' sections with shorter length and smaller mass. The natural frequencies and the mode shapes of the structures had already been compared in the previous study, but were briefly shown here as well to deviate, mainly due to manufacturing uncertainties, and potentially because of measurement errors. However, there were common patterns present. In order to explore those patterns in an SHM scheme, an added mass was used during testing in all of the structures and by using its effect on the FRFs, potential features were selected. Outlier analysis was the chosen method for 'damage' (added mass) detection and novelty detectors were built for the structures where the features were originally selected, but then subsequently tested with all the rest. It was shown that by choosing arbitrarily an FRF from a structure, without using the rest of the measurements from the other structures, then features can be used to identify the presence of mass successfully in all of them. Future work will address several of the testing inconsistencies with the help of a scanning laser vibrometer and also alter the structures more by actually damaging them or modify certain parts, e.g. stiffeners or replace rivets, to further explore the use of features in a population-based scheme.

**Acknowledgements** The support of the UK Engineering and Physical Sciences Research Council through grant reference EP/J016942/1 is greatly acknowledged. The authors would also like to thank Mr Jamie Park for his help during the experimental tests.

## References

1. Lombardo JS, Buckeridge DL (2007) Disease surveillance, a public health informatics approach. Wiley, New York
2. Deering S, Manson G, Worden K, Allen DW, Farrar CR, Lombardo JS (2008) Syndromic surveillance as a paradigm for shm data fusion. In: Proceedings of the 4th European workshop on structural health monitoring, Crakow, Poland, 4–6 July 2008
3. Papatheou E, Rahman TAZ, Barthorpe RJ, Park J, Worden K (2014) An experimental investigation of feature complexity and diversity in nominally similar test structures. In: Proceedings of ISMA2014, Leuven, Belgium
4. Farrar CR, Worden K (2013) Structural health monitoring, a machine learning perspective. Wiley, New York
5. Ewins DJ (2000) Modal testing: theory, practice and application, 2nd edn. Wiley, New York
6. Peeters B, Van Der Auweraer H, Guillaume P, Leuridan J (2004) The polymax frequency-domain method: a new standard for modal parameter estimation? Shock Vib 11(3–4):395–409
7. Maia NMM, Silva JMM (1998) Theoretical and experimental modal analysis. Research Studies Press LTD, Baldock, Hertfordshire, England
8. Papatheou E, Manson G, Barthorpe RJ, Worden K (2010) The use of pseudo-faults for novelty detection in shm. J Sound Vib 329(12): 2349–2366
9. Papatheou E, Manson G, Barthorpe RJ, Worden E (2014) The use of pseudo-faults for damage location in shm: an experimental investigation on a piper tomahawk aircraft wing. J Sound Vib 333(3):971–990
10. Worden K, Manson G, Fieller NRJ (2000) Damage detection using outlier analysis. J Sound Vib 229(3):647–667

MIX
Papier aus verantwortungsvollen Quellen
Paper from responsible sources
FSC® C105338

If you have any concerns about our products,
you can contact us on
**ProductSafety@springernature.com**

In case Publisher is established outside the EU,
the EU authorized representative is:
**Springer Nature Customer Service Center GmbH**
**Europaplatz 3, 69115 Heidelberg, Germany**

Printed by Libri Plureos GmbH
in Hamburg, Germany